Primate Life Histories and Socioecology

Primate Life Histories and Socioecology

EDITED BY PETER M. KAPPELER
AND MICHAEL E. PEREIRA

THE UNIVERSITY OF CHICAGO PRESS / CHICAGO AND LONDON

PETER M. KAPPELER is a research scientist at the German Primate Center, a lecturer at the University of Würzburg, and editor of *Primate Males: Causes and Consequences of Variation in Group Composition* (2000). MICHAEL E. PEREIRA is a science instructor at the Latin School of Chicago, a research associate at the Lincoln Park Zoo, and co-editor of *Juvenile Primates: Life History, Development, and Behavior* (with Lynn A. Fairbanks, 1993, reprinted 2002).

The University of Chicago Press, Chicago 60637
The University of Chicago Press, Ltd., London
© 2003 by The University of Chicago
All rights reserved. Published 2003
Printed in the United States of America

12 11 10 09 08 07 06 05 04 03   1 2 3 4 5

ISBN: 0-226-42463-4 (cloth)
ISBN: 0-226-42464-2 (paper)

Library of Congress Cataloging-in-Publication Data

Primate life histories and socioecology / edited by Peter M. Kappeler and Michael E. Pereira.
      p.   cm.
      Includes bibliographical references (p.   ) and index.
      ISBN 0-226-42463-4 (cloth : alk. paper)—ISBN 0-226-42464-2 (pbk. : alk. paper)
      1. Primates—Behavior—Evolution. 2. Primates—Ecology. I. Kappeler, Peter M.
      II. Pereira, Michael Eric, 1956–

  QL737.P9 P67253 2002
  599.815—dc21   2002010920

We dedicate this book to everyone who contributes to the preservation of biodiversity and our fascinating primate relatives in particular.

Complex interactions among genetics, development, and
ecology must be considered to truly understand evolution.
Our goal is to grasp the complexity of nature, not just
to imagine a world simple enough to be comprehended.
    —Paraphrased from Schlichting and Pigliucci 1998, p. 50

Contents

Foreword

ROBERT D. MARTIN

It has become increasingly apparent over recent decades that the individual components (traits) that contribute to the life history pattern typical of any mammalian species are in some way coordinated within an adaptive framework. The combined effects of these individual components —most notably gestation period, size and number of offspring, lactation period, size and age at weaning, pattern of postnatal growth, size and age at sexual maturity, interbirth interval and longevity—determine the pace of population growth and turnover. Within a species, these traits show typical values with varying degrees of flexibility and together yield the characteristic life history pattern. Various lines of evidence indicate that any species-typical life history represents an adaptive response to past or present environmental influences. Although there is now an increasing tendency to avoid the term "life history strategy" (with its unfortunate teleological flavor), it is generally accepted that natural selection has forged the basic pattern found in any given species. As part of this general acceptance, the concept of trade-offs between individual components of the life history has come to occupy a focal position.

Much of the work that has been conducted on mammalian life histories in general, and on primate life histories in particular, has involved broad interspecific comparisons of various kinds. Such comparisons have revealed a spectrum of life histories ranging from very fast to very slow reproductive turnover. It has also emerged that a fundamental factor influencing the rate of reproduction is body size, with a clear overall trend for life histories to show a slower pace with increasing size. For this reason, allometric analysis—which explicitly takes into account the scaling effect of body size—has played a prominent role in interspecific comparative investigations of life histories. Allometric analysis permits recognition of a general scaling trend and hence the identification of positive or negative departures (residual values) of individual species from that trend. As a result, it is possible to recognize species that have relatively slow or fast life history patterns at any given body size. In principle, any comprehensive theory of the evolution of life histories should be able to account both for the general trend toward slower reproduction with increasing body size and for the positive or negative deviations of individual species from that trend.

One important outcome of interspecific analyses—in fact, one of their major advantages—is the recognition of general principles. Identification of regular patterns in the variation of life history components relative to body size can provide valuable clues to underlying relationships. In some cases, it may be possible to identify invariant relationships among life history components that apply across a particular group of species at a higher or lower taxonomic level. The simplest rule concerns the general trend with body size: "The most obvious generalization about life history in mammals is that larger species lead slower lives" (Purvis et al., chap. 2, this volume). Beyond this, there are several indications that externally imposed mortality schedules play a major part in the evolution of life history components and help to explain differences in life history traits relative to body size. In this context, one important finding is that the product of age at sexual maturity and the instantaneous adult mortality rate is invariant across a given group of species. It has also emerged as a fairly consistent finding that life expectancy at maturity is strongly correlated with age at sexual maturity in mammals. At a lower taxonomic level, one interesting finding (yet to be explained) is that primate infants are fairly consistently weaned when they have attained a body mass that is about one-third of the adult value.

As comparative research has progressed, it has become increasingly apparent that there are fundamental methodological problems that must be effectively tackled for allometric analyses to be successful. It is now possible to identify at least four areas in which errors may arise: (1) choice of a line-fitting method, (2) recognition of grades, (3) allowance for the effects of phylogenetic inertia, and (4) inference of causality from empirical correlations. All of these potential sources of error have received some attention, and there have been a number of methodological developments specifically designed to tackle them one at a time. However, it has not been sufficiently recognized that these sources of error interact in complex ways, and we still have some way to go before achieving a synthetic approach that accounts for them all simultaneously.

The first step in identifying an overall trend in a bivariate data set for allometric analysis is determination of an appropriate best-fit line. In the majority of cases, least-squares regression has been used, but it can be argued that this method is inappropriate because the x variable (usually body mass) cannot be measured without error and because it is uncertain whether the y variable is directly dependent on the x variable in such data sets. For this reason, some authors have suggested that alternative best-fit lines avoiding these two requirements (e.g., reduced major axis or major axis) should be used instead. However, there is an additional problem in that the x and y variables in interspecific analyses (usually expressed in logarithmic form) are rarely normally distributed, thus violating a prerequisite of any parametric line-fitting procedure. It might therefore be preferable to use a nonparametric line-fitting procedure for allometric analyses. Discussion of the choice of an appropriate best-fit line has generally taken a back seat in recent years because attention has focused heavily on the problem of phylogenetic inertia (see below). However, the fact remains that the use of an inappropriate line will lead directly to errors of interpretation, so this issue should not simply be forgotten or passed over. As a general rule, a least-squares regression will have a lower slope than any other best-fit line that may be used. Hence, if least-squares regression is in fact inappropriate, residual values calculated for small-bodied species will be too small, while residual values for large-bodied species will be too large. As a minimal response to this potential source of error, it would be preferable to apply different line-fitting techniques and examine their effects on results obtained.

A different kind of error can arise when it is inappropriate to determine a single best-fit line for a particular data set. In allometric analyses, it is commonly observed that the data are in fact divided into two or more subsets that show a similar scaling trend (slope value) but are vertically separated in a bivariate plot (i.e., show different intercepts). Such subsets can be referred to as "grades," and the vertical separation between them can be termed a "grade shift." If the overall data set does not constitute a single distribution, but in fact contains two or more overlapping distributions (grades), calculation of a single best-fit line would be inappropriate. Instead, an individual best-fit line should be calculated for each grade. Interpretation of a single best-fit line for a data set containing distinct allometric grades ("grade confusion") can lead to erroneous conclusions. Clearly, an objective analytical procedure that would permit reliable recognition of grades within a data set would be valuable. As a first step, however, it is always advisable to test taxonomic subgroups within a data set in order to determine whether grade differences exist among them.

Such an approach led directly to the recognition of one of the clearest grade shifts so far reported for primates, which has direct implications for life history patterns. When neonatal body mass is plotted against adult body mass for primates, it emerges that strepsirrhine primates (lemurs and lorises) constitute a grade with markedly smaller values than haplorhine primates (tarsiers, monkeys, apes, and humans). It turns out that, for any given adult mass, the neonate of a haplorhine primate is almost three times heavier than that of a strepsirrhine primate. As there is no matching grade separation between strepsirrhines and haplorhines in a plot of gestation period against adult body mass, it follows that daily investment in fetal growth is markedly higher in haplorhines than in strepsirrhines. After birth, the situation is reversed, with strepsirrhines growing faster than haplorhines up to maturity. This fundamental distinction between strepsirrhine and haplorhine primates with respect to prenatal and postnatal development represents a major discovery that requires explanation within the overall framework of life history theory as applied to primates. Yet the initial finding underlying this discovery would not have seen the light of day if a single best-fit line had been determined for the plot of neonatal body mass against adult body mass in primates!

A striking example of a grade shift in a life history component of mammals is provided by a bivariate plot of gestation length against adult body mass. If a single best-fit line is determined for all placental mammals, the slope value is found to be 0.25. Some authors have discussed the potential significance of this value with respect to the scaling of longevity and other life history variables. However, placental mammals can be divided fairly clearly into altricial and precocial species, and this division has direct implications for the scaling of gestation periods. When the placental mammal sample is divided for allometric analysis, two separate grades are identified, each with a slope close to 0.10 rather than 0.25. Hence, a radically different scaling relationship emerges. Furthermore, the difference between the two grades is substantial. At any given adult mass, a precocial species typically has a gestation period about three times longer than an altricial species. This marked difference clearly bears major implications for life history patterns, and socioecology (e.g., Kappeler, Pereira, and van Schaik, chap. 1, this volume), in placental mammals. Thus it is arguably mandatory to examine altricial and precocial mammals separately when attempting to identify general principles.

Indeed, the occurrence of two grades in the scaling of mammalian gestation periods is of interest in its own right. When scaling is examined in relation to a single line fitted with a fixed slope of 0.10, the residual values are clearly bimodal, with very little overlap (in itself, evidence that two grades are present). This pattern is somewhat puzzling. In principle, it should be possible to find placental mammals with medium-length gestation periods and medium-sized litters; in practice, there is a virtual dichotomy between altricial mammals with large litters and short gestations and precocial mammals with small litters and long gestations. For some reason, the intermediate solution seems to be ruled out by the selective factors that determine life history patterns in mammals. (Such divergence between alternative solutions has been found in various areas of life history research, as shown by several chapters in this book.) Given that there is good evidence that the altricial condition is primitive for placental mammals, and that the precocial condition has apparently emerged several times in independent lineages (e.g., primates, ungulates, cetaceans, and elephants), the virtual lack of surviving intermediate forms cannot easily be attributed to historical accident. Accordingly, an explanation for the dichotomy between altricial and precocial mammals represents a major challenge for the further development of life history theory as applied to mammals. As noted more generally by Purvis and colleagues (chap. 2, this volume), "grade shifts are easy to demonstrate, but hard to explain convincingly."

Grade confusion has presented a common obstacle to sound interpretation of allometric scaling analyses, both with respect to life history patterns and in other contexts. There is, for instance, some confusion about the reported relationship between weaning mass and adult mass, with some authors claiming that the ratio is approximately constant and others stating that it increases with increasing adult mass. The reason for this discrepancy seems to reside in grade relationships. If the analysis is confined to primates, the ratio seems to be essentially constant, but if it is conducted for mammals generally, there seems to be confusion of a number of grades, resulting in an overall scaling trend when all grades are confounded.

The third potential source of error in allometric analyses—namely, phylogenetic inertia—was recognized relatively recently and has attracted a considerable amount of attention. Indeed, the extensive attention devoted to this single issue in recent years has probably diverted attention away from the other major issues. The essential point with respect to phylogenetic inertia is that all species are linked in a phylogenetic tree and it may therefore be inappropriate to treat data from closely related species as if they were statistically independent. As noted by Purvis and colleagues (chap. 2, this volume), the inclusion of such data in a comparative analysis may amount to pseudoreplication. The basic challenge here is reliable statistical testing, permitting meaningful assessment of significance. One simple approach to the problem is to make a suitable adjustment to the degrees of freedom calculated for a particular variable, as done by Godfrey and colleagues in this volume (chap. 8). With an original sample of forty species, these investigators found effective sample sizes between eight and twenty-two, depending on the particular variable being examined. Hence, the potential problem of pseudoreplication in data sets for individual species translates into the need for strong underlying relationships to obtain significant results.

The potential problem of phylogenetic inertia in comparative analyses is not, however, confined to the issue of statistical significance. In a bivariate allometric analysis, it is also necessary to know whether pseudoreplication might bias the allometric relationship determined, along with the corresponding residual values. For this reason, an alternative approach involving calculation of "independent contrasts" has been developed and is now readily available for application, thanks to generous distribution of the computer program CAIC. The notion underlying the calculation of independent contrasts is that differences between values for pairs of taxa in a phylogeny ("contrast values") are statistically independent and can therefore be used instead of raw values for individual species in comparative analyses. Starting with an original data set for N species, it is possible to calculate $N - 1$ contrast values by working down through dichotomous nodes to the base of the tree, assuming that it is valid to take average values for the two descendant taxa from any node as representative of the node itself. The resulting contrast values can be subjected to the same kinds of analysis as the raw data, with the advantage that any effects of phylogenetic inertia have hopefully been eliminated.

It may be thought that the calculation of independent contrasts, following the procedure now recommended with CAIC, might also eliminate problems associated with the choice of a best-fit line and with the presence of grades in a data set. In effect, the computation procedure requires the identification of one variable (usually body mass) as the "independent variable" for subsequent analyses. Furthermore, the prescription that the best-fit line in a bivariate plot of contrast values should pass through the origin alters the approach to line fitting. However, if it is true that least-squares regression (with its requirement for a clearly independent variable measured without error) is inappropriate for the raw data, it is difficult to see how this problem should simply disappear during the calculation and analysis of contrast values. It has also been stated that calculation of independent contrasts will eliminate the problem of grade shifts and, indeed, that the very existence of grades in itself provides a major reason for not treating species as statistically independent points for analysis (Purvis, et al., chap. 2, this volume). In principle, it might be expected that a single grade shift at a basal node in the tree might generate a single aberrant contrast value, with all other values fitting a single scaling relationship. In practice, however, the occurrence of several grade shifts higher up in the tree (as with mammalian gestation lengths) will produce a very confusing picture. Because the starting values for individual dichotomous nodes are calculated as the means of their two descendant taxa, any grade shift will be "smeared" through lower nodes of the tree by this averaging process and hence obscured. Until it is demonstrated with worked examples that the application of CAIC can lead to clear recognition and resolution of undoubted grade shifts, this potential problem must remain at issue. In fact, there is a worrying possibility that "grade smearing" generated by the method of inferring values for nodes in a tree may lead to erroneous conclusions when CAIC is applied to comparative data.

The fourth, and potentially most damaging, source of error in interspecific allometric analyses arises in the attempt to derive conclusions regarding causal relationships from the observed empirical correlations. It cannot be said often enough that correlation should not be equated with causality, particularly because it is so tempting to jump to conclusions when a strong correlation is found in the direction predicted by a particular hypothesis. Until numerous additional tests and exhaustive cross-checks have been conducted, it is simply unwarranted to draw any firm conclusions with respect to causal relationships. For this reason alone, comparative analyses of interspecific data must be treated with great caution in the absence of other kinds of evidence. Even if the procedures for analysis have been optimized such that the potential problems arising from line fitting, grade shifts, and phylogenetic inertia have all been effectively resolved, we are still faced with the major difficulty of advancing from empirical correlations to reliable causal inferences.

It is against this challenging methodological landscape that the undoubted achievements of the individual contributions to this volume must be assessed. As is noted by Stearns, Pereira, and Kappeler (chap. 13, this volume), primates are a particularly difficult group in which to study life history patterns. They are characterized by long life spans and especially slow reproductive turnover and, indeed, show grade shifts relative to other placental mammals in both respects. Consequently, collection of reliable data on primate life histories is a very long-term undertaking, and experimental approaches to test particular hypotheses regarding life history dynamics are effectively ruled out. On the other hand, primates are of particular interest in terms of life history theory precisely because they live so long and breed so slowly. In addition, the direct relevance of their life histories to human life history makes them particularly rewarding to study, as superbly exemplified

by the discussion of human menopause by Hawkes, O'Connell, and Blurton
Jones in chapter 9, which builds upon deviations of certain human life history traits from the typical primate pattern. Despite the many problems and
pitfalls, then, it is certainly worthwhile to explore primate life history patterns and attempt to identify general principles that may help to illuminate
the evolution of primates. Furthermore, it is eminently worthwhile to explore potential links between diversity in life histories and diversity in social
systems, which is the primary aim of this book.

Thus far, I have mentioned only interspecific comparisons, as these
comparisons are needed to identify general rules that apply to life history
patterns, in mammals generally or in primates in particular. However, it is
intraspecific variation that is subject to natural selection, so this is a topic,
hitherto greatly neglected with respect to primates, that deserves special attention. It is therefore particularly gratifying to see detailed treatment of
intraspecific variation in several chapters of this volume. In the first of these
(Lee and Kappeler, chap. 3), attention is directed to the phenotypic plasticity of certain life history variables (interbirth interval, age at weaning, fertility, and age at first reproduction) and their potential links with socioecological variables. To achieve this, comparisons are made between wild
and captive populations, on the assumption that the latter are close to the
maximal values for reaction norms. Apart from yielding the finding that
species with high plasticity in female body mass and age at weaning tend to
have a shorter relative gestation length, a lower relative weaning weight, and
younger relative weaning ages, this analysis also provides some evidence for
a link between variation in social behavior and variation in life history. In an
entirely different approach, Alberts and Altmann (chap. 4, this volume) use
demographic matrix models to assess the implications of intraspecific variation in life history components in savanna baboons. This approach has the
special advantage of making it possible to conduct perturbation analyses,
using either empirical data or manipulation of matrix entries, to assess the
effects of individual variables. Overall, the theory underlying this approach
indicates that age-specific changes in fitness are of great importance in the
evolution of life histories.

It is reasonable to expect that studies of intraspecific variation will eventually complement comparative interspecific studies in generating a body of
theoretical principles governing life history patterns and (hopefully) their
connections with patterns of social organization. We are, however, still some
way from achieving such a synthesis. Furthermore, on the way to this goal, it
will be necessary to resolve the apparent paradox that correlations for life
history components sometimes apply in opposite directions in intraspecific

and interspecific studies. For instance, within a species, it may be found that a neonate born after a longer than normal gestation will reach maturity relatively early. By contrast, interspecific studies consistently show that species with relatively long gestation periods have relatively late ages of sexual maturation. It is therefore difficult to see how selection for an extended gestation period within a species can lead to a positive correlation between gestation length and age at maturation across species. In chapter 5, Janson also remarks on striking differences between intraspecific and interspecific life history analyses.

For the time being, then, we are largely confined to identifying general scaling principles and intriguing patterns of correlation among life history variables across species, in the hope that these findings will ultimately lead to a full-blown theoretical framework. But even within this straitjacket it is possible to make significant advances. Pereira and Leigh (chap. 7, this volume) have, for example, addressed the neglected issue of differential modes of development in primates, rightly emphasizing the great significance of plasticity, which may itself have been a target of selection. Given the very long maturation periods of primates, the scope for plasticity is especially marked, and this again makes primates particularly interesting subjects of study.

In a wide-ranging interspecific study, Godfrey and colleagues (chap. 8, this volume) also break new ground, both by applying an impressive array of methods and by taking a closer look at the association between dental development, brain size, and life histories. Their analysis builds on the intriguing finding by Smith (1989) of a tight link between brain size and dental eruption patterns, which is confirmed. The authors show that dental development, as measured by dental precocity at different ages and dental endowment at weaning, relates to life history (age at weaning) and diet (foliage consumption), two factors believed to influence social organization. Overall, folivorous primates are convincingly shown to exhibit faster dental development than frugivorous primates.

On another tack, Janson (chap. 5, this volume) examines apparent puzzles concerned with the relationship between predation risk and group size in primates. The analysis explores the possibility that life history differences correlated with body size influence the relationship between intrinsic and current predation risk. As throughout this volume, there is laudable attention to methodological problems, and the allometric scaling of life history variables to body size is explicitly taken into account. The interspecific comparison addresses the potential dependence of the fitness effect of predation risk on life history traits and satisfactorily resolves the apparent puzzles noted at the outset. In the process, an explanation is provided for the fact that social group size tends to increase up to a certain body size in primates and then declines in the largest-bodies species (great apes). In fact, this particular case illustrates a further potential pitfall in comparative analyses, which is the assumption that there is necessarily a linear relationship between two variables. In initial analyses of the relationship between social group size and body mass in primates, it was simply noted that there was a positive correlation between the two variables, implying a consistent upward trend. Only later was it noted that group sizes decline again in the largest-bodied primates, such that fitting a single straight line to the data is inappropriate. It is, in fact, important to examine any bivariate data set to ensure that the relationship is likely to be linear before proceeding with an analysis based on that assumption.

In conclusion, this volume clearly shows that great progress has already been made in the study of primate life histories and in the investigation of potential links with patterns of social organization. Progress in the future will continue to rely upon improvement in methods of analysis in tandem with novel approaches to data, and further collection of data will surely play an important role. Sometimes, in the rush to conduct analyses, insufficient attention is given to data quality, and a lot will be gained through improvements in this area. Last but not least, real progress will depend on continued accumulation of long-term data from primate field studies, and those who have made such data collection possible under demanding conditions deserve our admiration, respect, and gratitude.

Preface

Life histories encapsulate the essence of life for organismal and evolutionary biologists. They summarize basic phenotypic causes of variation in fitness, thereby providing opportunities to illuminate the action of natural selection, the nature of adaptations, and constraints on development and evolution. Above all else, however, our fascination with life histories derives from their overwhelming diversity. Organisms have evolved myriad ways of expressing, modulating, and combining principal life history traits, such as size and number of offspring, growth pattern, size and age at maturity, reproductive effort, and longevity. Additionally, as contemporary research continues to reveal in ever-growing detail, these basic traits are linked in different ways, and to varying extents, across taxa by physiology (trade-offs) and genetics (pleiotropy, epistasis).

Variation in life histories, partly summarized by the relative speed of living, is encountered at many taxonomic levels. Mammalian orders vary systematically in key life history traits, for example, with most rodents and insectivores exhibiting relatively fast life histories and bats and primates relatively slow ones. Causes of life history variation at various taxonomic levels can include lineage-specific effects, divergent genotypes among individuals, and variation among pivotal ecological factors, such as the abundance or distribution of staple foods or major predators. Moreover, variation in life history traits, particularly among species, genera, and higher taxa, interacts importantly with major aspects of behavior, including social systems. This volume offers an overview of life history variation among primates as it is currently known.

Our undertaking embraces two specific objectives. First, we aim to provide an up-to-date summary of knowledge of life history diversity among primates, including evaluations of causes, constraints, and consequences.

Over the last two decades, many new details about the life histories of these slowly developing, long-lived mammals have accumulated alongside important theoretical and methodological advances in life history research. Second, this project offers the first systematic attempt to identify links between the diversity of primate life histories and the diversity of their social systems. Until now, the evolution of primate social systems has been examined within the socioecological paradigm, a theoretical model relating the distribution and relationships of individuals to ecological factors such as risk of predation and distribution of resources. The many links between life history and behavioral traits identified by contributors to this volume argue for an extension of the socioecological model to include life history considerations. This book, in other words, should help to launch still more integrative approaches in research on primate socioecology with a fresh perspective offering many new questions.

The volume is divided into three main parts. Part 1 examines the patterns and causes of variation in primate life histories and identifies links with social systems. The chapters in part 2 explore the unusually slow development and long life spans of primates. These chapters focus on immatures, the details of their development, and implications of their intense and extended dependencies for themselves, their adult caregivers, and other social partners. Part 3 addresses the single most intriguing aspect of primates: their very large brains. That the primate brain merits special consideration cannot be denied. This unusually large organ has long been suspected of not only processing information about social relations differently than do nonprimate brains but also constraining rates of development—that is, the speed of life histories. Without question, the primate brain plays central roles in the development and functioning of individual primates and their social systems. What remains is the excitement of elucidating these roles in relation to their mechanisms, and these final core chapters indicate many avenues fairly shouting for further investigation.

This project was conceived in early 1999, when P. M. K. began organizing a conference to explore relations between primate life histories and socioecology. It assumed intriguing shapes in December 1999, when seventeen invited speakers met in Göttingen, Germany, with scores of other primate researchers to discuss this issue for four days. Converting into a volume, it grew slowly over the next year, with chapters manifesting remarkable variation in relative speed of development. But, by November 2001, the present volume emerged as an integrated whole. We hope the final product is sufficiently mature to help fertilize research on life histories, ecology, and behavior for years to come.

Acknowledgments

This volume is based on contributions to a conference examining relations between primate life histories and socioecology (2. Göttinger Freilandtage) that took place 14–17 December 1999 at the German Primate Center and was made possible through the generous financial support of the Deutsche Forschungsgemeinschaft, the Wenner-Gren Foundation for Anthropological Research, and the German Primate Center. We thank all of these institutions for their support.

All chapters in this book benefited greatly from scientific review by at least one other contributor and one or more external referees. We wish to thank Tim Clutton-Brock, Tim Coulson, John Eisenberg, Lynn Fairbanks, Barbara Finlay, and Michael Power, in particular, for their helpful reviews. Thanks to Ulrike Walbaum for editorial help, especially with the reference lists. We thank Christie Henry for inviting us to submit the manuscript for review at the University of Chicago Press and for encouragement and support throughout its subsequent development. Two anonymous referees read the entire volume at an early stage and conveyed much-appreciated constructive criticism. We thank Norma Roche for providing finishing touches throughout the manuscript with thoughtful and careful copyediting. Finally, we are extremely grateful to Bob Martin for his multiple effects on our scientific work and for writing our foreword at short notice at an extremely busy time.

Our families and friends provided invaluable support throughout the development and preparation of this volume. P. M. K. is particularly grateful for the friendship of Claudia Fichtel, Theresa and Jakob Schmalzriedt, Dietmar Zinner, and Carel van Schaik. M. E. P. thanks The Latin School of Chicago, Tiana and Torin Pyer-Pereira, and, especially, Terri-Jean Pyer for making available time and other essential resources without which his work on this project could not have gone smoothly.

1 Primate Life Histories and Socioecology

PETER M. KAPPELER, MICHAEL E. PEREIRA,
AND CAREL P. VAN SCHAIK

One major goal of socioecology is to explain the evolution of diversity among social systems, which are defined in relation to both grouping characteristics and the nature of social and mating relationships within groups. The socioecological model, a theoretical framework that has guided much research in this area (Crook 1964; Crook and Gartlan 1966; Crook 1968, 1970; Crook, Ellis, and Goss-Custard 1976; Emlen and Oring 1977; Terborgh and Janson 1986; Standen and Foley 1989), has traditionally emphasized the effects of major ecological factors, especially resource distribution and risk of predation (Jarman 1974; Wrangham 1980a; van Schaik 1983; van Schaik and van Hooff 1983; Rubenstein and Wrangham 1986; Wrangham 1987; van Schaik 1989; Janson and Goldsmith 1995; van Schaik 1996; Sterck, Watts, and van Schaik 1997; Janson 1998).

During the past decade, however, this perspective has been broadened by the elucidation of sex-typical reproductive strategies and influences of sexual conflict on social relationships (Clutton-Brock 1989; Davies 1991; Clutton-Brock and Parker 1992; Smuts and Smuts 1993; Brereton 1995; Clutton-Brock and Parker 1995; Dunbar 1995a; van Schaik 1996; van Schaik and Kappeler 1997; Treves 1998; Kappeler 1999; Linklater et al. 1999; van Schaik, van Noordwijk, and Nunn 1999; Widemo and Owens 1999; Davies 2000). Since individual and taxon-specific life histories may constrain sexual strategies and other behavioral adaptations, earlier assertions that life history limits on behavioral evolution "are not surprising enough to catalyse interest" (Stearns 1992, p. 210) merit reconsideration. Indeed, considerable work has demonstrated that nontrivial relationships among life history, ecology, and behavior are shaped by ecological conditions (Caswell 1983; Stearns and Koella 1986; Stearns 1989a; Reznick et al. 1990; Janson and van Schaik 1993; Via et al. 1995; Przybylo, Sheldon, and Merilä 2000), indicating avenues for further expansion of the socioecological model (see also Stearns 2000). The main goals of this chapter are, therefore, (1) to formally expand the socioecological model by describing these additional interrelationships, (2) to identify potential links of causality among them, and (3) to illustrate these links using work on primates as examples from a diverse taxon with well-described life histories, ecologies, and social systems.

Evidence for Interrelationships

To substantiate the value of expanding the socioecological model, we first present some recent nonprimate examples of interactions among life history, ecology, and behavior. The first case provides a fresh perspective on the evolution of cooperative breeding, which occurs in about 3% of extant bird species, by illustrating how a change in life history traits can have consequences for individual behavior and, therefore, for an entire social system.

Whereas current explanations for the absence of independent breeding in helper individuals focus largely on ecological constraints (e.g., Emlen 1994), a comparison of avian orders by Arnold and Owens (1999) suggests that the evolution of slow life histories has also played an important role. The high adult survival rate characteristic of slow life histories retards the rate of territory turnover, thereby limiting opportunities for dispersal and independent breeding by young adults. Shifts from fast to slow life histories, in other words, have apparently augmented the potential advantages of helping to raise one's parents' next clutches of young, promoting the expression of helping behavior under particular ecological conditions.

A second example suggests effects in the opposite direction; that is, a life history trait modified in response to changes in a behavioral variable. Specifically, some females of a semelparous spider *(Stegodyphus lineatus)* delay reproduction to reduce the risk of infanticide by males (Schneider 1999). Female *Stegodyphus* typically guard a single egg sac until their young hatch and begin eating their mother. Sometimes males succeed in destroying a female's egg sac, however, thereby stimulating her to produce a second clutch, which is typically much smaller. Whereas males generally mature early, with an expected adult life span of only two to three weeks, females vary by a factor of four in the time taken to reach sexual maturity. The number of males, and thus the risk of infanticide, consequently decreases as the reproductive season progresses. Intersexual conflict thus contributes to plasticity in female reproductive strategy in this species, with early-maturing females reducing the risk of infanticide by delaying oviposition.

A final example shows that life history and behavioral strategies are generally finely tuned to each other, potentially linked genetically (i.e., pleiotropic), and conjointly responsive to ecological conditions. This case features growth rate as a life history parameter influencing foraging decisions in relation to predation risk (Werner and Anholt 1993). Exposing both transgenic salmon *(Salmo salar)* with enhanced growth rates and unmanipulated control individuals to predators, Abrahams and Sutterlin (1999) showed that the transgenic animals continued to express higher rates of feeding. The genetic manipulation's simultaneous effects on life history, foraging, and antipredator behavior suggest that the evolution of all three traits has been constrained by the ecological variable of predation risk (see also Arenz and Leger 2000).

A review of the relevant literature provides many more examples of direct interactions among features of life history, ecology, and behavior. The conceptual integration of these interrelationships into the socioecological model, however, is still in its infancy. We advocate the expansion of this theoretical framework by reviewing the new aspects suggested by primate studies, just as primate studies have previously helped to elaborate and refine the model. Because life histories are the new component we wish to add, we begin with an overview of life history concepts and the general characteristics of primate life histories. In subsequent sections, we examine interactions between life histories and social systems. The limited available information on modulation of life histories in response to ecological variation is summarized in other chapters (Ganzhorn et al., chap. 6; Janson, chap. 5; Lee and Kappeler, chap. 3; Pereira and Leigh, chap. 7; all this volume).

Life Histories
Basic Concepts
The life history of an organism is defined by features of its life cycle pertaining to developmental and reproductive rates as well as reproductive effort (Roff 1992; Stearns 1992). In mammals, gestation length, size, number, and sex of offspring, interbirth interval, age and size at weaning, age and size at first breeding, and life span are the most important life history traits (Charnov 1991).

Actual life histories vary tremendously among individuals of the same species. An elephant, for instance, can die as a fetus or soon after birth, but she can also live to be 80 years old. Nevertheless, each life history trait shows characteristic central tendencies when averaged for all individuals of a species, and there appear to be invariant relations between life history traits assessed across higher taxonomic levels (Charnov 1993; Hawkes, O'Connell, and Blurton Jones, chap. 9, this volume; Purvis et al., chap. 2, this volume). The elephant will endure for a maximum of almost 100 years—not 10 or 1000 years—no matter how well she is protected from predators, pathogens, the elements, malnourishment, poisons, and free radicals. Gestation length will show virtually no variation, and she will always give birth to a single, precocial infant, rather than litters of several small altricial ones. Life history traits are thus species-typical characters best defined as predispositions toward certain ranges of potential values among individuals in a given population (Roff 1992; Stearns 1992; Lee and Kappeler, chap. 3, this volume; Pereira and Leigh, chap. 7, this volume).

Variation in life history characteristics among mammals is enormous (Boyce 1988; Charnov 1991). A mouse is unlikely to live much beyond a year, almost two orders of magnitude less than the elephant, and some murid rodents have litters of ten or more, an order of magnitude greater than the elephants' singletons. Similar variation is found in the degree of development of young at birth, which varies from precocial to altricial (cf. Ricklefs and Starck 1998). Moreover, this interspecific variation in life history traits is not random; rather, life history characters are tightly intercorrelated across species (Portmann 1939; Eisenberg 1981; Harvey and Zammuto 1985; Harvey, Read, and Promislow 1989; Read and Harvey 1989; Charnov 1991; Harvey, Pagel, and Rees 1991), rendering a spectrum of life history "syndromes," from species with fast life histories (early maturation, large litters, short lives, etc.) to others with slow ones. Although larger animals inevitably tend toward slower life histories than smaller ones (Pagel and Harvey 1993), the size relationship is not entirely explanatory (Promislow and Harvey 1990). Significant variation in life history speed remains after controlling for size effects, especially in relation to grade shifts (see Purvis et al., chap. 2, this volume). Ultimately, particular life histories are considered to be adaptations to the rate of unavoidable mortality in the natural environment (Promislow and Harvey 1990; Charnov 1993; Harvey and Purvis 1999). How interspecific variability in traits defining life history rates is predictive of behavior remains to be examined in detail.

Primate Life Histories

Primate life histories are among the slowest among mammals (Harvey and Clutton-Brock 1985; Harvey, Martin, and Clutton-Brock 1987; Ross 1988; Charnov and Berrigan 1993; Lee 1996; Ross 1998). Primates' birth, growth, and death rates, in particular, are substantially lower than those of other mammals after controlling for differences in body size. Specifically, primates have relatively long gestation lengths, large neonates, low reproductive rates, slow postnatal growth rates, late ages at maturity, and long life spans in comparison to other mammals (Martin and MacLarnon 1988; Charnov 1991; Harvey and Nee 1991; Lee, Majluf, and Gordon 1991; Ross 1992a; Charnov and Berrigan 1993).

Whereas ultimate causes for the low developmental and reproductive rates of primates remain poorly understood (Charnov and Berrigan 1993), three main determinants have been suggested. First, the energetic costs of growing and maintaining primates' large brains have been suggested as constraints on these rates (Sacher and Staffeldt 1974; Sacher 1975; Armstrong 1983; Allman, McLaughlin, and Hakeem 1993a; Martin 1996). Theoretical objections and comparative data make this link unlikely, however (Harvey, Promislow, and Read 1989; Charnov and Berrigan 1993; but see modified view of Deaner, Barton, and van Schaik, chap. 10, this volume). Second, high juvenile mortality risk has been proposed as an ecological explanation for the comparatively slow growth of primates (Janson and van Schaik 1993), but the limited tests of this hypothesis to date are not consistently favorable (Ross and Jones 1999b; Godfrey et al., chap. 8, this volume). Finally, the arboreal lifestyle characterizing primates may have permitted the evolution of slow life histories (Eisenberg 1981; Martin 1995); recent tests both within primates and among mammalian orders do suggest a systematic link between arboreality and slow life histories (van Schaik and Deaner 2002).

The pronounced slow-fast continuum of life speed among mammalian orders is also found among primates. Small prosimians are sexually mature after less than one year of rapid growth, ultimately giving birth to two to four young once or twice annually, whereas great apes begin producing slow-growing singleton infants at four- to eight-year intervals after reaching seven to fourteen years of age (fig. 1.1). Such life history contrasts are pronounced between the prosimian and anthropoid grades and persist at lower taxonomic levels (e.g., among lemur and New World monkey families and genera: Harvey and Clutton-Brock 1985; Martin and MacLarnon 1985, 1988; Ross 1991; Kappeler 1995, 1996). Other traits more or less directly linked to life history variables, such as body size, brain size, metabolic rate, mode of infant care, habitat use, and diet, are also extremely diverse within the primate order (Leigh 1992c; Ross 1992a; Allman, McLaughlin, and Hakeem 1993a; Kappeler and Heymann 1996; Smith and Jungers 1997; Kappeler 1998a). This diversity in life history, in addition to their well-known diversity in social systems, makes primates an important taxon with which to examine relationships among life histories, ecologies and social behavior.

In sum, most primates have slow life histories, and thus long lives, and

FIG. 1.1. The order Primates encompasses tremendous diversity in life history and behavior, relating fundamentally to body masses distributed across four orders of magnitude. While adult female pygmy mouse lemurs *(Microcebus berthae)* weigh 30 g on average, adult female gorillas *(Gorilla gorilla)* typically weigh about 80,000 g. (Illustration by S. Nash; © by J. Fleagle.)

tend to produce, at long intervals, small litters of slow-growing young with long infant dependencies and also long juvenile periods of partial dependency. What relationships exist among these life history traits, ecological patterns, and social behavior?

Life History Effects on Socioecology

Life history traits should relate to social and ecological aspects of primate behavior in a variety of important ways. Progressing through the life cycle, we provide an overview of some demonstrated and potential links.

Prenatal and Postnatal Development

Durations of gestation and lactation among primates vary in duration between two and nine months and between two months and several years, respectively (Harvey, Martin, and Clutton-Brock 1987; Lee 1996). Maternal body size and litter size explain much, but not all, of this variation (Martin and MacLarnon 1985, 1988). The duration of postnatal maternal care is relatively difficult to quantify because weaning in primates is such a gradual process. Nonetheless, with appreciable variability within and among taxa, weaning appears generally to occur around the time when infants have reached about one-third of adult body mass (Lee, Majluf, and Gordon 1991; Lee 1996).

The relative lengths of prenatal and postnatal maternal investment have important consequences for behavior, especially for male reproductive strategies. Specifically, the ratio of lactation time to gestation time is a strong predictor of vulnerability to infanticide by males among eutherian mammals, including primates (van Schaik and Kappeler 1997; van Schaik 2000b), because females with relatively long lactations tend to undergo postpartum amenorrhea (i.e., lack postpartum estrus) to avoid concurrent gestation and lactation (van Schaik 2000b). Loss of an infant is unlikely to accelerate females' resumption of ovulatory cycling when lactation is brief relative to gestation. But when lactation is relatively long, infant loss generally brings females back into receptivity sooner, except in cases in which breeding is highly seasonal (van Schaik 2000a; cf. Jolly et al. 2000). Thus, males who have not sired current offspring but are in a good position to mate with females benefit from killing infants. Because weaning of primate infants occurs relatively late (Lee, Majluf, and Gordon 1991), primates have some of the highest lactation/gestation ratios among mammals and are therefore especially prone to sexually selected male infanticide (van Schaik 2000a), even in some taxa with seasonal reproduction (Jolly et al. 2000).

The basic set of conditions under which male infanticide is adaptive has been known for many years (Hrdy 1979), but the evolutionary consequences of infanticide for social behavior have been examined only recently. For example, risk of infanticide has been suggested to have selected for female gregariousness (Brereton 1995), year-round male-female association (van Schaik and Kappeler 1997), male-female friendships (Palombit, Seyfarth, and Cheney 1997), complex male-infant relationships (Paul, Preuschoft, and van Schaik 2000), multi-male group membership (Crockett and Janson 2000), female transfer and choice of target group (Steenbeck 2000; Sterck and Korstjens 2000; Watts 2000), female social and sexual behavior, and other aspects of female reproductive biology (van Noordwijk and van Schaik 2000; van Schaik, Hodges, and Nunn 2000). Although many of these particular hypotheses need to be tested more extensively, this unfolding view of the effect of infanticide risk on primate socioecology is based on a few basic life history features.

Litter Size and Degree of Infant Development

Litter size varies between one and four among primates, with larger litters very uncommon (Leutenegger 1979). The majority of primates has single infants; twins are common in callitrichids and some strepsirrhines, and triplets and quadruplets occur rarely in a few lemur taxa, at least in captivity. Mean litter size for wild primates therefore rarely exceeds two (Leutenegger 1979; Chapman, Walker, and Lefebvre 1990; Kappeler 1998a).

Litter size is broadly correlated with degree of infant development at birth, ranging from altricial, where infants are born with closed eyes and are unable to grip or climb for days or weeks, to precocial, where infants are alert and able to cling from birth (Portmann, 1939; Martin and MacLarnon 1985, 1988; Martin 1990). Altricial infants tend to be born in litters of two or more and spend the first days, weeks, or even months of their lives in nests, tree holes, or other shelters that protect them against inclement weather, predators, and hostile conspecifics, whereas most singletons are carried from birth by an adult, typically the mother (Kappeler 1998a). Only some prosimian singletons, as well as older infants from prosimian litters, are regularly parked in vegetation while their mothers forage.

Different modes of infant care have important consequences for the behavior of caregivers and, in turn, social organization. First, infants parked or left in a shelter force a mother to return to the same site repeatedly to nurse or move the infants. Dependent young should therefore constrain the ranging of lactating females. Some tree shrews, rabbits, and seals, which nurse their young only at intervals of one or more days (Martin 1968, 1990), demonstrate that this constraint can be overcome by producing large quantities of rich milk, and *Varecia* milk has been shown to be richer than that of other lemurids that carry their infants (Tilden and Oftedal 1997). Because detailed documentation of the movements of female primates that park young, during and apart from lactation, are still lacking (but see Morland 1990), however, it is difficult to evaluate fully the factors associated with changes in maternal behavior or milk qualities in primates.

Second, females that leave infants behind have more difficulty synchronizing their activities and movements with those of other females; therefore,

the presence of cached young ultimately diminishes the chances for formation of cohesive groups. Indeed, among living primates, we find a nearly per-
fect correlation between mode of infant care and female gregariousness
(van Schaik and Kappeler 1997). The sole group-forming primate that bears
litters *(Varecia variegata)* forms only small groups, whose members characteristically spend much time alone (Morland 1990) and exhibit unelaborated
social behaviors (Pereira, Seeligson, and Macedonia 1988). It is rare for two
females to reproduce simultaneously in *Varecia* groups (Morland 1990), and
males and nonreproductive females in this species take turns with mothers
to guard nests and parked young (Pereira, Klepper, and Simons 1987; Morland 1990). On the other hand, some other mammals, including many carnivores, appear to be less constrained in their ability to form permanent
groups despite the presence of litters of altricial young, which are often kept
in dens (Gittleman 1986b; Geffen et al. 1996). Reliance on transportable
high-energy food and cooperative infant care by several mothers and other
helpers may explain sociality in these taxa (Bekoff, Diamond, and Mitton
1981; Bekoff, Daniels, and Gittleman 1984; Pusey and Packer 1994). Taken
together, the extant patterns suggest that singleton young may have been
necessary for the evolution of permanent groups in primates.

A related observation in this context is that female mammals with altricial young show territoriality, whereas those with precocial young generally
do not (Wolff 1993, 1997). This link may have evolved because, based on resource competition, strange females would benefit from committing infanticide in order to usurp the nest site containing the altricial young. Although
this hypothesis is not uncontested (Tuomi, Agrell, and Mappes 1997) and is
yet untested for primates, it is another example of a possible social effect of
life history differences; namely, differences in litter size and developmental
rate.

Reproductive Rates

Among mammals with slow life histories, reproductive rates are lower because intervals between subsequent reproductive events are longer. As a
result, the number of females in a group or population that are ready to mate
at any time is greatly reduced. The deceleration of female reproductive rates
therefore has a direct effect on the operational sex ratio, and thus on the
mating system.

Operational sex ratio (OSR) determines opportunity for sexual selection (Clutton-Brock and Parker 1992); specifically, higher OSRs impart
greater opportunities for sexual selection on males. Although OSR is sim-

ply defined, in relation to the numbers of males and females ready to mate, the measure is not easily estimated. Mitani, Gros-Louis, and Richards (1996) used the following equation for primates:

$$OSR = m/f * B * d * (1/\Sigma c)$$

where m = number of potentially reproductively active males; f = number of potentially reproductively active females; B = interbirth interval in years; d = length of mating season (365 days in nonseasonal breeders); Σc = summed estrous periods before a conception (mean number of days per cycle that there is mating times number of cycles before conception).

Species with longer interbirth intervals have higher OSRs. Moreover, the opportunity for sexual selection depends on the number of males relative to the number of females, as well as on the number and length of estrous periods. Both of these features may represent the consequences of sexual selection in the past, however, and may therefore not be the best estimate of the current potential for sexual selection. For instance, the low OSRs of species that form one-male groups result partly from the violent exclusion of other males from the group by a single dominant individual (e.g., Rajpurohit, Sommer, and Mohnot 1995; Watts 2000). Although OSR is documented strictly at the population level (Kvarnemo and Ahnesjö 1996), these considerations help to illustrate challenges pertaining to OSR in structured populations.

Another factor biasing OSR is male maturation time. Many primates living in one-male groups are highly sexually dimorphic, which sometimes entails longer male maturation times (see Pereira and Leigh, chap. 7, this volume), thus biasing the sex ratio toward females compared with more monomorphic species (Alberts and Altmann 1995; Leigh 1995). Furthermore, a low OSR would arise from many mating days per conception. Pair-living gibbons, for example, have a few short estrous periods per conception, whereas baboons, which live in multi-male societies with intensive male-male competition, have about as many cycles, but each cycle has many mating days (Hrdy and Whitten 1987). Each pattern reflects female mating tactics, which evolve partly in relation to infanticide risk (van Schaik, van Noordwijk, and Nunn 1999).

Thus, several of the variables used to estimate the opportunity for selection on male traits are problematic, sometimes even showing correlations opposite to expectations, because their current values represent the consequences of selection. Hence, the best predictor of the intensity of male-male competition may be female reproductive rate. Large females, and otherwise slow-living ones, have low reproductive rates; therefore, their males are likely to be relatively large and polygynous. And, all else being equal, the potential for male sexual coercion should be high (Smuts and Smuts 1993). Future comparative research will reveal how well these predictions explain variation in male reproductive strategies among primates and other large mammals.

Extended Juvenility

Whereas many mammals essentially lack a juvenile phase of life history (from weaning until sexual maturation: Pereira 1993b), primates have life histories that include months or years during which individuals are independent but infertile members of their "neighborhoods" (nongregarious taxa) or social groups. Because primate somatic growth is slow (Case 1978; Kirkwood 1985) and because juveniles depend on experience to refine their basic behaviors (Poirier and Smith 1974; Janson and van Schaik 1993), primates spend large parts of their lives as small and significantly inexperienced individuals unable to reproduce.

One behavioral consequence of primates' slow growth and dependence on learning seems immediately apparent in the gregariousness of these mammals. As vulnerable as many adult primates are to predators, their slow-developing offspring are even more vulnerable, and primate group living may have evolved principally as a mechanism to safeguard vulnerable youngsters (Pereira and Fairbanks 1993a; see esp. Janson, chap. 5, this volume; also van Schaik and Kappeler 1997).

In turn, slow development and gregariousness have influenced primate behavior in myriad ways. Many anthropoids seem to "bootstrap" their development behaviorally; young juvenile males and females, for example, commonly prefer to affiliate with older groupmates of their own sex, facilitating the development of sex-typical behavior (Pereira 1988b; Edwards 1993). Also, juveniles target and are targeted by particular adults for the development of affiliative or adversarial social relationships (Silk et al. 1981; Pereira 1988a; Fairbanks 1993; Pereira 1995). For juveniles, certain alliances enhance their chances of reaching adulthood (e.g., Pereira 1988b), while others seem to be initiated to promote eventual adult success (Fairbanks 1990, 1993).

Many large anthropoids not only grow slowly, but actually postpone much somatic growth until puberty, whereupon sharp accelerations lead to rapid attainment of adult size (Pereira and Leigh, chap. 7, this volume). In some taxa, males and females both show this pattern, while in others, only males do. The significance of these prereproductive growth rate changes has hardly begun to be explored, but slow growth would help youngsters of a given species resolve whatever challenges may characterize particular developmental phases by maximizing the resources available to support their efforts. In populations in which both sexes defer somatic growth, individuals' large brains may be protected by ensuring the availability of resources to support their delicate development (Deaner, Barton, and van Schaik, chap. 10, this volume). If only male somatic growth is deferred, this may, instead, reflect primarily the advantages of a "crypticism" that helps small males to avoid sexually selected aggression from fully established males (Jarman 1983; Leigh, 1995; Maggioncalda et al. 2000).

It will be necessary to interweave, within and across taxa, detailed research on behavior, neurophysiology, metabolism, and ecology to evaluate such ideas formally. This work will be important, however, as developmental schedules can be expected to coevolve with patterns of juvenile-adult social relationship, and thus should have far-reaching consequences for primate social systems. Pereira (1995), for example, suggested how different timings and extents of pubertal growth spurts may have affected social relations between the sexes among papionin primates. In baboons and certain macaques, males rapidly increase in body mass 75–110% around puberty. Whereas male baboons begin rapid growth after having attained about 80% of adult female size, male macaques begin this growth spurt at smaller relative sizes. Also, female baboons do not undergo adolescent growth spurts, whereas some female macaques do. Rapid growth, size-dependent reproductive success, and correlations between adult size and prior rates of food intake (J. Altmann and S. C. Alberts, unpub.) all make food supply paramount for maturing males. And, because growth spurts begin in baboons when males are relatively large, male baboons can compete effectively with females for food. One consequence seems to have been that female baboons, unlike females in several macaques, have evolved a general disinclination to compete agonistically with males, and this, in turn, has probably contributed to divergences in male-female social relations, including friendships or "special relationships," among these species (Pereira 1995).

Species-typical behavioral tendencies turn our attention to brain evolution. Primates' large brains clearly relate to their slow development and related behavior, probably as cause and as effect. Juveniles' single most prominent behavior, play, appears pivotally linked to aspects of neuronal development, for example (Pereira and Leigh, chap. 7, this volume). Synaptogenesis can be modified by experience only soon after birth, and, across mammals, schedules for play behavior correlate closely with windows of opportunity in the development of both the cerebrum and cerebellum (Byers and Walker 1995; Fairbanks 2000). Expanding schedules for play from prosimians, through monkeys, to the apes suggest that the coupling of play and neuronal development contributes to the development of phyletic differences in behavior (Fairbanks 2000). Future research can be expected to reveal how a wide variety of juvenile behavioral initiative and other experiences help to sculpt neuronal circuitry in accord with species-typical life histories.

Finally, primates' long juvenilities and large groups enhance prospects for the evolution of modular plasticity of physiology and behavior by ensuring that individuals within populations encounter wide ranges of possible conditions during development (Pereira and Leigh, chap. 7, this volume). Variability among juveniles, adolescents, and young adults should be investigated longitudinally as possible differential expressions of reaction norms (Roff 1992; Stearns 1992; Schlichting and Pigliucci 1998). In return, information on taxonomic variation in such plasticity (Lee and Kappeler, chap. 3, this volume) will shed light on questions about primate phylogeny (see Pereira and Leigh, chap. 7, this volume).

Adult Life Span

A final unusual aspect of primate life histories is their relatively long life spans, compared with most other mammals, varying between five years in mouse lemurs to more than one hundred years in humans (Austad and Fischer 1992; Allman, McLaughlin, and Hakeem 1993a). The consequences of extended adult life for primate behavior have not been thoroughly explored, despite a variety of possible links.

Long lives should be associated with some social complexity because members of long-lived species have more time to differentiate their social relationships. Also, escalated fights should be less common because gravely negative outcomes represent greater penalties for the long-lived (Janson, chap. 5, this volume), who also enjoy greater chances to queue for reproductive opportunities and to reciprocate altruism and aggression over extended time periods (de Waal and Luttrell 1988; de Waal 1989; Aureli et al. 1992; Harcourt 1992). Increased social complexity and the need to manage social relations over many years may have selected for increased cognitive capacities (Cheney, Seyfarth, and Smuts 1986; de Waal 1991; Byrne 1995a; Dunbar 1998; Dunbar, chap. 12, this volume). In turn, unusually long lives and large brains may facilitate the evolution of other components of behavioral complexity, such as compensatory maternal adjustments across reproductive efforts (Fairbanks 1988a), grandmothering (Fairbanks and McGuire 1986; Fairbanks 1988b; Hawkes et al. 1998; Hawkes, O'Connell, and Blurton Jones, chap. 9, this volume), tradition, and culture (McGrew 1998; van

Schaik, Deaner, and Merrill 1999; Whiten et al. 1999; Tomasello 2000). Most of these potential consequences of long lives need further substantiation, but none of them are expected in animals with fast life histories.

Consequences of Socioecology for Life History

Behavioral variation is, theoretically, expected to affect life histories at two levels. First, individuals can modulate life history traits in relation to changes in their own behavior, social status, social organization, or the mating system (e.g., Rubenstein 1993). Second, these and other links can become canalized over evolutionary time, leading to relatively firm coupling of particular trait combinations. In general, both levels of this relationship remain poorly explored compared with effects in the opposite direction, such as those discussed above.

At the level of reaction norms, changes in behavioral or social variables can have qualitative or quantitative effects on life history traits. First, individual differences or changes in ranging, foraging, or antipredator behavior can have pronounced effects on reproductive effort or survival. At this level, however, behavior is typically the mechanism mediating the effects of ecological variations, such as food availability and predation risk, on life history traits, and not the ultimate cause itself. The kinds of primate life history traits affected by such changes, as well as their reaction norms, have been summarized by Lee and Kappeler (chap. 3, this volume).

Second, reproductive activity is often affected by social dominance, and is therefore subject to change along with social status across an individual's lifetime. Apart from the indirect effects of social status on reproductive success, such as rank-dependent access to resources, direct effects are also known. Reproductive inhibition of subordinates is an important example because reproductive rates in such cases are qualitative responses to social status. Among primates, both male and female inhibition, in which subordinate individuals develop reproductive capacity as soon as a dominant breeding individual has disappeared (e.g., Abbott 1989; Perret 1992; Maggioncalda, Sapolsky, and Czekala 1999), are known. Also, rank-dependent variation in reproductive success has been documented among females of several group-living species (e.g., Wolfe 1984; van Noordwijk and van Schaik 1999) and among males in many species (e.g., Bercovitch 1986; Cowlishaw and Dunbar 1991; de Ruiter and van Hooff 1993). Finally, infant sex ratio is often biased as a function of maternal social status. Overproduction of either males or females may be adaptive for high-ranking or low-ranking females under different competitive scenarios (Clark 1978; Silk 1983; Perret

1990; van Schaik and Hrdy 1991; Dittus 1998; Nunn and Pereira 2000; Pereira and Leigh, chap. 7, this volume).

Third, primate life histories can be modulated by group size and composition, which varies due to chance, habitat structure, natality, mortality, migration, and group fission or fusion. Group size can affect reproduction, especially in low-ranking individuals, because of its effects on feeding competition; reduced net food intake in larger groups slows individual or average reproductive rates (Dunbar 1988; Sterck, Watts, and van Schaik 1997). In some species, female group size or matriline size per se can also affect offspring sex ratios (Perret 1990; Nunn and Pereira 2000; Pereira and Leigh, chap. 7, this volume). A qualitative change in group composition can be achieved by modifying the number of adult males, and such a change can alter the mating system. Such changes can induce, or be induced by, changes in female reproductive parameters and behavior (Hamilton and Bulger 1992; van Schaik, van Noordwijk, and Nunn 1999), and in some cases, additional males demonstrably improve infant survival (Heymann and Soini 1999; Heymann 2000; Watts 2000).

A particular behavioral tendency or aspect of a social system can promote corresponding evolutionary changes in life history traits. Because only a few phylogenetically based analyses demonstrating such links have been conducted, the following examples are only suggestive, offered in the hope of stimulating further investigations.

First, females' inclinations to allow nonmaternal caregivers access to their dependent offspring (Goldizen 1987; Fairbanks 1990; Stanford 1992) affect several life history parameters. Species showing substantial allocare show higher postnatal growth rates, and thus attain weaning earlier, than species in which mothers alone care for their offspring (Ross and MacLarnon 1995; Mitani and Watts 1997; Ross 1998; Ross, chap. 11, this volume). Earlier weaning decreases interbirth intervals, increasing female reproductive rates.

Second, social factors such as mode of affiliation and discrete antipredator behaviors such as alarm calling and communal defense should affect levels of unavoidable mortality and, consequently, life history. Age-specific mortality schedules are thought to provide the strongest selective effects on speed of life history (Promislow and Harvey 1990). Predation, as a major source of extrinsic mortality, should therefore be an important link between behavior and life histories, potentially explaining other correlations of life history traits with size and activity period (Janson, chap. 5, this volume).

A final example is provided by the energetically most costly form of paternal care among primates (Sanchez et al. 1999): infant carrying in twinning callitrichids. An initial transition from nongregariousness, with roving males and polygamy, to monogamy has been argued to have occurred in the Callitrichidae, perhaps to decrease the risk of infanticide (Dunbar 1995a). Pair-bonded males, now investing in young, permitted the evolution of twinning, additionally augmenting rates of reproduction. The subsequent loss of lactational amenorrhea contributed further to these maximal reproductive rates among primates, completing an evolutionary cascade that began with a change in the social system (see also fig. 4.8 in van Schaik 2000a).

Coevolution of Life History and Behavior

The foregoing examples provide substantial evidence for proximate and ultimate links between behavioral and life history traits. The general nature of such links remains to be illuminated, however, alongside the mechanisms of specific relationships. For clarity's sake, we have presented case studies as though changes in particular behavioral traits caused changes in corresponding life history traits, or vice versa. However, we do not know whether behavior drives life history evolution or whether it is typically the other way around. Alternatively, traits inevitably influence one another reciprocally, and cohesive combinations result from whatever selective regime, favoring particular adjustments of life history or behavior, affects a taxon at a particular juncture. The view that changes in one category typically precede and cause changes in the other begs evidence that, generally, traits in the former class are evolutionarily more labile.

Behavioral traits are often considered evolutionarily more labile than morphological or physiological ones (e.g., Mayr 1963; Lorenz 1965; West-Eberhard 1989), with behavioral evolution preceding physical evolution (Wcislo 1989; Basolo 1990; Arnold 1992; Gariépy, Lewis, and Cairns 1996). Also, behavioral mechanisms have been identified that could mediate physical change (Bateson 1988). On the other hand, rapid evolutionary change in some morphological and physiological traits is well documented (James 1983; Grant 1986), and in some cases, physical changes may have been constrained by behavioral inflexibility (Huey 1991). So, broad qualitative comparisons are of limited help in deciding this issue.

Because morphological traits are much better preserved in the fossil record than behavioral ones, it has only recently become possible to address this question more directly, with the aid of newly developed molecular phylogenies and comparative tests (Brooks and McLennan 1991; Gittleman et al. 1996a). To our knowledge, however, only one study has used these methods to investigate the relative lability of behavioral, morphological, and life history traits across mammals. Evaluating eight higher mammalian taxa, Gittleman et al. (1996b) took group and home range sizes as behavioral traits, body and brain sizes as morphological traits, birth mass and gestation length as life history traits, and degree of correlation with phylogenetic distance as evolutionary lability. The results of their comparative analysis suggested that these behavioral traits were evolutionarily more labile than the life history traits, which in turn appeared more flexible than the morphological traits. Thus, "the net rate of evolution in the behavioral traits may be different and may change more randomly over time than the other traits" (Gittleman et al. 1996b).

More analyses, examining more trait relationships and using more methods, will be required before more definitive conclusions can be drawn in this regard, however. Beyond the small samples used, Gittleman et al.'s (1996b) interpretation is complicated by the fact that the degree to which group and home range sizes constitute *individuals'* behavioral traits is confounded with degree of gregariousness across taxa. Also, whereas the evidence suggests that behavioral and life history traits may show different evolutionary rates, this does not translate usefully into a concept of *random* behavioral change.

To determine whether one class of phenotype characteristically evolves at different rates than another, we will need to understand better what constitutes a meaningful unit in each phenotypic class, a challenge that has historically hounded students of behavioral sampling and of homology. Behaviorists, for example, examine everything from the ontogeny of simple movement patterns (e.g., scratching) to levels of locomotor activity, circadian rhythms, shyness-boldness spectra, learning patterns, patterns of dispersal, detailed aspects of social behavior, and classes of social systems. At which level, if any, across taxa does behavior most consistently respond in an integrated manner to selection pressure? What are the consequences, if any, of bias in our selections of morphological traits for study toward gross developmental end products?

Entertaining the notion that certain levels or domains of phenotypes serve more importantly than others as targets for selection risks losing sight of the fact that entire, integrated organisms, not body parts, are selectively favored or disfavored. Nonetheless, we seek major effects, at least in particular comparisons. A bit of our own work provides an example. Whereas ringtailed *(Lemur catta)* and brown lemurs *(Eulemur fulvus)* are closely related, partially sympatric, and extremely similar morphologically and physiologically (Kappeler 1997; Pereira et al. 1999), their social systems have diverged radically (Kappeler 1993; Pereira and Kappeler 1997). Ringtails are relatively diurnal, use dominance behavior, show female philopatry and bonding among close female kin, and form moderately large social groups. Brown lemurs, by contrast, are active during the night and the day, do not use dominance, show bisexual dispersal and male-female bonding, and form smaller social groups. Ringtailed lemurs probably shifted much of their activity to the daylight hours (van Schaik and Kappeler 1993) before evolving their system of ritualized social dominance (Pereira and Kappeler 1997). But in many other ways, ringtails seem to have remained fundamentally the same mid-sized primate that, like *Eulemur* spp., ravenously feeds, grows, and fattens during summer and fall and uses stored resources during winter (Pereira et al. 1999). Myriad neuronal and hormonal differences that support the divergent patterns of aggression, affiliation, and reciprocity (Kappeler 1993, 1998b; Pereira and Kappeler 1997; Pereira and McGlynn 1997) and fine aspects of nutritional ecology (Ganzhorn et al., chap. 6, this volume) in these two closely related lemurs doubtlessly remain to be discovered, however.

Work on animal domestication provides evidence of strong canalization of basic patterns of behavior and of their pleiotropic linkages with physiological and anatomical traits. In reviewing the history of domestication, Diamond (1997) addresses why only 10% of the world's large terrestrial herbivorous mammals have ever been domesticated successfully, even after extensive efforts, the most recent of which have applied modern practices of animal management and genetics. Three of Diamond's six reasons for these failures are distinctly "behavioral" (dangerous disposition, tendency to panic, obstructive social structure), while two others (diet and failure to breed) derive from close relations between behavior and physiology. Diamond's account makes clear that the *inflexibility* of fundamental behavioral patterns, at least those pertaining to interaction with humans, is what has precluded domestication for 90% of the world's large terrestrial herbivores.

Complementing Diamond's account is a remarkable but little-known 40-year experiment in selective breeding with wild canids (reviewed by Trut 1999). Beginning in 1959, Belyaev (1969) and co-workers investigated the history of *Canis familiaris* by asking whether selection solely for behavioral tractability in relationship with humans could have led to the sorts of workers and companions we have among domestic dogs today. The behavioral obstacles identified by Diamond loomed large at the outset of Belyaev's project, which used silver foxes *(Vulpes vulpes)*. Ninety percent of the wild individuals first brought into the project died from stress, unable to cope with captive conditions. Even the few survivors' descendants were not domestic dogs, remaining averse to human company and dangerous for humans to try to touch (Trut 1999).

Forty years later, full domestication has been achieved. Most remarkable in light of present concerns, however, is the host of changes in anatomy, physiology, and unselected patterns of behavior that derived from the strict regime of selection solely for docility. These included changes in endocrine glands, timings and levels of circulating hormones, pelage characteristics, aspects of ear and tail carriage, patterns of vocalization, and offspring sex ratios (Trut 1996, 1999). Belyaev and colleagues (1981; Trut 1999) noted that particular changes in several of these traits are characteristic of all mammalian species undergoing domestication, suggesting a basic pleiotropy across these vertebrate taxa. Belyaev suggested that genetic transformations deeply affecting behavior should be difficult to achieve and of paramount importance, entraining various other genetic events throughout development (Trut 1999). Genes controlling neuronal and hormonal functioning, he pointed out, help importantly to regulate not only behavioral ontogeny, but all ontogenesis, from its earliest phases.

In sum, the results of human efforts to domesticate animals suggest that animals' repertoires of behavior typically lack the variability or flexibility needed for success, whereas, given success, diverse anatomical and physiological changes typically accompany selection for particular behavioral traits. In addition, life history traits as fundamental as body size are generally easy to change, relatively independently, via selection (see also Deacon 2000).

Summary and Conclusions

The diversity of primate social systems has traditionally been explained as the result of adaptations to variable ecological factors, particularly the distribution of resources and the risk of predation. These links have been formalized by the socioecological model, and empirical and comparative tests have supported many of its specific predictions. In recent years, however, it became clear that several aspects of male-female relations, in particular, could not be satisfactorily explained within the existing theoretical framework. Incorporating sexual conflict and its underlying life history traits suggested a promising extension of the socioecological model. Using primates as an example, we have explored possible links between salient life history traits and aspects of the social system. Our review has identified numerous links between life history and social variation, including viable hypotheses on causality. These findings support our suggestion that life history traits be incorporated into the socioecological model, where they should stimulate experimental and comparative tests of specific links. Clearly, consideration of life history evolution will continue to provide insights into behavioral ecology (and vice versa), as well as into other disciplines of population and organismal biology (Stearns 2000).

Acknowledgments

We thank Dietmar Zinner for helpful comments on an earlier version of this chapter.

PART ONE

Life History and Socioecology

The chapters in this first section examine patterns and causes of variation in primate life histories. Purvis and colleagues (chap. 2) provide an up-to-date review of life history theory and examine the relationships between phylogeny and life history variation. Their review of current mammalian life history models indicates that neither slow growth nor low adult rates of mortality provide a complete explanation for primates' slow life histories. Using modern comparative methods and information on phylogenetic history, Purvis et al. go on to show that the rate of life history evolution has differed among the major primate clades and that different lineages have moved along the slow-fast life history continuum differently—that is, via adjustment of different life history traits. A look in the other direction—at whether characteristic life history strategies have affected species richness among major taxa (via effects on rates of speciation or extinction)—reveals no strong or consistent relationships.

Lee and Kappeler (chap. 3) focus on another aspect of life history variation; namely, variation within species. So-called phenotypic plasticity—environmentally induced flexibility of traits within a species—is useful for investigating links between ecology and life history. It is interesting to ask, for example, which external variables most influence life history traits and whether all traits within a species exhibit comparable flexibility. Given the difficulties of collecting, and compiling, the necessary data for several populations per species, Lee and Kappeler's results, while preliminary, are encouraging. They include, for example, evidence that some plasticity in life history traits may trade off against overall life speed in an unexpected way, resulting in certain traits being more flexible in relation to environmental variation in smaller than in larger species. Also, particular analyses suggest interesting roles for plasticity in early growth in response to ecology, social-

21

ity, and reproductive strategy. The observation that strepsirrhines and haplorhines appear to exhibit similar degrees of plasticity along the dimensions
of different life history traits supports a similar conclusion drawn in the earlier chapter by Purvis and colleagues.

A major remaining gap in the study of primate life histories concerns
the systematic and standardized quantification of major life history traits in
study populations. Alberts and Altmann (chap. 4) fill this gap by providing a step-by-step introduction to the demographic and statistical methods
that yield the necessary data. They advocate the use of matrix models because of certain theoretical advantages and because these models can deal
with the limits of the demographic data available from many primate field
sites. Moreover, selection pressures can be explored via perturbation procedures, making matrix methods excellent guides for further research on
sources of fitness variation. The authors' theoretical discussion is amplified
by examples from their own exemplary fieldwork on baboons and by references to important practical applications of these methods (e.g., in conservation biology).

The chapter by Janson (chap. 5) was inspired by the puzzling observation that primate species with putative adaptations to high predation risk,
such as large body size or large group size, tend to have relatively low predation rates. Janson uses a mathematical model to examine the possibility
that these puzzles may be explained by life history variables. By varying different life history variables and selective forces independently of each other,
he identifies interspecific variation in longevity as the trait that can explain
all of the observed puzzles, assuming that predation risk varies little across
living primates. Accordingly, if longevity increases with increasing body size,
larger species will pay the costs of predation over more years than will small
species and thus will gain a larger fitness benefit by reducing predation risk
(e.g., by increased sociality). The benefits to small species of reducing predation risk are limited by their short life spans. Thus, this model can also explain why primate group size increases with body size. Janson's work contributes not only important theoretical progress toward understanding the
roles of predation in primate evolution, but also another example of the in-
timate evolutionary interplay between life histories and social systems.

The final chapter in this section (chap. 6) reviews the growing evidence
of adaptations to environmental seasonality among primates. Although primates have a predominantly tropical distribution, many taxa are exposed to
significant seasonal variation in their habitat, which poses potential problems for basic maintenance and the timings of growth and reproduction.
Ganzhorn and colleagues begin at the community level, where previous

work suggests that primate populations are not constrained by seasons of food scarcity, contrary to long-standing assumptions. This inference is supported by comparative studies from dry and wet forests and results from methodologically sophisticated fieldwork examining the most extreme adaptation to seasonality: torpor. Ganzhorn et al. next sample anatomical and physiological mechanisms promoting survival of lean seasons in non-primates, underscoring that, while a variety of similar mechanisms presumably exist among primates, they remain virtually unexplored.

In sum, the chapters of part 1 introduce clearly a great variety of approaches to research pertaining to primate life histories. Along the way, the contributing authors highlight a cornucopia of major issues requiring further attention, providing many foci for upcoming research on interactions among factors of ecology, behavior, and life history.

2 Primate Life Histories and Phylogeny

Andy Purvis, Andrea J. Webster,
Paul-Michael Agapow, Kate E. Jones,
and Nick J. B. Isaac

Primates are a difficult group in which to study life history. They tend to live in inaccessible places, at low population densities, and (worst of all) for a long time. Good estimates of natural vital demographic rates therefore require long and careful field studies. Due to their undoubted charisma and their particular fascination for us, primates have nonetheless attracted a disproportionate number of such studies and so provide a rich seam of comparative data that has long given us insight into their evolutionary ecology (e.g., Harvey and Clutton-Brock 1985; Ross 1988; Kappeler 1996). Nonetheless, most of the early data came from more tractable mammalian groups such as rodents and ungulates (Millar 1977; Western 1979), providing the initial grist for the life history theorists' mill (Millar and Zammuto 1983; Harvey and Zammuto 1985; Sutherland, Grafen, and Harvey 1986) and the result that most of the patterns and issues that emerged in broad-scale overviews of primate life history evolution, such as the pervasive "fast-slow continuum" (Read and Harvey 1989), were already known from studies of mammals as a whole. We therefore start with a selective review of some of these patterns before outlining the most recent models designed to provide a comprehensive explanation of them (for a historical review of the interplay between theorists and empiricists in this area, see Harvey and Purvis 1999). We then consider how data from primates fit the models' predictions.

The "speed" of a species' life history is a central concept in these models. It is a combination of several parameters—age at first reproduction, gestation length, litter size, and interbirth interval (Stearns 1992). How have changes in the speed of life evolved? We report the preliminary results of an ongoing look at the rates of evolution of these traits and body size in primates and carnivores, which finds evidence of a complex mosaic pattern: different traits show different patterns, and rates have varied between and within major clades. Primate litter size has evolved especially slowly.

Last, we invert the usual phylogenetic comparative approach, which views phylogeny as a nuisance to be "got rid of" in order to make sense of trait covariation. Several workers have hypothesized that life history should be an important determinant of a key attribute of mammalian phylogenies: species-rich versus species-poor lineages. Some primate clades, notably the Old World monkeys, are very species-rich for their age (Purvis, Nee, and Harvey 1995), whereas others are very species-poor. Can life history differences predict these patterns? We give a progress report on our current work, first outlining how our phylogenetic approach is an improvement over previous tests of such hypotheses.

Life History Patterns in Mammals

The most obvious generalization about life history in mammals is that larger species live slower lives (fig. 2.1). Larger species mature later, have a longer gestation, have smaller litters, wean them later and at a larger size, and have longer gaps between births. These differences combine multiplicatively to produce prodigious overall differences in reproductive potential: a newly founded population of a typical small mammal could reach its carrying capacity within the time required for a single generation of a large species (Purvis and Harvey 1996).

Although body size correlates with much of the interspecies variation in

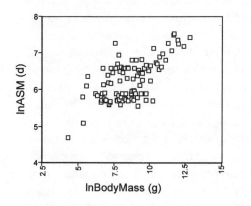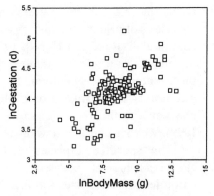

FIG. 2.1. The relationship between two life history variables—age at sexual maturity (ASM) and gestation length—and adult body mass across a range of mammalian species from several different orders. (Data from Purvis and Harvey 1995.)

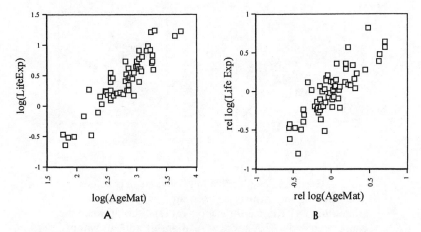

Fig. 2.2. The relationship between two components of the speed of life (life expectancy at maturity and age at maturity) across a range of mammalian species. *(A)* Species values. *(B)* Partial correlation controlling for body size effects. The relationship holds when phylogenetic comparative methods are used. (Data from Purvis and Harvey 1995.)

life history, it is not the whole story: components of the speed of life are still correlated among species when body size effects are removed (fig. 2.2). Various approaches have been taken to explain this "fast-slow continuum." Hypotheses based on allometric constraints, brain size, metabolic rate, and r and K selection have failed to stand up to close scrutiny (reviewed by Harvey and Purvis 1999; Stearns 1992). By far the most insightful approach so far has been to view life history decisions as adaptations to mortality schedules that are imposed by the environment, a line of thinking with a long history in life history theory (see Stearns 1992 for a review). The most recent models incorporating this approach will be outlined in the next section.

Another general pattern in mammalian life history evolution—one with major implications for how comparative studies must be conducted—is that relationships between variables differ among groups because of confounding variables that themselves take on different values in different groups. The groups might be phylogenetic clades (such as Old World monkeys), grades (such as prosimians), or ecological categories (such as nocturnal species). As an illustration, figure 2.3A shows how gestation length scales with adult body size in a set of mammalian species. The overall correlation is weak, with a great deal of scatter. However, if the species are grouped into orders and the relationship within each order is calculated separately, much

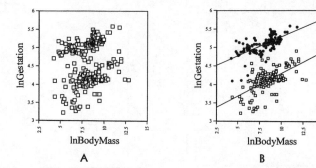

FIG. 2.3. Gestation length and adult body size in primates (solid circles) and carnivores (open squares), showing a grade shift. *(A)* There is little apparent relationship when the data are pooled. *(B)* A stronger relationship appears within each order when the orders are considered separately.

tighter correlations emerge (fig. 2.3B). The slope of the allometric relationship between the two traits is roughly the same in the two orders, but the primate line lies above the carnivore one. Such patterns are known as "grade shifts," and they are seen at all taxonomic levels. It is also possible for the slope relating two traits to differ among groups. In either case, fitting an overall line through the pooled species values is meaningless, as the slope of the line will reflect simply the degree to which the different groups are represented in the data set by accidents of sampling, rather than telling us anything about the true evolutionary relationship between the variables. If the confounding variables affecting the relationship between traits Y and X could be identified with certainty and included in the analysis, grade shifts would not be a problem. However, we can usually only guess what the confounding variables might be: grade shifts are easy to demonstrate, but hard to explain convincingly.

Grade shifts provide one compelling reason why comparative trends should not be assessed by treating species (or, indeed, any taxonomic unit) as statistically independent points for analysis. Even in the absence of obvious grade shifts, taxa cannot be treated as independent because similarities among related taxa, in both measured and unmeasured variables, often arise through inheritance from common ancestors rather than through independent evolution. Consequently, related species are pseudoreplicates, rather than true replicates, in statistical analysis (Felsenstein 1985; Grafen 1989; Harvey and Pagel 1991; Ridley 1992). Experimentalists rightly avoid experimental designs based on pseudoreplication; comparative biologists should likewise avoid pseudoreplicated comparisons (Rees 1995; Nee, Read, and Harvey 1996; Purvis and Webster 1999), which simulations show to have unacceptably high type I error rates (Grafen 1989; Martins and Garland 1991; Purvis, Gittleman, and Luh 1994).

All these arguments point to the same, now widely accepted, solution: phylogeny can be used to construct comparisons that are independent replicates for hypothesis testing (Felsenstein 1985; Burt 1989). Pairs of sister taxa automatically provide independent matched pairs for comparison: differences between sister taxa must have evolved since they shared a common ancestor, and must have evolved independently of differences elsewhere in the phylogeny. Felsenstein's (1985) method assesses the relationship between traits Y and X between the two members of each sister-taxon pair. When the Y differences are plotted against the X differences, grade shifts show up as marked outliers. This method provides an objective and rigorous way of identifying grade shifts (e.g., Garland et al. 1993; Barton, Purvis, and Harvey 1995). Differences in slope can also be tested objectively by assessing the significance of heterogeneity of relationship among contrasts from different major groups (Purvis and Rambaut 1995). Such principled ways of identifying heterogeneity of pattern in a data set are clearly preferable to the ad hoc procedures that are commonly used as part of nonphylogenetic analyses.

Comprehensive Models of Mammalian Life History Evolution
Charnov's Model

The first comprehensive model of mammalian life history evolution was produced by Charnov (1991, 1993). A conceptually simple yet potentially very powerful optimality model, Charnov's formalism (also described by Harvey and Nee 1991; Harvey and Purvis 1999) has led to what may well prove to be important insights into life history evolution in primates (e.g., the grandmother hypothesis for the evolution of menopause: Hawkes, O'Connell, and Blurton Jones, chap. 9, this volume) and elsewhere (Owens and Bennett 1995). It views externally imposed mortality schedules as the drivers of life history evolution, mediated by a growth law that constrains the relationship between age and size before maturity. Charnov assumes that all species grow like money in a bank account (Harvey and Purvis 1999); their body size is the capital, and the interest is the amount of energy that they get to spend on either growth or reproduction. Before maturity, all the interest is plowed back into growth, so mammals grow exponentially (specifically, they grow according to $dW/dt = AW^{0.75}$, though A varies among species). At maturity, they shift all their spending from growth to reproduction. Delaying maturity thus enhances reproductive effort (because it increases size at maturity), but at the cost of perhaps dying before reproducing at all: natural selection maximizes lifetime reproductive success and hence optimizes age and size at maturity. Because A, the coefficient in the growth law, varies among species, species maturing at the same age can be different sizes. Density-dependent mortality in the youngest age classes keeps population size constant, which in the long term it must be.

Charnov's model not only correctly recovers the size-independent fast-slow continuum, but also predicts correctly the allometric slopes of many life history traits on body size, and points out that certain dimensionless combinations of life history traits (e.g., the product of age at maturity a and the instantaneous adult mortality rate m —henceforth $a.m$) will be invariant across species. However, despite its generally impressive fit with known life history patterns across mammals as a whole, this model has been undermined by the recognition that one of its central assumptions—namely, that the ratio of weaning weight to adult weight is uncorrelated with adult weight —is wrong. This assumption is contradicted by the data (Purvis and Harvey, 1995, 1996). Furthermore, it is theoretically flawed: with the assumption in place, there is no trade-off between fecundity and survival to maturity because both are highest for the smallest species (Kozlowski and Weiner 1997). Additionally, Charnov's model assumes that growth rate scales with body size with the same allometric exponent (0.75) both within and between species, whereas the within-species trajectories in fact have lower exponents (Purvis and Harvey 1997). We hope it will prove possible to tweak the model to accommodate these observations without damaging the rest, but at present, Charnov's model must be placed in the category of an extremely constructive near-miss.

Kozlowski and Weiner's Model

More recently, a second comprehensive model of mammalian life history evolution has appeared. Kozlowski and Weiner's (1997) model is broadly similar to Charnov's, but makes fewer assumptions. Whereas Charnov assumes that, among adults of the same species, mortality rate is independent of size, Kozlowski and Weiner permit it to be independent of or to decrease with increasing size. Charnov incorrectly assumes that all species in a group follow the same growth law, with the same allometric exponent relating growth rate to size within each species; Kozlowski and Weiner permit the exponent to be the same or to vary.

In this model, the energy available for growth or reproduction is given by the difference between assimilation (A) and respiration (R), both of which scale allometrically with weight (w):

$$A(w) = aw^b$$
$$R(w) = hw^\beta$$

Mortality rate, m, also scales with w:

$$m(w) = gw^\lambda$$

The six parameters—three intercepts and three exponents—of these within-species allometric relationships are determined by the species' niche and vary among related species. In general, b and β are both positive ($\beta > b$, so maximum net productivity occurs at an intermediate size rather than increasing without end as size increases) and λ is negative. The trade-off is very similar to Charnov's: delaying reproduction risks dying before reproducing while increasing productivity, and natural selection optimizes age and size at maturity. Neither model assumes any mortality cost of reproduction (though Charnov's model is compatible with up to a 20% mortality cost of reproduction: Charnov 1993). When body size is optimized in many species with slightly different parameter values, the resulting comparative relationships among life history variables can look very like those observed in mammals. Three results are particularly noteworthy: First, the fast-slow continuum emerges independently of body size. Second, the among-species allometries relating traits such as age at maturity to body size are generally different from the within-species allometries. Third, one of Charnov's invariants—$a.m$—is also invariant in this model.

A program has been written that permits simulations of many species under Kozlowski and Weiner's full model and two simplified versions (Gawelczyk 1998). Each "species" is assigned a value for each of the model's six parameters (drawn from normal distributions), and body size is then optimized as outlined above. Plugging the resulting body size into the allometric relationships above gives each species' rates of assimilation, respiration, production, and mortality. As figure 2.4 shows, species evolved under the model lie on the fast-slow continuum.

The model is much less predictive than Charnov's because it makes fewer assumptions. For instance, the exponents of cross-species allometries depend on the means, and particularly the spread, of the within-species parameter distributions. Frustratingly, the parameters are extremely difficult to measure in natural populations, so it may never be possible to pin down exactly what the model predicts.

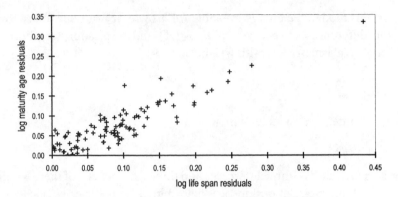

FIG. 2.4. Kozlowski and Weiner's model captures the fast-slow continuum, with life history variables intercorrelating even when body size effects are controlled for. The points represent "species" from simulations as described in the text.

Primates and the Models

How do primates fit with general mammalian patterns? In some ways they are like "ordinary" mammals, but in others they are very different. Primate species, like those of other orders, lie on a fast-slow continuum. For instance, species maturing late for their body size have low fecundity for their size (Jones and MacLarnon 2001). However, there is more than a single continuum: strepsirrhines are "faster" than haplorhines of the same size. Grade shifts are common among primates, either among clades (strepsirrhine/haplorhine) or among grades (e.g., altricial/precocial: Martin and MacLarnon 1985).

Primates are also grade-shifted compared with other mammals. They have unusually long lives and unusually low fecundities for their size (Charnov and Berrigan 1993), and lie very much at the slow end of the fast-slow continuum (as do bats) when size is factored out (Read and Harvey 1989). In keeping with this pattern of nested grade shifts, the supposed "invariant" $a.m$ differs significantly among the major mammalian clades in a sample of species for which it can be estimated. Rodents have a mean $a.m$ of 0.98; the mean is 0.84 in carnivores, 0.58 in ungulates, 0.44 in bats, and just 0.33 in primates (0.38 in strepsirrhines and 0.30 in haplorhines).

Why are primates so different? Charnov's model points the finger at their slow individual growth rates (Charnov and Berrigan 1993). In Kozlowski and Weiner's model, however, only changes in the within-species mortality allometry can make clades faster or slower for their body size; simple changes in the productivity equation can change the speed of life

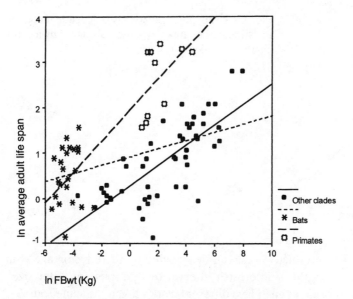

FIG. 2.5. Average adult life span and female body weight (FBwt) in a range of mammalian species. Bats and primates have longer life spans for their size than do most other mammals. (From Jones and MacLarnon 2001.)

and the body size, but not the size-independent speed of life. It is tempting to speculate that primates may be adapted to the low mortality rates prevalent in their ancestral habitat—tropical forest trees. Other arboreal mammals also have low mortality rates, as do bats (Jones and MacLarnon 2001); figure 2.5 shows that bats and primates have unusually long life spans for their body size. More detailed data on a range of forest mammals might clarify the issue.

It must be remembered that both the models outlined in this section rely on simplifying assumptions that are caricatures of reality. For instance, real primate growth curves are much more complex than they are modeled to be (see Pereira and Leigh, chap. 7, this volume), and the same is likely to be true of mortality patterns. The usefulness of the next generation of models will depend on the reasonableness of such assumptions. This, in turn, will depend on the availability of good empirical data on a range of species.

Rates of Life History Evolution

If the speed of life is indeed adapted to mortality rates, as Kozlowski and Weiner's model assumes, then which components of life history adapt most

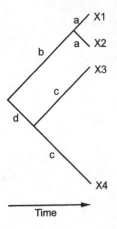

FIG. 2.6. Using phylogenetically independent comparisons to estimate the rate of evolution displayed by a continuous character.

readily? Reproductive rate can be speeded up by reducing age at maturity, gestation length, or interbirth interval or by increasing litter size. Which route has been taken? Have different clades taken the same route? One can envisage the whole suite of traits changing together, or most of the difference being due to changes in a single component. Equally, one can envisage all clades showing the same pattern, or different traits evolving most rapidly in different groups. Another question can be addressed here: Do rates of life history evolution correlate with rates of cladogenesis or of ecological divergence? In primates, the two possibilities make different predictions: Cercopithecidae show the highest rates of taxonomic diversification, whereas probably the greatest ecological radiation within the order has taken place in the Malagasy strepsirrhines (especially given that extant strepsirrhines represent only part of the ecological diversity present before recent extinctions: see, e.g., Kappeler and Heymann 1996; Godfrey et al. 1997b).

Each independent sister-taxon comparison can provide an estimate of the long-term average rate of evolution under an assumed model of character change, Brownian motion (Felsenstein 1985; Webster and Purvis 2002). Figure 2.6 illustrates the process. The difference $X_1 - X_2$ evolved along a total branch length of $2a$, where a is units of time (say, millions of years). The rate parameter of the assumed Brownian motion process is estimated by $(X_1 - X_2)^2/2a$, which is the square of the usual quantity used when testing hypotheses of correlated evolution using independent comparisons. Likewise, the difference $X_3 - X_4$ evolved at a rate estimated by $(X_3 - X_4)^2/2c$. A third estimate can be computed (though the formula is more complicated) for the evolution along branches b and d. Note that each estimate is of the long-term average rate: one clade might show a higher rate not because it changes faster when it is changing, but because it has changed more

Table 2.1 Rates of life history evolution in primates and carnivores

Trait	d.f.	Primates	Carnivores	t
Body mass	264	8.77	7.84	3.18
Age at maturity	135	7.94	5.91	4.90
Gestation length	170	6.19	4.04	6.86
Interbirth interval	135	8.50	4.69	7.78
Litter size	229	2.72	5.84	−8.05

often. The pair of branches descended from each node gives an independent estimate of the rate parameter, so rates can be compared among groups. With large samples, t tests or ANOVAs are valid; nonparametric alternatives can also be used. It must be borne in mind that such estimates assume the phylogeny, including the ages of nodes, to be correct. If there is bias such that ages of nodes in part of the phylogeny are systematically over- or underestimated, then the rate estimates for that part will be consistently too low or too high, respectively. More discussion of rate estimation and comparison can be found in Garland (1992), Martins (1994), and Webster and Purvis (2002).

A. J. Webster and A. Purvis (unpub.) compared rates of evolution of body mass, age at maturity, gestation length, interbirth interval, and litter size in primates and carnivores, and also within primates between strepsirrhines, platyrrhines, cercopithecines, colobines, and hominoids. The data (from Purvis et al. 2000) were log-transformed prior to analysis, so the comparison is between rates of proportional rather than absolute change. Tables 2.1 and 2.2 show some of the results. Four of the traits (all but litter size) have evolved significantly faster in primates than in carnivores (table 2.1). Litter size is a striking exception, having evolved very much faster in carnivores. It is clear that lineages in the two orders have followed different evolutionary pathways when adapting to changed mortality schedules. This heterogeneity of response is also clear within the primates: rates of evolution of two of the traits (body mass and litter size) have differed

Table 2.2 Rates of evolution of life history traits in different primate clades

	Body mass	Age at maturity	Gestation length	Interbirth interval	Litter size
Strepsirrhines	9.36	8.28	7.05	7.97	6.09
Platyrrhines	7.72	7.39	5.59	7.89	3.35
Cercopithecines	9.35	7.96	5.76	9.13	1.01
Colobines	9.27	9.92	6.98	—	0.00
Hominoids	8.15	7.29	6.31	8.91	1.89
p	.03	.37	.07	.22	.00

Note: Shaded cells indicate clades in which variables have evolved fastest.

significantly among the major primate clades (table 2.2), with litter size having evolved as rapidly in strepsirrhines as in carnivores. These results may reflect the fact that litter size cannot go below 1.0; any selection pressure for further slowing of life history must affect other parameters instead. They may also reflect differential lability among major groups in nesting behavior: evolution of larger litters is associated with the use of shelters (Kappeler 1998).

The pattern of rates may be a better match to the degree of ecological diversification than to rates of cladogenesis. Three of the traits have evolved most rapidly in the ecologically diverse strepsirrhines, and none in the most marked radiation (cercopithecines). The caveat about phylogenetic error is relevant here: an alternative explanation is that the ages of strepsirrhine nodes are underestimated relative to cercopithecine nodes. However, such errors cannot explain the mosaic pattern of rates described above.

Life History and Species Richness

In this section, we consider how life history may have shaped primate phylogeny. Marzluff and Dial (1991) found a significant association among mammalian and other taxa between a measure of "taxonomic dominance" (roughly, the species richness of a taxon compared with that of other related taxa) and both longevity and age at first reproduction. Earlier, they had shown that the dominant subtaxon within a higher taxon typically had a smaller than average body size (Dial and Marzluff 1988), a finding also reported by others for mammals (Van Valen, 1973a; Martin 1992). Their results form the basis of a model of taxonomic diversity that still provides much of the theoretical background for modern comparative tests. In their model, taxa with short generation times are able to track changing environments and recover quickly from population crashes. Both traits confer a reduced risk of extinction. The associated rapid evolution (Bromham, Rambaut, and Harvey 1996) enhances the speed of adaptation to new niches (Van Valen 1973b) and fosters a concomitant increase in the chance and rate of speciation. This is a classic "bigger-cake" clade selection mechanism (Purvis and Hector 2000): some lineages possess one or more traits that allow them to be more successful (attain larger biomass or metabolic turnover: Williams 1992) than other lineages.

However, the tests reported above, like the other nonphylogenetic tests referred to earlier in this chapter, are statistically flawed for at least two reasons. They treat taxa as being comparable, yet mammalian families vary greatly in age (Gardezi and da Silva 1999). So, genera of small-bodied species might be more species-rich simply because they are, on average, older. The tests also treat taxa as independent, again raising the prospect of pseudoreplication (see Purvis 1996 for a review). Murid genera, for instance, might tend to be atypically species-rich not because they are small-bodied, but because of any other attribute of murid rodents. Phylogenetic methods are required, therefore, for testing correlates of species richness. Most published tests to date have focused on hypotheses relating diversity to discrete characters, such as phytophagy in insects (Mitter, Farrell, and Wiegman 1988); methods for continuous characters have lagged behind. Typical independent contrasts approaches, such as CAIC (Purvis and Rambaut 1995), are not suitable—clade richness is not heritable in the same way as traits used in standard comparative analyses—but it is clear that the comparison of sister taxa will be central, as in CAIC.

Two sister-taxon comparison techniques are currently in use: non-nested and nested comparisons. Non-nested comparisons (Barraclough, Harvey, and Nee 1995) are made between sister taxa that differ in the trait of interest. Different rules can be applied to determine the precise set of comparisons made, although the usual rules are either to make all possible terminal comparisons or all terminal comparisons with at least a minimum difference. The results are usually analyzed nonparametrically (is an increase in X associated with more or fewer species?), although this need not be the case. The principal advantage of this method is that it makes no assumptions about the evolutionary processes underlying the observed differences other than that the taxonomic units at the tips of the phylogeny are monophyletic.

Nested comparisons have been implemented in a new program, Macro-CAIC (Agapow and Isaac 2002), which calculates contrasts in one or more independent variables in the same way as CAIC does. These contrasts are compared with clade richness contrasts, which are calculated using the species richness values of the two descendant clades (fig. 2.7): the contrast used in these analyses is the logarithm of the ratio of species richnesses of the two clades, with the numerator being the number of species in the clade having the larger value of X. Under the null hypothesis of no association between X and diversity, this quantity has an expectation of zero. Contrasts can be analyzed by regression through the origin. Nested comparisons run the risk of nonindependence if the evolutionary model being assumed is violated (Gittleman and Purvis 1998). However, simulations (N. J. B. Isaac et al., unpub.) have shown that type I error rates for the sorts of analyses presented here never exceed 8% under a wide range of models of clade growth and character change. MacroCAIC is freely available at <http://www.bio.ic.ac.uk/evolve/software/index.html>.

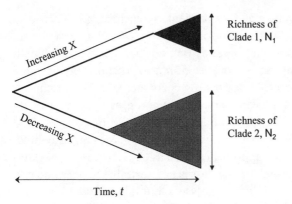

FIG. 2.7. Using phylogenetically independent comparisons to test proposed correlates of species richness. The contrast used is $\ln(N_1/N_2)$, with the numerator being the number of species in the clade having the larger value of X.

Does life history predict species richness in primates? In a preliminary assessment, we tested body mass and the same life history variables as used in the previous section. Despite reasonable sample sizes (averaging over 100 contrasts), no variable approached significance (all $p > .5$). (Macro-CAIC had already been used to show that body mass does not significantly predict species richness in primates: Gittleman and Purvis 1998.) We did, however, find suggestions that ecological predictors might be important: lineages living in larger groups were more species-rich (96 contrasts, $t = 2.07$, $p = .04$), and high ecological population density (i.e., density in ideal natural habitat: Damuth 1993) was a near-significant predictor overall (70 contrasts, $t = 1.93$, $p = .06$) and was significant within the haplorhines (56 contrasts, $t = 2.20$, $p = .03$) (fig. 2.8). We caution, however, that we tested several variables, leading to problems of interpreting multiple test results (Rice 1989), and we note that body size and life history do not appear to be consistent correlates of species richness in birds either (Owens, Bennett, and Harvey 1999).

Summary and Conclusions

The study of life history, both in primates and more generally, has progressed rapidly. There are now answers to some key questions: It seems that life history strategies evolve in response to mortality schedules, rather than other proposed "drivers." Body size seems to be an adaptive response to life history strategies, rather than the other way around. These recent developments at least suggest where we should look for the answer to the key question about primates—namely, why do they live their lives so slowly? Charnov's model points the finger at a slow growth rate; Kozlowski and Weiner's model implicates low adult mortality rates. We probably need new

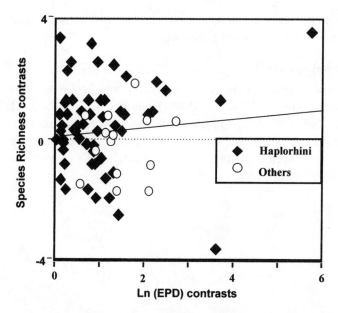

FIG. 2.8. Ecological population density (EPD) as a predictor
of primate species richness, as tested by regression through
the origin of independent contrasts. The overall relationship
is not quite significant ($p = .06$); within the haplorhines
(solid symbols), it is weak but significant ($p = .03$).

data at this point, especially data from nonprimates, to see how growth rates
and mortality schedules fit with both models' predictions and, indeed, their
assumptions.

In this chapter, we have shown two new ways in which phylogeny can be
used to study life histories in primates. Our analyses of rates of life history
evolution show that life history evolves in a mosaic fashion at two levels.
First, the rate of accumulation of life history variation has differed among
major primate clades, in a pattern that appears to correlate with their ecological diversity. Second, different life history traits show different patterns
of variation, suggesting that lineages are moving along the fast-slow continuum in different ways—some by changing litter size, others by varying age
at sexual maturity, and so on.

We have found no evidence that life history has played much of a role in
shaping primate phylogeny by influencing speciation or extinction rates.
The marked differences in species richness among primate clades do not
correlate with life history differences, but may correlate with ecology. It may

be (though we have no evidence) that a fast life history is necessary for radiation, but it is certainly not sufficient.

Acknowledgments

This work was funded by the Natural Environment Research Council, U.K., through grant GR3/11526 (A. P., P.-M. A., and N. J. B. I.), grant GR8/04371 (K. E. J.), and studentship GT4/96/164/T (A. J. W.). We thank Peter Kappeler for the invitation, and Tim Clutton-Brock, Rob Deaner, Peter Kappeler, Michael Pereira, and two anonymous referees for suggestions that improved the manuscript.

3 Socioecological Correlates of Phenotypic Plasticity of Primate Life Histories

PHYLLIS C. LEE AND PETER M. KAPPELER

Life history "invariants" now appear to be accepted as concepts relating reproduction to phylogeny, mass-specific traits, and selection (Pianka 1970; Stearns 1976; Blueweiss et al. 1978; Calder 1984; Harvey, Martin, and Clutton-Brock 1987; Partridge and Harvey 1988; Charnov 1991). However, such a perspective necessarily ignores plasticity and variability as a component of selection (Lande 1982b; Kozlowski 1993). The ecology and evolutionary history of most species is characterized by environmental heterogeneity, in the form of seasonal variation, geographic variation, or both (Boyce 1979; Boyce and Daley 1980; Berven and Gill 1983). This heterogeneity poses an evolutionary problem for adaptations in the form of life history traits directly involved in survival and reproduction, because it is unlikely that a single phenotype will confer high fitness in all situations (Schlichting 1986; Stearns and Koella 1986; Via et al. 1995; Brommer 2000).

Phenotypic plasticity is a common solution to this problem. It occurs when a single genotype produces a set of different phenotypes across a range of environmental conditions (Stearns 1989a). Traits that exhibit such variability have a certain norm of reaction, which provides the substrate for natural selection and thus evolutionary change (Caswell 1983; Via and Lande 1985; West-Eberhard 1989). Understanding the extent and limits of plasticity is therefore a useful and necessary starting point for examining the evolution of life histories with respect to patterns and processes within species and between individuals. This approach complements those that examine causes of interspecific variation in average species traits (Blueweiss et al. 1978; Western 1979; Hennemann 1983; Martin and MacLarnon 1985; Gittleman 1986b; Harvey, Promislow, and Read 1989; Read and Harvey 1989; Harvey and Keymer 1991).

Phenotypic responses to spatial and temporal variation in the environment have been documented for numerous traits in many plants and animals (Schlichting 1986; Stearns 1992; Via et al. 1995). These studies have demonstrated that plasticity can be variable between characters and that plasticity itself, rather than one particular phenotype, can be adaptive under certain environmental conditions (Via et al. 1995; Schlichting and Pigliucci 1998). Most theoretical and empirical progress on plasticity has been based on studies of short-lived organisms that can be easily subjected to experimental manipulations, including control of genotypes (Schlichting 1986; Stearns, de Jong, and Newman 1991; Newman 1992; Via et al. 1995; but see Stearns 2000). Less is known in this respect about taxa with relatively slow life histories, such as primates (see Rubenstein 1993; Holmes and Sherry 1997; Pereira and Leigh, chap. 7, this volume).

Primates represent a diverse but monophyletic group of species with an array of radiations of different time depths, occupying a range of niches. They thus make an interesting group for an exploration of phenotypic plasticity. Nonhuman primates exhibit considerable interspecific variation in key life history traits, so that principles and constraints shaping variability can be examined at different taxonomic levels and as a function of social and ecological variables (Harvey and Clutton-Brock 1985; Harvey, Martin, and Clutton-Brock 1987; Ross 1988, 1992b; Charnov and Berrigan 1993; Ross 1998). There is also remarkable variation in some life history traits among individuals, and explorations of the sources and causes of this variation have been undertaken in relation to individual reproductive success. For example, interbirth intervals, age at weaning, fertility, and age at first reproduction are known to vary among individuals, and adult body mass itself is a plastic trait (Lee, Majluf, and Gordon 1991; Altmann et al. 1993; Dietz, Baker, and Miglioretti 1994; Lee 1997; Smith and Jungers 1997; Schmid and Kappeler 1998; Richard et al. 2000). In relation to phenotypic plasticity, successive links need to be made between the variation seen at the level of individuals or populations, which may be genetic, developmental, or the result of local ecology and group structure, and the variation seen among species, and ultimately with the larger-scale phenomenon of phylogenetically related invariants.

There are a number of sources of variation in life history traits between individuals, some of which are due to genetic consequences of mate choice (Grob et al. 1998), others to social and ecological influences on early development (Altmann 1991; Lee 1999), and still others to social status or rank (Bercovitch and Berard 1993; Bercovitch and Strum 1993; Pusey, Williams, and Goodall 1997). In relation to local ecology, the sources of variation are well known: diets change both within a year in response to seasonality and over time (Dunbar 1990; Bronikowski and Altmann 1996; Pereira et al. 1999), group size varies among populations (van Schaik et al. 1983; Stacey 1986; Crockett and Eisenberg 1987; Wrangham, Gittleman, and Chapman 1993; Chapman, Wrangham, and Chapman 1995; Barton, Byrne, and Whiten 1996), and risks of predation are seldom constant (Cowlishaw 1994; van Schaik and Hörstermann 1994; Cowlishaw 1997; Janson and Goldsmith 1995; Stanford 1995; Hill and Lee 1998; Treves 1999; Janson, chap. 5, this volume). Even though ecological and social variation are well documented, we do not know whether they underlie a "need" for individuals of a species to remain facultatively responsive within their evolved life history strategies.

While primates as a group are relatively well studied, with many data now coming from long-term field studies of known individuals, individual life histories are still less well known than are social and behavioral adaptations. Attempts to examine phenotypic plasticity would ideally use experimental populations in which genetic variation, social variation (status and group size), and sampling error could be controlled. In a comparative study, population-level variance might be considered a surrogate for plasticity, but only when there are sufficient samples of different populations to take into account the potential sources of error. Population-specific life history data are available for few species and are highly biased toward the few long-term studies available (Ross 1998). These studies may neither reflect the range of variation possible nor be "typical" of the primary habitats for the species, since they are often accidents of history and funding. For the majority of life history traits, there is still little information on species averages, and many of the available means are based on very small samples. It is therefore not surprising that there are few comparative analyses of variability (Charnov and Berrigan 1993; Ross 1998).

Our first aim, therefore, was to describe some of the extent of variation in primate life history parameters in order to make comparisons between traits and across taxonomic levels, integrating from theoretical expectations of individual variance to phylogenetically stable traits. Because data from different populations of the same species in different habitats were rarely available, we used data from both wild and captive populations, assuming that the latter enjoy optimal access to resources so that their developmental and reproductive rates are near the maximum of their reaction norms. Our second goal was to initiate an exploration of the correlates, if any, of the observed level of variation with basic social and ecological factors. Our analyses focused on variance in age at first reproduction, age at weaning, growth rates, and adult body mass. Primates represent some of the slowest mammalian life histories (Harvey, Promislow, and Read 1989; Harvey, Read, and Promislow 1989; Martin 1990; Read and Harvey 1989; Promislow and Harvey 1990) and include many species with relatively large brains (Martin 1983; Armstrong 1985; Harvey and Krebs 1990; Allman, McLaughlin, and Hakeem 1993a; Deaner, Barton, and van Schaik, chap. 10, this volume). Thus, studies of correlates and patterns of primate life history plasticity can contribute important comparative information for the development or testing of more general principles (Promislow 1991; Austad and Fischer 1992; Barton 1996; Promislow 1996; Dunbar 1998). Finally, we hope to stimulate fellow primatologists to investigate and report individual- and population-based life history variation in understudied taxa, so that more detailed analyses of this fundamental aspect of primate life history evolution will be possible in the future.

Methods

Our data analysis was based on eight life history traits and four ecological variables compiled from the literature for 141 species. Unpublished personal measures contributed data points for some strepsirrhines. Table 3.1 presents the sample sizes for each taxonomic group for the different variables.

Mass and Life History Parameters

Body mass: We used adult female body mass wherever available ($n = 124$); if there were no differences in mass between males and females, then the overall mean was used. When data on other variables were specific to a population, then the weights for that population were used. Thus, there are some differences between the mean weights used here and those recently reported by Smith and Jungers (1997). We have also used mean captive weights ($n = 49$) and mean wild weights ($n = 68$) (Leigh 1994b; Terranova and Coffman 1997) as part of our attempt to explore intraspecific plasticity.

Age at first reproduction ($n = 47$): We used the average age reported in the literature for first birth. We have included the range of ages (minimum–maximum) for 29 samples.

Gestation length: Mean length of gestation is reported for 91 species. Although there is some variation in gestation length due to infant sex (Clutton-Brock, Albon, and Guinness 1989; Clutton-Brock 1991) or maternal nutrition (Silk 1986), this variation is on the order of a few days and is difficult to assess in comparative analyses. Yet unquantified factors such as maternal

Table 3.1 Available sample sizes (number of species) for life history and ecological measures by phylogenetic group

Family/subfamily	No. species in taxon	Body mass	Gestation length	Birth mass	Interbirth interval	Age at weaning	Mass at weaning	Age at first reproduction	Litter size	Residence strategy	Diet type	Group size
Cheirogaleidae	9	9	4	3	3	1	0	3	3	2	9	4
Lemuridae	11	11	8	8	6	4	1	6	8	10	11	11
Indriidae	8	6	3	2	4	3	0	2	1	6	6	5
Daubentoniidae	1	1	1	1	1	1	1	0	1	1	1	1
Lepilemuridae	7	4	1	0	0	1	0	1	0	0	7	1
Galaginae	17	11	6	5	6	2	1	6	4	0	16	5
Lorisinae	18	12	5	7	4	4	1	3	4	1	17	4
Tarsiidae	5	4	1	1	2	2	0	2	5	5	5	2
Callitrichinae	35	10	4	10	6	7	6	4	10	10	4	10
Aotinae	9	2	1	1	2	2	1	0	2	2	1	1
Cebinae	11	4	4	4	4	4	3	1	4	4	4	4
Atelinae	19	5	5	4	5	5	3	1	5	5	5	5
Pithecinae	22	2	1	2	1	1	0	0	2	1	2	1
Colobinae	40	7	6	6	7	6	5	1	8	8	8	6
Papioninae	34	19	18	19	19	16	14	7	19	19	15	15
Cercopithecinae	24	10	6	9	7	6	4	7	10	10	9	8
Hylobatidae	10	2	2	2	2	1	1	1	2	2	2	2
Hominidae	6	5	5	5	5	5	5	2	5	5	5	4

Note: See Appendix for values.

immunology, placental function, hormonal competence, and maternal nutrition at different stages of pregnancy may also affect fetal growth rate and thus birth mass and relative infant development at birth.

Birth mass ($n = 89$): As noted above, considerable variation in birth mass is possible, at least some of which is pathological or has negative consequences for infant survival when birth mass is low or when infants are born relatively immature neurologically or physically (Martorell and Gonzalez-Cossio 1987; Dang et al. 1992). However, few studies exist on fetal growth or on variation in birth mass in primates, and thus we have had available only the range of birth mass by species reported in the literature.

Interbirth interval ($n = 84$): We used the species average, as well as the minimum ($n = 42$) and maximum ($n = 46$) for reproductively active females. We have attempted to use minimum values for surviving infants only. Such data are highly biased toward long-term studies, and are probably also affected by interindividual variation due to maternal age, sex of infant, dominance status, nutritional status, and so forth. Interbirth interval was converted to a measure of *fertility* (using litter size; see below), calculated as number of infants/female/year. We analyze variation in fertility presented as maximum, minimum, and species average.

Age at weaning: Again, along with a species mean ($n = 71$), maximum and minimum ages for weaning were used when available ($n = 45$). Weaning is problematic to determine for those species in which the cessation of suckling is a gradual process. For these species, we have used the age at which suckling is reduced to levels that allow for the return of the mother's cycles (Lee, Majluf, and Gordon 1991). For species with a relatively abrupt transition from suckling to independent feeding, we have used the earliest and latest dates for this transition.

Mass at weaning: Mass at weaning was available for only 46 species, most of which were haplorhines.

Litter size ($n = 93$): Average litter size and maximum litter size were both used in analyses. For the catarrhine primates, maximum litter size was given as two for those species in which twinning was reported in the literature. Twinning is a problematic issue, since twins may occasionally be produced, but both infants are unlikely to survive. In humans, for example, maximum litter size can be as great as five or six, but without medical intervention both infants and mothers have a high probability of mortality. We constructed a categorical variable that considers both the average litter size and the litter maximum, coding litters as either habitual singletons with a low probability ($< 50\%$) of multiple births, twins as the main form of reproduction, and typical twins with larger litters common (maximum litter > 2).

Ecological Parameters

Group size: For 89 species, information on mean reproductive group size was available (Kappeler and Heymann 1996). While mean group size may differ from the size of the foraging group (Chapman, Wrangham, and Chapman 1995) or the size of a sleeping aggregation, flexibility in grouping was included in the form of a code for fission and fusion, which was entered into analyses of variance simultaneously with group size. Because group size is a response to local predation pressure (Terborgh and Janson 1986; van Schaik and Hörstermann 1994; Hill and Lee 1998) or other environmental factors (Crockett and Eisenberg 1987; Chapman 1990; Wrangham, Gittleman, and Chapman 1993; Janson and Goldsmith 1995; Sterck 1999), the use of a "species mean" group size remains problematic. In most analyses, group size was considered as a categorical variable based on the distribution of observed sizes: small (< 4), medium ($4–20$), or large (> 20).

Predation risk: We explored the effect of predation on life history through codes based on subjective assessments of predation risk. The ranking of low, medium, and high was based on responses to predators (vocalizations, flight, defensive acts) and on whether predators were likely to be encountered by groups or individuals. This ranking is not a measure of mortality, but rather of perceived risk or potential for response in an evolutionary sense (Hill and Lee 1998; Janson 1998; Janson, chap. 5, this volume). Sample sizes were small, and these analyses should be considered preliminary.

Diet type: A code for diet type, based on the majority food type in the annual diet, was used for 127 species (Kappeler and Heymann 1996). If two food types predominated, then diets were coded for these two major types. Again, the level of seasonal variation in diets may be more important than the modal diet, but such data are as yet unavailable for a sample large enough to be analyzed meaningfully (see also Ganzhorn et al., chap. 6, this volume).

Residence strategy: Finally, a code for residence strategy was included for 91 species. These strategies were determined by the predominant sex remaining in a natal breeding group and were coded as follows: (1) both sexes disperse prior to breeding, producing a group of unrelated residents of both sexes; (2) males disperse, producing female kin resident groups; and (3) females disperse, producing male kin resident groups.

Statistical Analyses

All continuous measures were log-transformed to meet the normality assumption of parametric tests. Repeated tests on the same set of data were rare, and for these we accepted only an alpha level of $< .001$. For other tests, we used an alpha of $< .05$, but given the tiny sample sizes involved, we have presented some trends as possibly informative with $p = .10$ (two-tailed).

Phylogeny and Allometry

One of the key issues in any life history analysis is the problem of the statistical contribution of closely related taxa to any variation observed (Felsenstein 1985; Harvey and Pagel 1991). Another, possibly more interesting question is whether phylogeny is a constraint on the potential for plasticity in life history traits (Harvey and Keymer 1991; Purvis et al., chap. 2, this volume).

We have not used phylogenetic subtraction techniques, such as independent contrasts, in these analyses because we are attempting to assess the level at which variation primarily resides, rather than to remove potentially confounding effects of phylogeny. CAIC (Purvis and Rambaut 1995) does not yet allow for comparisons between variance partitioned at different taxonomic levels. We have therefore used a series of codes for phylogeny as a variable in analyses rather than as correction factors. The first code relates to "grades" (Martin and MacLarnon 1988), and was used where analyses were done only on strepsirrhines and "anthropoids" (suborder Anthropoidea; following Kay, Ross, and Williams 1997) because of a lack of sufficient data for tarsiers within the semiorder of haplorhines. The second level used was the family, since there were too few representatives of many subfamilies to analyze these in any statistically meaningful sense. We also used a code for genus and species, but it should be noted that the number of species with available data in most genera was one. Thus, these analyses are subject to biases due to small samples of unequal size and variance, and due to unbalanced cells.

In many life history analyses, body mass has been shown to exert a major influence on the nature and extent of observed relationships. We used allometric slopes as a means of removing some confounding effects of mass between closely related groups, but we did not use the slopes as predictors of an evolutionary difference. We recognize that this attempt to explore the effects of phylogeny is limited by a lack of data, which also constrains methods relying on contrasts. We suggest that phylogenetic similarity is unlikely to be a more significant source of error than gaps in the data. We also recognize that error is probably a major contributor to questions about phylogenetic plasticity, since the determination of an average value is subject to observer error and measurement error as well as phylogenetic error.

Preliminary analyses were made to determine whether mass correlations existed within the interspecific sample used here. We used female body mass as the underlying metabolic limit on aspects of reproduction to reduce potential biases introduced by high levels of adult sexual dimorphism. Observed correlations, separated by "grade," are presented in table 3.2. We present these allometric results so that the relationships are made explicit for the sample used here. It is important to achieve consistency across different investigators because of the use of different data sets and different mean values.

Several differences between the two grades were worth noting. Gestation length is known as a phylogenetically conserved trait (Martin and MacLarnon 1988), and this pattern was replicated in this data set (see table 3.3). The relatively weak relationship with mass did not differ between the grades in either slope or intercept. Mass correlations where the intercepts differed but the slopes were similar between grades were found for group size and interbirth interval. For litter size and age at weaning, both intercepts and slopes differed. Although the sample sizes were small, it appeared that, while primates as a group maintain allometric similarity in gestation length and birth mass, results here for the strepsirrhines and anthropoids confirmed previously reported differences in mass relations with life history traits. These differences between grades were most marked for the suite of traits associated with postnatal reproductive investment. Regressions were then used to determine the residual values for species. If there were no differences between the grades, then the overall regression was used. For those traits where a difference in either slope or intercept was found, residuals were calculated separately for each grade. For the remainder of our analyses, we used these unstandardized residuals from the least-squares regression line at the level of the species.

Results
Phylogenetic Influences on Grade-Specific Residuals

Mass-controlled relative values were explored at the level of family, genus, and "remaining" variance (table 3.3). Whether the remaining variance is due to "error" or, as we would suggest, mostly to variation within species, within populations, and between individuals, and thus equivalent to plasticity, can be assessed only when a larger sample of different populations for each species becomes available. It is worth noting that there were only a few

Table 3.2 Correlations between female mass and various life history traits

Trait	N_s	r^2	Slope	SE	Intercept	SE	N_a	r^2	Slope	SE	Intercept	SE
Gestation length	28	**0.154**	0.103	0.042	4.160	0.286	52	**0.578**	0.112	0.013	4.224	0.116
Age at first reproduction	22	**0.788**	0.347	0.039	0.916	0.250	27	**0.703**	0.353	0.045	1.170	0.378
Interbirth interval	24	**0.429**	0.239	0.056	0.936	0.358	57	**0.641**	0.312	0.031	0.366	0.261
Age at weaning	15	**0.416**	0.287	0.083	3.136	0.534	53	**0.656**	0.466	0.047	1.880	0.392
Litter size	21	**0.177**	−0.098	0.043	0.888	0.272	66	**0.484**	−0.031	0.004	0.283	0.034
Birth mass	26	**0.912**	0.646	0.040	−0.473	0.257	62	**0.931**	0.797	0.028	−0.939	0.233
Mass at weaning[a]	4	**0.725**	1.145	0.383	1.636	1.405	42	**0.922**	1.197	0.054	0.402	0.316
Group size	35	0.014	0.106	0.088	0.615	0.563	61	**0.416**	0.287	0.083	3.136	0.534

Note: Correlations and residuals were calculated separately for varying numbers of strepsirrhines (N_s) and anthropoids (N_a). Regressions significant at $P < .05$ are noted in bold.

[a]Mass at weaning was correlated with birth mass as this has been shown to be a more informative relationship (Lee 1999).

Table 3.3 Percentage of variance in mass-corrected residuals accounted for by similarity at different phylogenetic levels, as determined by hierarchical ANOVA

Variable	Family	Genus	Remaining
Body mass	**85.2**	**12.6**	2.2
% mass difference	4.9	11.1	***84.0***
Ratio max/min mass	0.1	0.1	***99.8***
Relative gestation length	**76.0**	**19.6**	4.5
Relative interbirth interval	35.8	29.6	***34.6***
Relative min interbirth interval	36.6	49.8	13.6
Relative max interbirth interval	23.0	65.0	12.0
Ratio max/min interbirth interval	1.5	8.5	***90.0***
Relative age at weaning	38.8	32.8	28.4
Ratio max/min age at weaning	0.3	0.4	***99.3***
% difference age at weaning	12.3	8.5	***79.2***
Relative group size	**48.5**	**37.2**	14.3
Relative age at first reproduction	31.8	37.7	30.5
Relative mass at weaning	17.5	16.2	***66.3***
Relative birth mass	**29.9**	**44.4**	25.7
Relative litter size	**40.5**	**44.9**	14.6
Relative fertility	30.8	46.4	22.8
Relative min fertility	24.5	63.9	11.6
Relative max fertility	37.4	53.9	8.7

Note: The remaining variance resides at either the level of the species or, more importantly, at the level of the individual (phenotypic plasticity). Boldfaced type = $p < .001$ for each level. Boldfaced italic type in remaining variance notes those variables for which over one-third of the variation is unexplained by phylogeny.

variables for which either the family or the genus was the predominant explanatory level of variance. Absolute mass was strongly phylogenetically conserved, but we found no significant association with phylogeny for the measures of mass plasticity. Relative gestation length and litter size exhibited significant variation among families; the litter size effect is probably a feature of the small-bodied callitrichids and the strepsirrhine grade. One surprising result was that relative group size was only weakly related to mass (see table 3.2), suggesting that it is only weakly affected by scaling effects and more responsive to local ecology and social strategies.

Overall, there was a trend for 25–65% (mean = 36%) of the variation in mass-corrected traits to be unexplained by either familial or generic similarity. Since over a third of variation was relatively independent of phylogeny, that variation might be indicative of an evolved potential for plasticity, a possibility that is explored below. Furthermore, more than 70% of the variance resided at the level of the species for measures of range or variability. Thus, the extent of differences within measures (maximum to minimum) may be most suggestive of the potential for plasticity in response to local socioecological conditions.

Intercorrelations of Grade-Specific Residuals

Residual values are known to be interrelated as suites of nonindependent traits. Here, few correlations between mass-controlled residuals, either across the sample as a whole or within grades, were found (table 3.4). Relative age at weaning was positively correlated with relative age at first reproduction and relative gestation length. Relative interbirth interval was also correlated with age at first reproduction. Slow or fast rates of reproduction appeared as typical robust patterns, and these patterns were similar between the grades. Litter size was negatively related to gestation length for strepsirrhines and to age at weaning for all groups. Species with larger litters were those with relatively shorter gestations and earlier ages at weaning.

Relative birth mass was correlated with litter size and relative group size (table 3.4), but this latter effect was due to the anthropoids alone, and there were grade differences in relative birth mass ($F = 6.35, p = .014$), but not in

Table 3.4 Correlations between residual values of time and mass

	Age at first reproduction	Interbirth interval	Age at weaning	Litter size	Birth mass	Mass at weaning	Group size
Gestation	.208	.075	**.513**	**−.575**	.147	.108	−.098
Strepsirrhines	−.022	−.352	**.751**	**−.743**	.223	−.543	−.144
Anthropoids	**.680**	.527	.493	.078	.159	.008	−.084
Age at first reproduction		.562	.662	−.155	.065	.066	−.242
Strepsirrhines		.542	.539	−.328	−.365	.997	−.474
Anthropoids		.583	**.748**	−.039	.221	−.169	−.138
Interbirth interval			.356	.075	−.049	.127	−.231
Strepsirrhines			−.092	.156	−.129	.290	−.025
Anthropoids			**.474**	−.015	−.049	.159	−.303
Age at weaning				−.337	.302	.236	.087
Strepsirrhines				**−.848**	−.053	.221	−.402
Anthropoids				−.178	.352	.229	.191
Litter size					−.334	−.025	.051
Strepsirrhines					−.429	−.683	.290
Anthropoids					**−.523**	.229	−.271
Birth mass						−.236	**.397**
Strepsirrhines						−.157	.085
Anthropoids						−.307	**.521**
Mass at weaning							−.234
Strepsirrhines							−.358
Anthropoids							−.230

Note: Boldfaced type indicates R values significant at $p < .001$; italic type at $p < .01$ (other significance levels are not included to account for repeated tests). The first correlation is for all species combined, the second for strepsirrhines only, and the last for anthropoids only (excluding tarsiers, $n = 2$). For sample sizes, see table 3.1.

group size. Mass at weaning was negatively correlated with birth mass, but again, this was probably due either to an anthropoid effect ($F = 3.99, p = .052$) or simply a lack of data for strepsirrhines ($n = 4$).

Ecological and Social Influences on Relative Traits

The potential contributions of socioecology to relative life history traits were examined through analysis of variance. As noted above, relative group size was associated with relative reproductive rates and mass. Fission-fusion, diet type, and residence strategy were initially entered together with group size category to test for main effects and interactions. No significant effects of diet type were found, suggesting that once mass is removed, dietary strategies are unrelated to relative rates of offspring production for species in general. There was one suggestive difference between the grades of primates, however: relative ages at weaning showed opposing trends in relation to diet type for strepsirrhines and anthropoids (interaction, $F = 5.18, p = .008$). Among anthropoids, the highest relative ages at weaning were found among frugivores, while for strepsirrhines these were greatest for folivores.

Both group size and residence strategy affected some measures of relative reproductive output. Significant effects were found for group size with respect to relative reproductive mass, but not relative reproductive time. Relative birth mass ($F = 5.66$, d.f. $= 2,43, p = .007$) and relative mass at weaning ($F = 8.98$, d.f. $= 2,82, p < .001$) both varied in relation to group size categories (fig. 3.1), with large groups having the highest relative birth mass, but the lowest mass at weaning (post hoc tests, $p < .05$). This significant difference in mass at weaning between large and small groups was largely due to anthropoids, as there are no strepsirrhines living in large groups. There were significant differences in relative gestation length ($F = 7.08$, d.f. $= 2,62, p = .002$), age at first reproduction ($F = 3.80$, d.f. $= 2,37, p = .033$), interbirth interval ($F = 5.05$, d.f. $= 2,67, p = .009$) and age at weaning ($F = 7.40$, d.f. $= 2,61, p = .001$) as a function of residence strategy. Relative reproductive times tended to be longer for groups in which females were resident as kin and shortest for male kin residents (all post hoc, $p < .05$; fig. 3.2). As the majority of strepsirrhines fall into a category in which neither sex remains in the natal group, these differences were not simply a function of a "strepsirrhine bias."

The analysis of interbirth intervals highlighted the classic pattern of species that breed annually (IBI = 12 months) and those that are non-annual breeders (IBI < or > 12 months). Since relative litter size was phylogenetically conserved, we used the categorical variable constructed from average and maximum litter size. These two tactics of reproduction, defined by the

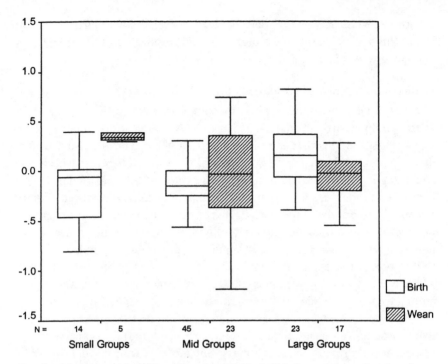

Fig. 3.1. Median and interquartile range of relative birth mass (open bars) and relative mass at weaning (shaded bars) by group size categories.

combination of breeding pattern and litter size, might be expected to determine relative reproductive rates. However, there were few effects of annual breeding on reproductive rates, although there was an expected trend for relatively short gestations to be associated with annual reproduction ($F = 6.06$, d.f. $= 1,80$, $p = .016$). Litter size (fig. 3.3) was associated with reproductive time in that both relative gestation length ($F = 6.04$, d.f. $= 2,71$, $p = .004$) and relative age at weaning ($F = 3.34$, d.f. $= 2,66$, $p = .042$) were shorter when litter size was large (post hoc, $p < .005$ for gestation length, $p = .05$ for age at weaning). Large litter size was associated with reproductive rate as measured by relative fertility ($F = 3.13$, d.f. $= 2,75$, $p = .05$, post hoc $p = .06$), but there was no association between IBI or relative age at first reproduction and litter size.

There was a trend for birth mass to vary with respect to litter size ($F = 2.78$, d.f. $= 2,60$, $p = .07$), and the effect here was most marked for twin births, rather than singletons or multiple births (post hoc, $p = .1$). However, relative mass at weaning did not vary as a function of litter size category. This finding suggests a separation of prenatal investment (via gestation length

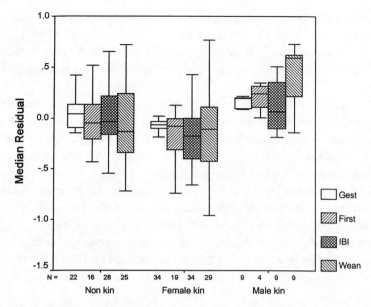

FIG. 3.2. Median and range of relative reproductive times by residence strategy. Gest = gestation, First = age at first reproduction, IBI = interbirth interval, Wean = age at weaning.

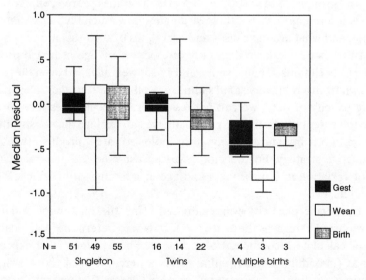

FIG. 3.3. Median and interquartile range of relative gestation length (Gest), age at weaning (Wean), and birth mass (Birth) by litter size category.

and birth mass) from the time and energy strategies used postnatally between species with small and large litters (see also Tardif 1994; Hartwig 1996; Pereira et al. 1999). It would be helpful to have more data on mass at weaning for species with large litter sizes to explore this relationship in more detail.

Predation risk was expected to influence rates of reproduction, as high reproductive output is more likely under higher probabilities of mortality. However, there were few relationships between predation risk and any of the measures of relative reproductive output. Among strepsirrhines, relatively short gestations were associated with high risks of predation (fig. 3.4A) (gestation length: $F = 4.33$, d.f. $= 2,27, p = .025$; age at weaning: NS), but short gestations relative to mass did not produce greater reproductive output, either through shorter interbirth intervals or higher potential fertility. Among anthropoid primates, reproductive input via birth mass and mass at weaning varied as a function of predation risk (fig. 3.4B) (mass at weaning: $F = 6.20$, d.f. $= 2,56, p = .005$; birth mass: $F = 3.93$, d.f. $= 2,40$, $p = .025$).

Phenotypic Plasticity

The general patterns in relative reproductive parameters noted above are consistent with previous analyses: the classic differences between "slow" and "fast" groups, be they species, families, or grades, were seen. However, the question remains: within these strategies, how much further variation is possible, and what are the causes and correlates of such variation?

At this stage, we have data only on body mass, age at first reproduction, fertility (infants/female/year), and age at weaning, and even these data are drawn from a relatively small sample of well-studied species. Despite the relative paucity of data, we can address several specific questions in relation to the extent of plasticity. Specifically, we can explore the constraints on variation in fertility and ask why anthropoids so rarely produce twins as a reproductive strategy. Furthermore, we can examine the general associations of variation in female mass and age at weaning with socioecological parameters.

Intraspecific plasticity was examined using the maximum and minimum reported values, as there are too few data to determine statistical variance or calculate a coefficient of variation. These values were expressed either as (maximum minus minimum) as a percentage of the average or as a ratio of maximum to minimum. As noted above, there was little consistent taxonomic influence on the pattern of variation (fig. 3.5; see table 3.3). There were no significant correlations between the measures of variation

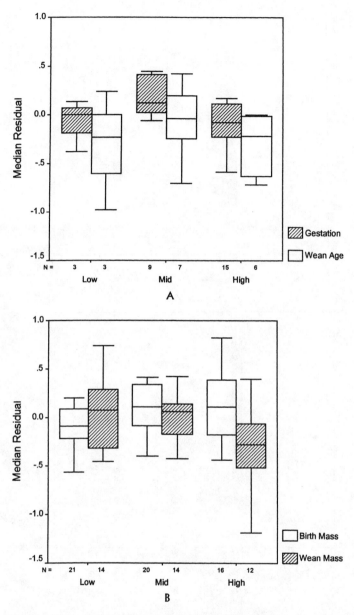

FIG. 3.4. Effects of predation risk. *(A)* Effects of predation risk on relative reproductive times (gestation length and age at weaning) for strepsirrhines only. *(B)* Effects of predation risk on relative reproductive mass (birth mass, mass at weaning) for anthropoids only.

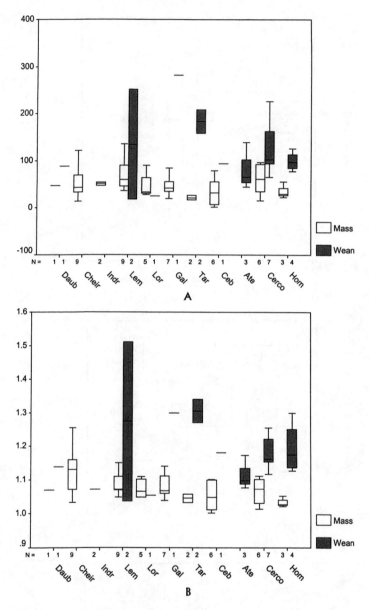

Fig. 3.5. Plasticity measures for families as (A) percentage of difference and (B) ratio of minimum/maximum for body mass and age at weaning. For family names, see Table 3.1.

Table 3.5 Correlations between measures of plasticity and relative reproductive time and body mass

	Relative mass		Relative time				Other
Plasticity	Birth mass	Mass at weaning	Gestation length	Age at first repro-duction	Interbirth interval	Age at weaning	Litter size
Ratio body mass	−.267	.039	**−.508**	.165	.243	−.164	**.352**
% difference body mass	−.167	−.056	**−.378**	−.043	−.012	−.112	.010
Ratio age at weaning	.127	**−.578**	**−.516**	−.419	−.284	**−.641**	**−.868**
% difference age at weaning	.197	**−.723**	**−.448**	−.507	−.300	**−.612**	**−.676**
Ratio interbirth interval	−.102	−.065	−.147	−.123	−.239	−.212	.112
Ratio fertility	−.102	−.065	−.150	−.077	−.200	−.198	.112

Note: Correlations significant at $p < .05$ indicated in boldfaced type.

themselves, which was unexpected in that plasticity could be predicted to be a generalized trait.

As the ratio of maximum to minimum body mass increased and as the percentage of difference increased, there tended to be a negative relationship to relative gestation length (table 3.5). The extent of variation in age at weaning was also negatively correlated with relative mass at weaning, gestation length, and age at weaning. In other words, species with higher plasticity in female mass and age at weaning tended to be those with shorter relative gestations, lower relative mass at weaning, and younger relative weaning ages. Such species also tended to have younger relative ages at first reproduction, higher relative group sizes, and relatively larger litters, although not significantly so. Variability in fertility had no association with either relative mass or relative reproductive time measures.

Socioecology and plasticity were interrelated. Variation in age at weaning was affected by seasonality of breeding ($F = 17.23$, d.f. $= 1,21, p < .001$), showing greater potential for plasticity when reproduction is not constrained to an annual cycle. Group size as a continuous variable was negatively associated with mass variation, but positively associated with variation in age at weaning, although none of these correlations was significant. There was a weak effect of group size on mass variation ($F = 2.51$, d.f. $= 2,38, p = .088$), and mid-sized groups tended to have the greatest variation in mass. The relationships between group size and variation in relative reproductive time or female body mass appeared to be nonlinear (fig. 3.6). Finally, although the *range* of variation in fertility was unrelated to any ecological or

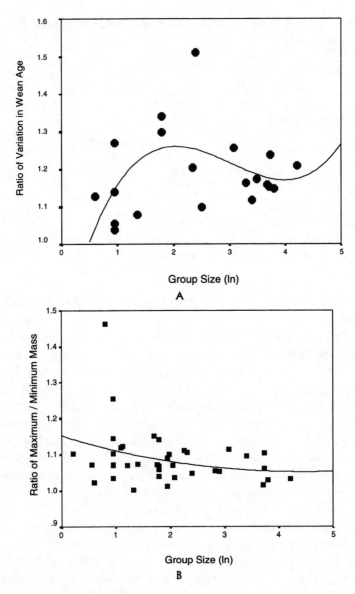

FIG. 3.6. Nonlinear relationship between group size and plasticity in *(A)* age at weaning and *(B)* body mass. Lines are fitted from quadratic equations.

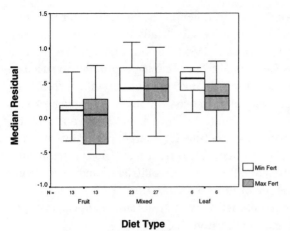

FIG. 3.7. Relative maximum (Max Fert) and minimum (Min Fert) fertility by diet type. Diet types have been combined into frugivores (Fruit), folivores (Leaf), and Mixed.

social variable, and indeed, was uncorrelated with other measures of plasticity, both relative maximum fertility (F = 4.42, d.f. = 2,42, p = .018) and relative minimum fertility (F = 5.20, d.f. = 2,41, p = .01) were associated with diet type. Thus, the potential for extremes in rates of reproduction appeared to be less marked for committed frugivores and folivores, but was associated with mixed diets (fig. 3.7).

Discussion

Variation among primate species in life history traits remains a topic of major interest in that both variation in evolved features and individual variation are marked with respect to rates of reproduction. The preceding analyses confirmed previous findings of grade differences between strepsirrhines and anthropoids in a larger sample and highlighted some novel relationships between the classic r–K continuum and socioecological variables. The questions of interest here are how sensitive rates of reproduction are to general trends in ecology and social systems; how much variation is exhibited, or indeed possible, as an adaptive or plastic response to the environment; and what might constrain the potential for variation.

Our analyses suggest that one major way in which females respond flexibly to varying ecological and social conditions is by altering litter size. We can posit two strategies, which appear to differ between the two grades. For strepsirrhine primates, reproductive success (at the level of the individual) is most constrained by the rate of reproduction, and thus plasticity in fertility (infants/female/year) should be expected, as variation in relative age at first reproduction (which is typically on the order of months) will have less of an effect than varying litter size. Females are predicted to maximize prenatal investment and respond to environmental variation via litter manipulation. Thus, we see plasticity in fertility, specifically in response to energy availability (Pereira 1993c; Pereira et al. 1999). For the anthropoid primates, reproductive life span may be critical to maximizing reproductive success, but a greater contribution may come via investment in infant survival, and for this group of species, plasticity in birth mass and mass at weaning, as well as avoidance of extrinsic sources of mortality such as predation and infanticide, may be crucial. Anthropoids trade time for energetic investment in infant growth, while strepsirrhines effectively opt for numbers at a significant mortality cost. It is interesting that the twinning callitrichids may have replicated the strepsirrhine strategy of litter maximization, but we suggest that they can afford this only in combination with an intensive allocare system (e.g., Dunbar 1995a; Ross and MacLarnon 1995).

Does Social Variance Relate to Life History Variance?

Group size for primates is both a social and an ecological feature. Because individuals are embedded within the social structure and composition of their groups, this aspect of group size may affect their reproductive performance (Abbott 1989). Similarly, because group size is a facultative solution to problems of contest and scramble competition for resources, as well as a strategy for reducing predation risk, these ecological determinants of group size will also affect individual reproductive performance (van Schaik and van Hooff 1983). As it is difficult to disassociate these two aspects of group size, the "social" dimension, in terms of numbers (e.g., Dunbar 1998), will be discussed here.

When group size is relatively large, relative birth mass is small, but relative mass at weaning is large. While we do not have a large sample on phenotypic variation, there was a suggestion that group size was associated in complex ways with variability in both reproductive time and reproductive mass. Larger groups were associated with reduced plasticity in mass, while mid-sized groups were associated with greater variation in age at weaning. Why should relative or absolute group size relate to fast or slow life histories and the extent of variation? If group size functions to reduce infant or juvenile mortality—for example, due to predation (cercopithecoids: Hill and Lee 1998; platyrrhines: Janson and Goldsmith 1995) or infanticide (Janson and van Schaik 2000), then an association with postnatal investment might be expected (Lee 1999). If, however, group size reflects the extent of within- or between-group food competition, then associations with female reproductive rates are likely (van Schaik and van Hoof 1983; Isbell 1991). In all likelihood, both these factors are operating, and they are impossible to separate in the small sample analyzed here.

Male kin residence was associated with longer relative gestation times, a later age at first reproduction, and a later age at weaning. In addition, relative interbirth intervals were long, and this trend was marked for large groups. Given the distribution of male-kin sociality within the primates, these associations are unlikely to be due to phylogenetic similarity. We can only speculate that male-kin sociality may be associated with lower rates of extrinsic infant mortality, thus facilitating extensive maternal investment. However, residence and dispersal strategy showed no evidence of covarying with any aspect of phenotypic plasticity, suggesting that these are phenotypically conserved traits acting at the level of invariants rather than individual life history strategies.

Ecological Variation

Local ecological variation typically affects mortality rates and energy acquisition. As these factors are known to influence reproductive output, it was expected that both species-level life history traits and plasticity in those traits should be related to local ecology. Predation risk was found to be associated with both birth mass and mass at weaning in anthropoids, such that under high predation risk infants were born large but weaned small, thereby minimizing the risks immediately after birth, but also minimizing investment costs during the high-risk postnatal growth phase (see also Lee 1999). Predation risk in strepsirrhines showed no association with reproductive mass, possibly due to the small sample of mass at weaning, but we suggest that predation should relate to plasticity measures as it is a major element of mortality. Indeed, predation risk among strepsirrhines did appear to interact with reproductive times, especially that of relative gestation length. Low- and high-risk species tended to have the shortest relative gestation times, with a similar trend for age at weaning. The partitioning of investment in relation to litter size was also explored. Younger relative ages at weaning are found with large litters, and birth mass is smaller, although here the effect is probably due to strepsirrhines alone. However, twinning or "littering" might be a straightforward way of facultatively exploiting favorable conditions. We can thus ask why haplorhines, with the exception of callitrichids and humans, have not taken up this option with greater regularity.

Mating seasonality, as expressed in birth seasons, appeared to result in high plasticity in age at weaning, but also was associated with relatively smaller weanlings. Strict seasonal breeding may force statistical variance, in that females who miss a breeding period must then wait until the subsequent season, so that the variance will be a fixed rather than a plastic response. However, the evolution of a breeding season itself represents an adaptation—either to predation levels (Boinski 1987; Ims 1990; Saether and Gordon 1994), to optimal conditions for the maintenance of ovulation and pregnancy (Follett 1984), or to conditions necessary for sustaining peak lactation and weaning (Lee 1987; Meyers and Wright 1993; Pereira 1993c). Thus, the observed plasticity in response to mating and birth seasonality might simply be a consequence of the primary adaptation to a seasonal pattern of reproduction in response to energy availability for lactation and weaning.

There were surprisingly few significant associations between diet type and any measures of relative reproductive rates or mass (c.f. Clutton-Brock and Harvey 1978; Chapman, Wrangham, and Chapman 1995). Only relative fertility extremes differed by diet type, and mixed diets appeared to be associated with higher relative maximum fertility. We suggest that while the primary relationship with diet is mass-specific, phenotypic variation in energy balance does appear to be related to the maximum potential reproductive output.

Summary and Conclusions

Within the primates, plasticity may be less a phenotypic trait than it is a function of local ecology or individual experiences during development and over a series of reproductive events. However, we suggest that plasticity may be traded off against absolute time in life history strategies. While this argument runs counter to predictions from general life history theory, according to which slow strategies should be less constrained, being larger and taking longer to reproduce may effectively reduce the extent of plasticity possible, as time delays have greater potential energy and mortality costs. Alternatively, plasticity may be a reasonable response to environmental uncertainty, which quickens reproductive rates and reduces the investment in individual offspring (see also Richard et al. 2000). This strategy may be characterized as "Live fast, at variable mass, wean early, have high variation in weaning age and litter production, and suffer greater potential infant mortality." We also suggest that the scale of temporal variability may be key, in that short-term variation in energy intake can be responded to only in the context of the current reproductive event. Thus, seasonal extremes may promote plasticity in rates of reproduction and fertility (e.g., litter size) rather than in those features that occur over a longer time scale, such as age at first reproduction or even age at weaning.

It is clear that our exploration of variation has been preliminary, and that we need far more data on the extent of variation possible. Despite this, we suggest that species life history traits are not necessarily invariants, but rather selective outcomes of compromises at the level of the individual between energy expenditure (costs of growth, activity, and reproduction), local environments (changes in energy balance and mortality risks), and social solutions to foraging and reproductive problems. Even in this limited analysis, we have found interesting suggestions about phenotypic plasticity in early growth (both via time and mass) in response to ecology, sociality, and reproductive strategy. We also suggest that mass itself represents a selected aspect of plasticity among primates, as well as being the product of the interaction between early growth, nutrition, and disease. However, plasticity in mass may require trade-offs with mortality during the juvenile growth phase (e.g., Abrams et al. 1996), and these might underlie some of the patterns of growth to weaning seen here. Furthermore, variation in body composition (fat, fat-free mass, skeletal mass) may prove more informative than simple overall mass measures, especially in relation to primate reproduction (Altmann et al. 1993; Pereira and Pond 1995). Such data, however, are not yet available for comparative analyses.

In summary, we have analyzed general life history traits relating to reproduction, but to a large extent we have focused on early infant growth, as this appears to have secondary consequences for fertility within and between individuals (Altmann 1991; Lee 1999). Despite obvious grade differences in overall mass (Fleagle 1978; Smith and Jungers 1997), in brain size (Martin 1996), and in socioecology (van Schaik and Kappeler 1996), both strepsirrhines and haplorhines appear to exhibit similar degrees of plasticity in the life history traits examined here, although which specific traits are plastic varies. In part, our lack of results may simply be due to assessing extremes of reaction norms for a species, rather than being able to directly link population or individual variance to local conditions. However, if "evolutionary rules" (e.g., Clutton-Brock and Harvey 1976) operate across primate species, then plasticity should be evident at this level. More sensitive measures are needed, and far more data are required to further test the potential and causation of primate phenotypic plasticity.

Acknowledgments

We thank Caroline Ross, Andy Purvis, and especially Michael Pereira for their constructive comments on an earlier version of this chapter. We also thank all those who contributed to the data set and critiqued the presentations, and we thank the British Academy for funding (P. C. L.).

4 Matrix Models for Primate Life History Analysis

Susan C. Alberts and Jeanne Altmann

A major theme of this book is the analysis of primate life histories through broad interspecific comparisons of selected life history traits. In this chapter we present a complementary approach using demographic matrix models, which allow for a detailed analysis of the life history (the schedule of survival and reproduction across the life span) of a single species. This approach has not yet been used extensively for nonhuman primates, but as demographic and life history data accumulate on an increasing number of species, matrix models will offer a powerful means of exploring life history variation within species as well as alternative ways of exploring interspecific differences.

Demographic matrix models produce two results of major interest. The first is λ, an estimate of the population growth rate, which is also analytically equivalent to the relative fitness of the mean phenotype in the population (van Groenendael, de Kroon, and Caswell 1988; McDonald and Caswell 1993; Caswell 2001). The second is elasticity (or sensitivity), which estimates the effect of perturbations in life history parameters on λ. In an ecological context, elasticity analyses reveal how population dynamics change as individual life history parameters change. In an evolutionary context, elasticities measure the relative strength of selection on life history parameters. Thus, demographic models can provide evolutionary as well as ecological insights.

Demographic matrix models have several applications. First, they provide a method for evaluating the viability of populations that are threatened or endangered and for assessing management strategies for such populations. For example, Heppell, Walters, and Crowder (1994) used a matrix model and elasticity analysis to examine a declining population of endangered red-cockaded woodpeckers. They determined that a critical factor affecting the population growth rate (λ) was the probability that a nonbreeding male woodpecker would become a breeder; perturbing their model by increasing the proportion of nonbreeding males that became breeders resulted in a relatively large increase in λ. They noted that the rate at which nonbreeders become breeders in this species is dependent on the availability of nesting cavities, and proposed that management efforts should focus on increasing the number of potential nesting cavities for woodpecker populations.

Second, the demonstration that λ is equivalent to fitness and that sensitivities are equivalent to selection gradients (Lande 1982a; McDonald and Caswell 1993; Caswell 2001) means that the strength of selection on life history parameters can be estimated. For example, McDonald (1993) used elasticity analysis to demonstrate that, for male long-tailed manakins, selection on survival is an order of magnitude stronger than selection on fertility, a somewhat counterintuitive result for a bird species with elaborate male displays and extreme variance in male reproductive success. McDonald also demonstrated that selection on prereproductive survival is much stronger for male manakins than it is for females and that male generation times are more than double those of females. The consequences of such sexual differences in demographic parameters have been little explored.

Third, while most uses of matrix demography have focused on within-population analysis, matrix methods can also be used to examine patterns across populations and even between species. For example, Pfister (1998) performed a broad interspecific comparison of phenotypic variance in life history traits and found that over a wide variety of taxa, ranging from annual plants to long-lived vertebrates, phenotypic variance tends to be lowest in traits with high elasticities (traits that have a large effect on fitness). This finding suggests that natural selection has resulted in the evolution of a suite of traits that minimizes temporal variation in fitness, which in turn supports the notion of bet hedging (i.e., that reduced phenotypic variance in fitness is selected for because it increases lifetime fitness: Stearns 1992; Benton and Grant 1999).

The purpose of this chapter is to provide an introduction to demographic matrix methods for the analysis of primate life histories, using examples from an analysis of baboon life histories. The wide applicability of matrix models makes them a useful tool for life history analyses from several perspectives. Their usefulness for conservation applications in particular is unequaled by other methods. Their use for analyzing fitness differences among individuals and for identifying selection pressures overlaps with that of the multivariate methods developed by Arnold and Wade

(1984a,b) and Brown (1988), and we therefore conclude this chapter with a brief comparison of matrix methods with these multivariate methods. For additional introductions to matrix methodology, the reader is referred to van Groenendael, de Kroon, and Caswell (1988), McDonald and Caswell (1993), and Morris and Doak (2002). For more detailed discussions of matrix models, the reader is referred to Caswell (2001).

Background

Life history theory describes the distribution of mortality and reproductive effort over the life span (Roff 1992; Stearns 1992). Life history traits include size at birth, pattern of growth, age at first reproduction, age-specific fertility, age-specific mortality, longevity, and number, size, and sex ratio of offspring. Life history theory asserts that these traits have evolved as a suite, with the target of selection being fitness over the lifetime rather than instantaneous fitness or maximization of any single trait.

The general approach for examining the life history of a particular species is to construct a model of the average life history in a given population, using age-specific rates of survival and reproduction. Collectively, these are known as vital rates, and in a real sense they define a life history: they encompass the probability of surviving for any given time period, the age at first reproduction, the rates of reproduction thereafter, and the average longevity. The model then generates a measure of population growth rate, as well as projected estimates of a population's size and its age distribution over specified time periods. The model can also serve as a point of comparison for variants of the mean life history, making it possible to compare the mean vital rates with a range of alternatives.

Historically, most demographic models have taken the form of life tables, or $l_x m_x$ tables (see Sade et al. 1976; Altmann et al. 1977; Ricklefs 1983; Stearns 1992, chap. 2; Charlesworth 1994, chap. 1, for examples and discussions). In the last two decades, however, developments in demographic analysis have greatly extended classic life table analysis, overcoming some of its limitations and advancing methods for studying variation in vital rates (for excellent introductions, see van Groenendael, de Kroon, and Caswell 1988; McDonald and Caswell 1993). These developments have yielded several important results, three of which are particularly relevant to this chapter. First, life histories that are difficult to describe in terms of age-specific fertilities and mortalities (such as those in which size or social status is a better predictor of vital rates than age) can be analyzed using stage-specific vital rates. Second, perturbation analyses of demographic models provide a simple method for examining the consequences of changes in vital rates, as exemplified by the red-cockaded woodpecker example (Heppell, Walters, and Crowder 1994). Finally, the analytic demonstration (Lande 1982a,b; see also McDonald and Caswell 1993; Caswell 2001) that λ, the measure of population growth rate, is equivalent to the relative fitness of a given life history (i.e., of a set of vital rates) means that matrix approaches can be used to describe the effects of life history changes at the level of the individual (in terms of the relative fitness of different phenotypes) and at the level of the population (in terms of effects on population growth).

Constructing the Model

Projection Matrices and Life Cycle Graphs

A demographic analysis of life history can be formulated as a projection matrix (so called because it allows one to project estimates of population size) or as a life cycle graph. These are exactly equivalent formulations: the matrix representation is more traditional, while the life cycle graph is considered by many to be more intuitive. In the life cycle graph (fig. 4.1A), each age class is represented by a node, and arrows between the nodes represent the probabilities of moving from one age class to the next. Arrows back to the first age class represent age-specific fertilities. In the matrix formulation (fig. 4.1B), the vital rates are represented by the elements, a_{ij}, of the matrix. The first row of the matrix, with elements a_{1j}, represents the expected production of newborns by each age class (known as fertilities; see appendix 4.1), and corresponds to the arrows back to the first age class in the life cycle graph. Subsequent rows represent the probabilities of moving from one class to another. In general, rows can be thought of as representing demographic input to the corresponding age classes, with columns representing their demographic output (fig. 4.1B).

Multiplying the projection matrix by a vector representing the current numbers of animals in each age class (the census vector; see fig. 4.1B) yields an estimate of the population size and age distribution in the next time period (see Caswell 2001, appendix A, for a clear introduction to the rules of matrix manipulation). Repeated multiplication, then, gives the population size and age distribution after any arbitrary number of time units. The time unit over which population size is calculated is known as the projection interval.

Determining the Projection Interval

The first step in constructing a demographic model is determining the projection interval—the period over which one will measure vital rates and take

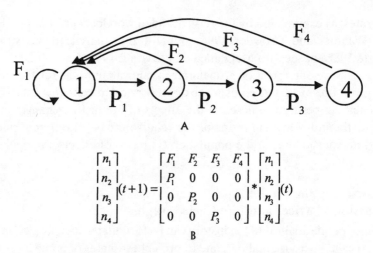

A

$$\begin{bmatrix} n_1 \\ n_2 \\ n_3 \\ n_4 \end{bmatrix}(t+1) = \begin{bmatrix} F_1 & F_2 & F_3 & F_4 \\ P_1 & 0 & 0 & 0 \\ 0 & P_2 & 0 & 0 \\ 0 & 0 & P_3 & 0 \end{bmatrix} * \begin{bmatrix} n_1 \\ n_2 \\ n_3 \\ n_4 \end{bmatrix}(t)$$

B

FIG. 4.1. *(A)* Life cycle graph for a population with four age classes, in which reproduction occurs in each age class. *(B)* Projection matrix with census vectors corresponding to the life cycle graph in part A. Rows can be thought of as representing demographic input to the corresponding age classes, with columns representing their demographic output. For instance, the cell at the intersection of the first row and the third column (a_{13}) designates production of newborns by animals in the third age class (input to the first age class, output from the third). The cell at the intersection of the fourth row and the third column (a_{43}) designates the probability of surviving from the third age class to the fourth (input to the fourth class, output from the third).

censuses. To some extent this is an arbitrary choice, but the interval must be long enough to yield meaningful vital rates (for instance, daily survival and fertility measures will not yield reasonable values for vertebrates). Further, to ensure accurate estimates of vital rates, the interval should not be longer than the duration of the age classes (for instance, if the projection interval were two years, the resulting census data would allow estimation of survival from two to four years of age, but not from two to three years of age; thus, the age classes for such a model must be two years in duration). For primates and many other large-bodied animals, a projection interval of one year is convenient because it often encompasses a single birth season. Estimates of yearly survival and birth rate for each age class are then retrieved from yearly censuses or from continuous observations.

Age-Structured versus Stage-Structured Models

The next step is to determine whether the life history in question is best described by an age-structured model or a stage-structured model. Age-structured models are appropriate if mortality or fertility changes with age, and if age can be accurately measured. For instance, a matrix representation of a population of female baboons in Amboseli, Kenya, is a 27 × 27 matrix describing survival and fertility for 27 age classes (fig. 4.2A); the 21 × 21 matrix for Amboseli males reflects the fact that males in this population have shorter life spans on average than females do (fig. 4.2B). Note that some life history characteristics, such as age at first reproduction and life span, are apparent in the matrix and that others, such as life expectancy, can readily be calculated by simple multiplication of the diagonal elements.

In many cases, however, such fine-grained age classifications are not possible. Instead, researchers identify individuals only as infants, young juveniles, older juveniles, young adults, and so forth. In such cases, stage-based models may be employed, as illustrated by a stage-based life cycle graph for female elk in Yellowstone National Park, USA (fig. 4.3; Dixon et al. 1997). Some cautions are required in constructing stage-based models of this sort, and these will be described in more detail below.

Finally, for some species, an individual's reproductive status and survival probabilities depend more on its position in the social group than on its age (among primates, callitrichids are the best example). In these cases, a stage-based model may be most appropriate, as illustrated by a stage-based model for male red-cockaded woodpeckers, cooperatively breeding North American birds (fig. 4.4; Heppell, Walters, and Crowder 1994).

Cautions for Constructing Stage-Based Models

While age-based matrices (Leslie matrices; after Leslie 1945) are historically most common, stage-based matrices (Lefkovitch matrices; after Lefkovitch 1965) are increasingly popular because their flexibility makes them applicable to a wide range of species and data sets. We have described two types of stage-based models: those in which vital rates are better predicted by social status than by age (as in red-cockaded woodpeckers), and those in which vital rates are age-specific but the data are not sufficiently fine-grained to generate age-based models (as in Yellowstone elk).

In constructing stage-based models, two points bear emphasis. First, the duration of the stage class is independent of the length of projection interval, the period over which vital rates are measured. Regardless of the length of the projection interval and the duration of the stage classes, vital rates are calculated as the number of events per interval (Caswell 2001). For instance, in the model of Yellowstone elk (see fig. 4.3), some classes cover one-year durations while others cover several years (e.g., class 4 comprises three- to seven-year-olds). In both cases, survival and fertility are calculated yearly

Fig. 4.2. Demographic matrices for wild-foraging baboons in Amboseli, Kenya (birth-flow model): (A) females, (B) males.

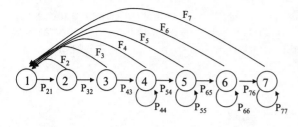

FIG. 4.3. Life cycle graph for female elk in Yellowstone National Park. Stage classes: 1, newborns; 2, yearlings; 3, 2-year-olds; 4, 3–7 years; 5, 8–15 years; 6, 16–20 years; 7, 20+ years. (After Dixon et al. 1997.)

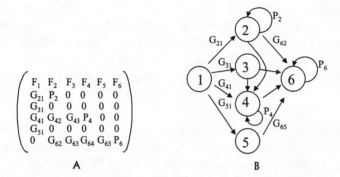

$$\begin{pmatrix} F_1 & F_2 & F_3 & F_4 & F_5 & F_6 \\ G_{21} & P_2 & 0 & 0 & 0 & 0 \\ G_{31} & 0 & 0 & 0 & 0 & 0 \\ G_{41} & G_{42} & G_{43} & P_4 & 0 & 0 \\ G_{51} & 0 & 0 & 0 & 0 & 0 \\ 0 & G_{62} & G_{63} & G_{64} & G_{65} & P_6 \end{pmatrix}$$

A   B

FIG. 4.4. *(A)* Projection matrix for male red-cockaded woodpeckers. Stages: 1, fledgling; 2, helper; 3, floater; 4, solitary; 5, 1-year-old breeder; 6, older breeder. *(B)* Life cycle graph for male red-cockaded woodpeckers; no fertilities and only some transition probabilities are shown. *P*s represent survival probabilities; *G*s represent probabilities of transition from one stage to another. (After Heppell, Walters, and Crowder 1994.)

(i.e., the projection interval is one year). The projection interval is independent of the duration of stages, and must be the same for all classes.

Second, choosing the durations of the stages is a critical step. If stages of long duration are chosen, one runs the risk of pooling together ages that have very different vital rates (Vandermeer 1978; Benton and Grant 1999). For instance, if three-year-old elk have much lower survival rates than four- to seven-year-old elk, then grouping them into a single class of three- to seven-year-olds will inflate their survival, and will consequently inflate λ — the model will, in essence, project that three-year-olds will remain alive and reproduce for longer than they really do. A good rule of thumb is to avoid grouping the early age classes, which often have high and rapidly decreasing mortality rates. Employment of this rule often coincides with available data, as researchers typically have more fine-grained age estimates (and thus survival data) for infants and young juveniles than they do for adults. In the case of Yellowstone elk, the first three classes each represent one-year intervals, and grouping is not employed until well into adulthood, when mortalities are likely to be less variable over larger age spans (see fig. 4.3, Dixon et al. 1997).

Obtaining Vital Rates
Initial Decisions about the Matrix

Vital rates, the elements a_{ij} of the projection matrix, may be obtained from continuous observational data or from periodic census data. In constructing a matrix model, several key decisions must be made prior to beginning. These decisions include (1) which sex will be modeled, (2) whether the population will be modeled as a birth-flow or a birth-pulse population, and (3) whether the censuses (either actual or taken from a long-term database) will occur before the birth season (a prebreeding census) or after (a postbreeding census). The implications of each decision are described below.

Males versus females. Unless females and males exhibit the same vital rates (generally not the case for primates), separate models will be constructed for the two sexes. Within any population, the sexes will vary in their age distributions, in their reproductive patterns, and in how changes in these parameters affect fitness. However, models for each sex should yield roughly equivalent values of λ; this must be the case unless one sex is increasing in frequency relative to the other. For questions of general population dynamics, females are typically modeled (e.g., Yellowstone elk: Dixon et al. 1997), both because of the relative ease of measuring female fertility rather than male fertility and because, particularly among mammals, females are often the nondispersing sex, leading to better data on female survival rates. In some cases, conservation issues may be highly sex-specific, so that one sex rather than the other becomes the focus of demographic models (e.g., male red-cockaded woodpeckers: Heppell, Walters, and Crowder 1994). In evolutionary studies in which selection pressures on life history are likely to be different for the two sexes, both sexes are modeled if possible (e.g., McDonald 1993). Reproductive data for males are often difficult to obtain, but if paternity data come from genetic studies or behavioral data have been validated with genetic data (both of which are being accomplished for a number of primate populations; e.g., de Ruiter, van Hooff, and Scheffrahn 1994; Altmann et al. 1996; Bercovitch and Nürnberg 1996; Borries et al. 1999), the task is somewhat less onerous.

Birth-pulse versus birth-flow populations. Many primate species reproduce seasonally, so that all births occur within a fairly short period. Such species are termed birth-pulse populations. Species such as baboons and the great apes, which reproduce throughout the year, are termed birth-flow populations. In both cases, the matrix model makes the simplifying assumption that age classes are discrete and that population growth is a discrete rather than a continuous process. This assumption is less troublesome for birth-pulse populations than it is for birth-flow populations. Birth-flow methods for calculating matrix elements are designed to mitigate the effects of assuming discrete population processes, and so are more complex than birth-pulse methods. Below we present a birth-pulse model for a hypothetical population of seasonally breeding monkeys, as well as birth-pulse and birth-flow models for female baboons in Amboseli. In some cases, the results of birth-flow and birth-pulse models for the same population may be quite different (Caswell 2001). However, the results of our birth-flow and birth-pulse models for Amboseli baboons are very similar, suggesting that in some cases a birth-pulse model provides an adequate description of population processes, even for birth-flow populations.

Prebreeding versus postbreeding censuses. For a birth-pulse population, if the census takes place before the birth season each year (a prebreeding census), it includes pregnant females and nearly-one-year-olds, but not newborns. Hence, survival during the first year of life is not observed directly, although it can be inferred by taking the difference between the number of pregnant females each year and the number of one-year-olds each subsequent year. In contrast, if the census takes place immediately after the birth season (a postbreeding census), newborns are counted each year, and survival during the first year of life can be observed directly by censusing each newborn cohort both immediately after birth and immediately after its first birthday. Prebreeding and postbreeding censuses yield exactly equivalent estimates of population growth (Caswell 2001). In the birth-pulse models presented here, we employ postbreeding censuses (i.e., we count newborns). We do not cover methods for employing prebreeding censuses, as many of the principles are the same. However, care must be taken to make explicit which type of census is being taken, as it affects calculations of both survival and fertility. We recommend Morris and Doak (2000) and Caswell (2001) to interested readers.

A Birth-Pulse Model for a Hypothetical Population of Seasonally Breeding Monkeys

Here we present a birth-pulse model that employs a postbreeding census for females in a hypothetical monkey population that reproduces seasonally (table 4.1 and fig. 4.5). We have made some simplifying assumptions to make

Table 4.1 Female matrix data for a hypothetical seasonally breeding monkey population

A	B	C	D	E	F	G	H	I	J[a]	K[b]	L[c]
Age	Age class (i)	Census took place when?	Females in this age class gave birth when?	Number of females that entered the age class	Number of age class members that died during the age class (i.e., did not give birth)	Number of age class members that survived to the next age class but did not give birth in the current age class	Number of female offspring born to the age class	$l(i)$ (survivorship)	$P_i = l(i)/l(i-1)$	Birth rate, m_i	$F_i = P_i m_i$
Newborn				1000	—	—	—	1.00	—	—	—
Newborn to 1	1	Just after birth	—	1000	200	800	0	0.80	0.80	0.00	0.00
1 to 2	2	Just after 1st birthday	—	800	300	500	0	0.50	0.63	0.00	0.00
2 to 3	3	Just after 2nd birthday	—	500	100	400	0	0.40	0.80	0.00	0.00
3 to 4	4	Just after 3rd birthday	On 4th birthday	400	80	60	260	0.32	0.80	0.81	0.65
4 to 5	5	Just after 4th birthday	On 5th birthday	320	60	30	230	0.26	0.81	0.88	0.72
5 to 6	6	Just after 5th birthday	On 6th birthday	260	110	40	110	0.15	0.58	0.73	0.42
6 to 7	7	Just after 6th birthday	On 7th birthday	150	100	10	40	0.05	0.33	0.80	0.27
7 to 8	8	Just after 7th birthday	On 8th birthday	50	50	0	0	0.00	0.00	0.00	0.00

[a] P_i is also equal to 1 − hazard, where hazard = (column F/column E).
[b] Birth rate, m_i = column H/(column G + column H).
[c] F_i is also equivalent to column H/column E (births/entering females).

$$\begin{pmatrix} 0.00 & 0.00 & 0.00 & 0.65 & 0.72 & 0.42 & 0.27 & 0.00 \\ 0.80 & 0.00 & 0.00 & 0.00 & 0.00 & 0.00 & 0.00 & 0.00 \\ 0.00 & 0.63 & 0.00 & 0.00 & 0.00 & 0.00 & 0.00 & 0.00 \\ 0.00 & 0.00 & 0.80 & 0.00 & 0.00 & 0.00 & 0.00 & 0.00 \\ 0.00 & 0.00 & 0.00 & 0.80 & 0.00 & 0.00 & 0.00 & 0.00 \\ 0.00 & 0.00 & 0.00 & 0.00 & 0.81 & 0.00 & 0.00 & 0.00 \\ 0.00 & 0.00 & 0.00 & 0.00 & 0.00 & 0.58 & 0.00 & 0.00 \\ 0.00 & 0.00 & 0.00 & 0.00 & 0.00 & 0.00 & 0.33 & 0.00 \end{pmatrix}$$

FIG. 4.5. Demographic matrix for the hypothetical seasonally breeding monkey population in table 4.1.

the process of calculating vital rates clear. In particular, we assume (1) that all females gave birth to a single offspring at the same time each year (and hence gave birth on or near their own birthdays), (2) that yearly census data are available, (3) that the censuses were taken immediately after the births occurred, (4) that all newborns were counted, and (5) that the population is now extinct, so that no population processes were ongoing when the model was constructed. In table 4.1, the numbers of individuals in each age class represent census data that are pooled over a number of years.

Survival (P_i). In the case of a postbreeding census, survival, P_i, is the probability of surviving from age class $(i - 1)$ to age class i. For each age class, the fate of all animals that ever entered that class must be determined; that is, animals must be designated as having survived through the age class or died in it. Survival for each age class can be calculated as simply $(1 - \text{hazard})$, where the hazard is defined, in this simple case, as the proportion dying in the age class (table 4.1, column J).

It is also valuable to calculate survivorship, $l(i)$, which is the probability of surviving from birth to the ith birthday (i.e., to the end of age class i in the case of a postbreeding census: table 4.1, column I). Matrix models begin with age class 1 (there is no 0 age class), but survivorship of newborns, $l(0)$, is retained as a placeholder, and is set to 1.00 (see tables 4.1 and 4.2). Survivorship, $l(i)$, is calculated as the number of females that survive to the ith birthday divided by the number of females ever born. Thus, survivorship is a cumulative measure of survival (the proportion of all animals born that survive to the ith birthday), while survival, P_i, is a conditional one (the proportion that survive age class i, given that they entered age class i). In most presentations of matrix methodology, survival is presented as

$$P_i = \frac{l(i)}{l(i - 1)},$$

and it can be seen in table 4.1 that this yields the same values for P_i as does $(1 - \text{hazard})$.

Birth rate (m_i) and fertility (F_i). The calculation of fertility can be a source of confusion in matrix models, even though it may seem intuitively straightforward. The desired fertility rate for age class i is the expected number of female offspring born to a female *entering* the ith age class. In the case of the birth-pulse population presented in table 4.1, F_i can be calculated directly by dividing the number of female offspring produced by females in age class i by the number of females that ever entered age class i (column H/column E; see notes to table 4.1).

Care must be taken here to distinguish fertility, F_i, from the birth rate, m_i, which is the average number of offspring produced by a female who *reaches her ith birthday* (Caswell 2001, p. 27). More females *enter* the ith age class than reach their ith birthday unless survival for the age class is 1.00 (which will be rare). Thus, the age-specific birth rate, m_i, will be higher than the age-specific fertility, F_i. In table 4.1, age-specific birth rate, m_i, is calculated as the number of female offspring born to females in age class i (column H), divided by the number of females that gave birth in age class i (column H again, assuming all births are singletons) *plus* the number of females that survived through age class i but did not give birth (column G).

In most descriptions of how to produce vital rates, fertility for postbreeding censuses is described as

$$F_i = P_i m_i.$$

In table 4.1 it can be seen that this equation produces exactly the same value of F_i that is produced by dividing number of births by number of females entering the age class (see notes to table 4.1).

A Birth-Flow Model for Female Baboons in Amboseli

Three things distinguish the birth-flow model we constructed for baboons from the birth-pulse model presented above. First, the Amboseli baboons reproduce year-round, so that no clear birth season occurs. This reproductive pattern both makes the population better suited to a birth-flow than a birth-pulse model and makes the calculation of the vital rates less obvious.

Second, for the baboons, we had continuous data rather than yearly censuses. In principle, we could have easily constructed, post hoc, the equivalent of yearly censuses from the continuous data. This would be a reasonable approach to the use of continuous data. However, we wanted the more accurate survival estimates obtainable from continuous data.

Finally, the population was still extant at the time we constructed the matrix, so our analysis included many incomplete life histories (i.e., many females were still alive and contributing to population growth at the time of the last census). Data on incomplete life histories are termed censored data, since the fate of each animal alive at the time of the last census (will it die in its current age class? survive to the next age class? reproduce in the current or future age classes?) is unknown. Including censored data is important, particularly in studies of long-lived species, because the sample of completed life histories is often biased toward animals that died young.

There are several methods for handling censored data, some of which can be used with either yearly census data or continuous data. For instance, Proc Lifetest, a computational program in SAS, produces age-specific survival estimates using numbers of animals that died or were censored in each age class. (For a clear description of how to produce survival estimates manually or using SAS, see Kalbfleisch and Prentice 1980; Lee 1992; Allison 1995.)

Birth-flow survival using continuous data. Survival, P_i, is the probability that an individual in age class i will survive from time t to $t + 1$. To calculate this probability for birth flow populations, the first step, for each age class, is to designate all animals that ever entered that age class as having survived, died, or been censored in that age class. For studies with continuous data on survival, we recommend the following method, which we employed for the Amboseli baboons. We calculated the age-specific hazard rate, H_i, as

$$H_i = N_i/T_i,$$

where N_i = the number of individuals that died in the ith age class and T_i = the total "exposure" time in the ith age class—the cumulative length of time that all individuals that entered the ith age class, including those that died, survived, or were censored in it, spent in it. For instance, in Amboseli, five females died while they were between eighteen and nineteen years of age (i.e., while they were in the nineteenth age class), and exposure time in the nineteenth age class totaled 4091 female-days (11.2 female-years). Some of this exposure time was contributed by the five females that died during the age class, some by females that survived through it, and some by females that were censored during it. The resulting hazard rate for the nineteenth age class is thus 5/11.2, or 0.4464 (table 4.2).

In the case of the birth-pulse model, it was helpful, but not critical, to calculate survivorship, $l(i)$, but for the birth-flow model, both survival and fertility values depend on estimates of survivorship, the probability of surviving from birth to the ith birthday (Caswell 2001, chap. 2). We estimated survivorship, $l(i)$, as follows:

$$l(i) = (1 - H_i)l(i - 1)$$

Table 4.2 Matrix data for female baboons in Amboseli

A	B	C	D	E	F	G	I	J	K	L	M	N	O	P
		Fe-males that en-tered the age class	Fe-males that died in the age class	Fe-males born to mothers in age class = B_i	De-nomi-nator for m_i^a	$m_i =$ B_i/ColF	Exposure time (days)	Hazard[b]	1 − Hazard	$l(i)$	Birth-flow survival	Birth-flow fertility	Birth-pulse survival	Birth-pulse fertility
Age	Age class													
Newborn	(0)									1.0000				
0–1	1	274	59	0	195	0.0000	81536	0.2643	0.7357	0.7357	0.7938	0.0000	0.7357	0.0000
1–2	2	195	22	0	164	0.0000	63146	0.1273	0.8727	0.6421	0.8884	0.0000	0.8727	0.0000
2–3	3	164	14	0	138	0.0000	54577	0.0937	0.9063	0.5819	0.9366	0.0000	0.9063	0.0000
3–4	4	138	4	0	132	0.0000	48684	0.0300	0.9700	0.5645	0.9688	0.0000[c]	0.9700	0.0000
4–5	5	132	4	1	120	0.0083	45106	0.0324	0.9676	0.5462	0.9529	0.1284	0.9676	0.0081
5–6	6	120	7	33	108	0.3056	41033	0.0623	0.9377	0.5121	0.9439	0.2727	0.9377	0.2865
6–7	7	107	5	35	100	0.3500	36843	0.0496	0.9504	0.4868	0.9481	0.2755	0.9504	0.3327
7–8	8	96	5	29	94	0.3085	33539	0.0545	0.9455	0.4603	0.9483	0.2647	0.9455	0.2917
8–9	9	89	4	28	86	0.3256	29906	0.0489	0.9511	0.4378	0.9427	0.2598	0.9511	0.3097
9–10	10	83	5	22	74	0.2973	27636	0.0661	0.9339	0.4088	0.9233	0.2237	0.9339	0.2777
10–11	11	74	6	17	70	0.2429	24900	0.0880	0.9120	0.3729	0.8910	0.1981	0.9120	0.2215
11–12	12	68	8	15	61	0.2459	22135	0.1320	0.8680	0.3236	0.8943	0.2310	0.8680	0.2134
12–13	13	58	4	18	55	0.3273	19402	0.0753	0.9247	0.2993	0.9419	0.2615	0.9247	0.3026
13–14	14	54	2	15	50	0.3000	18469	0.0396	0.9604	0.2874	0.9160	0.2596	0.9604	0.2881
14–15	15	50	6	14	42	0.3333	16819	0.1303	0.8697	0.2500	0.8701	0.2842	0.8697	0.2899
15–16	16	42	5	14	37	0.3784	14113	0.1294	0.8706	0.2176	0.8731	0.3010	0.8706	0.3294
16–17	17	37	4	10	27	0.3704	11785	0.1240	0.8760	0.1907	0.7880	0.2656	0.8760	0.3245

17–18	18	27	7	6	19	0.3158	8181	0.3125	0.6875	0.1311	0.6329	0.2169	0.6875	0.2171
18–19	19	15	5	3	10	0.3000	4091	0.4464	0.5536	0.0726	0.6278	0.2440	0.5536	0.1661
19–20	20	9	2	3	7	0.4286	3066	0.2383	0.7617	0.0553	0.8648	0.1838	0.7617	0.3265
20–21	21	7	0	0	7	0.0000	2562	0.0000	1.0000	0.0553	0.7455	0.2398	1.0000	0.0000
21–22	22	7	3	3	4	0.7500	2153	0.5089	0.4911	0.0271	0.5517	0.3216	0.4911	0.3683
22–23	23	4	1	0	3	0.0000	1124	0.3250	0.6750	0.0183	0.6090	0.1306	0.6750	0.0000
23–24	24	3	1	1	2	0.5000	747	0.4890	0.5110	0.0094	0.4174	0.2144	0.5110	0.2555
24–25	25	2	1	0	1	0.0000	477	0.7657	0.2343	0.0022	0.3796	0.0000	0.2343	0.0000
25–26	26	1	0	0	1	0.0000	365	0.0000	1.0000	0.0022	0.5000	0.0000	1.0000	0.0000
26–27	27	1	0	0	1	0.0000	193	0.0000	0.0000^c	0.0000	0.0000^c	0.0000	0.0000^c	0.0000

[a] Denominator for m_i = females that gave birth in the age class plus those that survived the age class without giving birth.

[b] Hazard = column D/(column J/365.25).

[c] An asterisk next to the value means the value was forced to zero. In the case of F_4, this was because no wild-feeding female below the 5th age class had ever been observed to become pregnant. In the case of P_{27}, this was because, even though our oldest age class was not the oldest possible, it is both close to it and associated with zero fertility, suggesting that it does not contribute to population growth.

Thus, $l(19) = (1 - H_{19})l(18)$, or $(0.5536)(0.1311) = 0.7256$ (table 4.2). Note that $l(0)$, the survivorship of newborns, is by definition 1.0 (table 4.2).

Survival values, P_i, for birth-flow models (Caswell 2001, chap. 2) can then be estimated as

$$P_i = \frac{l(i) + l(i + 1)}{l(i - 1) + l(i)}.$$

(For other possible estimates, see Caswell 2001, chap. 2.)

Birth-flow birth rates (m_i) and fertilities (F_i). Recall that the birth rate, m_i, is the average number of offspring produced by a female who *reaches her* ith *birthday* (Caswell 2001, p. 27), or

$$m_i = \frac{B_i}{N_i},$$

where B_i is the number of births to females in age class i and N_i is the number of females that survived to their ith birthday without giving birth plus the number that gave birth in the ith age class (i.e., on their ith birthday in the hypothetical case presented above). Because females in birth-flow populations do not reproduce on their birthdays, it is less obvious in this case which animals should be included in the denominator.

For the Amboseli birth-flow model, we included four classes of females in the denominator for m_i: (1) females that survived through the ith age class without giving birth in that age class, (2) females that survived through the ith age class and gave birth in that age class, (3) females that gave birth in the ith age class and then died without surviving through the entire age class, and (4) females that gave birth in the ith age class and then were censored before their fate was known (i.e., on 31 December 1999, the last census date we included in the analysis, these females had produced an infant and were still alive in the age class). Females that were censored without giving birth in the ith age class might eventually contribute to the birth rate in that age class, but we deemed it incorrect to include them in the denominator of m_i if they had not yet done so. For instance, in Amboseli, three female offspring were born to females in the nineteenth age class, and ten females either gave birth in that age class or survived childless through it, yielding a value of $m_{19} = 3/10$, or 0.30 (table 4.2).

Fertility in birth-flow models may be estimated as

$$F_i = l(0.5)\left(\frac{m_i + P_i m_{i+1}}{2}\right)$$

(Caswell 2001). This formula reflects the fact that in forming age classes from a continuous age distribution, we have given up all knowledge of age within each age class. Thus, to account for fertility over the entire projection interval, we must take into account the fact that some reproductive females in age class i will transition to age class $i + 1$ with probability P_i. Further, their offspring may be produced at any time during the projection interval, and must survive for varying lengths of time to be included in the next census; on average, they must survive half the projection interval. Thus, birth-flow fertilities depend on an estimate of $l(0.5)$ as well as estimates of m_i. $l(0.5)$ may be estimated directly from the continuous data or by interpolation from the values of $l(0)$ and $l(1)$ as follows (Caswell 2001, chap. 2):

$$l(0.5) = l(0)\sqrt{l(1)}.$$

We recommend that those interested in employing birth-flow estimates of survival and fertility read Caswell (2001, chap. 2) for a full description of the logic behind them.

Female Baboons in Amboseli Modeled as a Birth-Pulse Population

In order to explore the differences between modeling a continuously breeding population as a birth-pulse population and modeling it as a birth-flow population, we developed a birth-pulse model for the Amboseli females, assuming postbreeding censuses.

Birth-pulse survivals were taken directly from the hazard rate, $P_i = (1 - H_i)$, as described for the hypothetical population in table 4.1. Similarly, we estimated birth-pulse fertility, F_i, as for the hypothetical population in table 4.1, as

$$F_i = P_i m_i.$$

We estimated m_i in the same manner for the birth-pulse model as for the birth-flow model.

Our resulting birth-pulse estimates of survival and fertility for the Amboseli population are presented in table 4.2 for easy comparison with the birth-flow estimates. Birth-flow estimates of vital rates show less variation over age classes, which reflects the fact that they are like moving averages. That is, in the birth-flow model, the vital rates for each age class include contributions from the birth rate and survivorship estimates of previous and successive age classes.

Survival and Fertility for Males

Survival and fertility calculations for male baboons (and males of many mammal species) are less straightforward than those for females. Survival estimates are complicated by the facts that males disperse from their natal group and that some dispersing animals have unknown fates, so that the duration of their lives is uncertain. (In species with female dispersal, this problem will apply to females rather than, or in addition to, males, but the principles are the same.) Fertility estimates are complicated by the difficulty of establishing paternity. Here we present methods for estimating both survival and fertility for males. Using these methods, we constructed a birth-flow model for male baboons in Amboseli so that we could compare its output to that of the female model.

Male survival. In the case of baboon males, we estimated P_i for the predispersal age classes (1–6) in the same manner as for females (table 4.3). We estimated survival after the age of dispersal by examining the age distribution. The age distribution gives exact measures of survival from one age class to the next, assuming (1) that the population produces and receives migrants at the same rates, (2) that if the population includes age-biased groups (such as all-male bands in which old or juvenile males are overrepresented), such groups are included in the census in proportion to their representation in the population, and (3) that the population is at equilibrium and is neither growing nor shrinking. If these assumptions are met, then any drop in numbers from one age class to the next will represent mortality in a stable population. In a growing population, however, the age distribution will underestimate survival. That is, in a growing population, the drop in numbers between age class 1 and age class 2 will reflect not only mortality from age class 1 to 2, but also the fact that the cohort represented by age class 1 was larger to begin with than that represented by age class 2. Before equilibrium is reached, population growth will be reflected in the age distribution at successive time periods as a wave that is moving through the population from younger to older age classes over time. Similarly, in a declining population, the age distribution will overestimate survival.

Our analysis of female baboons suggested that the population in Amboseli was growing. Consequently, when we constructed survival estimates for males based on their age distribution, we adjusted those estimates by examining the relationship between age distribution and survival for females. We reasoned that the female and male populations were growing at the same rate and, because both age distribution and survival were known exactly for females, that the female data would provide us with a measure of the extent to which the age distribution overestimated mortality. In particular, we regressed observed female survival (l_x) on the observed female age distribution and used the slope and intercept of the resulting line as a correction factor for our male survival estimates (fig. 4.6, table 4.3). We did not employ this correction for the oldest age classes (16–20), as both

Table 4.3 Matrix data for male baboons in Amboseli

Age	Age class	Exposure time (days)	Males that died in the age class	Hazard	Survivorship ($l(i)$), based on observed hazard	Survivorship ($l(i)$), based on age distribution	Flow survival for predispersal classes	Flow survival for dispersing classes	Birth rate (m_i)	Flow Fertility
Newborn					1.0000				0.00000	0.0000
0–1	1	62735	47	0.2736	0.7264		0.7825		0.00000	0.0000
1–2	2	49480	19	0.1403	0.6245		0.9122		0.00000	0.0000
2–3	3	40963	3	0.0267	0.6078		0.9337		0.00000	0.0000
3–4	4	34146	10	0.1070	0.5428		0.9167		0.00000	0.0000
4–5	5	32108	5	0.0569	0.5119		0.9588		0.00000	0.0000
5–6	6	29717	2	0.0246	0.4993	0.5217	0.9427		0.00000	0.3734
6–7	7	28170	7	0.0908	0.4540	0.4849		0.9390	0.00000	0.7759
7–8	8					0.4603		0.9416	0.06669	0.7791
8–9	9					0.4297		0.9311	0.57542	0.7663
9–10	10					0.3990		0.9112	0.81552	0.7423
10–11	11					0.3561		0.8944	0.64931	0.7221
11–12	12					0.3193		0.8820	0.81757	0.7074
12–13	13					0.2764		0.8456	0.55429	0.6652
13–14	14					0.2273		0.8295	0.49193	0.6467
14–15	15					0.1906		0.7816	0.30070	0.5950
15–16	16					0.1361		0.6665	0.33257	0.4747
16–17	17					0.0816		0.6251	0.07760	0.4334
17–18	18					0.0544		0.8002	0.06467	0.6252
18–19	19					0.0544		0.7500	0.04850	0.5859
19–20	20					0.0272		0.3333	0.07760	0.2131
20–21	21					0.0000		0.0000	0.00000	0.0000

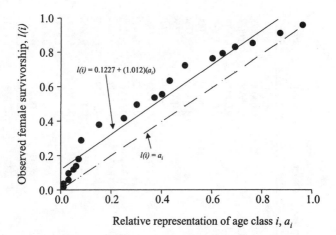

FIG. 4.6. Relationship between survivorship, $l(i)$, and age distribution, a_i, for wild-foraging female baboons in Amboseli. The x-axis represents the relative age distribution, normalized to 1.0 for age class 1. In the ideal case of an equilibrium population with $\lambda = 1$, $l(i) = a_i$ (dashed line). In Amboseli, the observed relationship between $l(i)$ and a_i (represented by the solid circles) is described by the solid line $l(i) = 0.1227 + (1.012)(a_i)$.

sample sizes and survival in those age classes were low for both males and females; thus, this detailed correction was unlikely to improve those estimates very much.

This procedure for estimating survival may be employed for either age-based or stage-based models as long as animals progress from one class to the next sequentially (e.g., the method would not work for red-cockaded woodpeckers, where animals in a given class may move to or come from a number of other classes). In either case, accurate assignment to age or stage classes is critical. In the case of Amboseli males, our sample included 218 natal males with known birth dates. Of 114 immigrant males, 78 were assigned ages based on an estimation process developed for the Amboseli population (see Alberts and Altmann 1995 for details); 36 males with unassigned ages were excluded from the analysis.

Male birth rates. For males, some approximation to birth rate must be identified. Among Amboseli males, behavioral observations of mate guarding (consortships) correspond well to genetic paternity assignments (Altmann et al. 1996). Therefore, we used observational measures of mating success to approximate birth rates. We measured the proportion of available female consort hours obtained by males of each age class, then distributed live births of male newborns ($n = 194$) across those age classes accordingly. Birth rates, m_i, were then calculated as the number of live male births per age class divided by the total number of male-years per age class (table 4.3). Fertility, F_i, was calculated from P_i and m_i as a birth-flow estimate, in the manner described for females.

Model Output

Here we focus on a subset of four results yielded by matrix models (see Caswell 2001 for a complete discussion). These results were all obtained from fairly straightforward manipulations of the matrix, which can be accomplished using computational programs such as Mathematica (Shuchat and Shultz 2000), Maple (Kamerich 1999), or Matlab (Pratap 1998). In appendix 4.2 we present a Matlab program that calculates these parameters.

Population Growth Rate

Population growth rate, λ, is the dominant eigenvalue of the demographic matrix (see Caswell 2001, appendix A, for a clear description of eigenvalues and eigenvectors). It is a direct measure of the rate at which the population is growing or shrinking, given the set of vital rates used in the model. Hence it is important for assessing any population deemed in need of management, and it is a key result for estimating the viability of threatened or endangered populations. λ also represents the mean fitness of a population (Lande 1982a; McDonald and Caswell 1993; Caswell 2001), and as such is the means by which alternative phenotypes (alternative sets of vital rates) are judged vis-à-vis the likelihood that they will spread in the population once introduced (for an excellent comparison of various measures of population increase and fitness, including λ, see Stearns 1992, chap. 2).

Values of λ for the hypothetical population shown in table 4.1, as well as for the birth-flow and birth-pulse models for Amboseli, are shown in table 4.4. As noted earlier, separate models for males and females in a population should yield roughly equivalent values of λ unless one sex is increasing in frequency relative to the other. In the case of the wild-foraging Amboseli baboons, the female model yields $\lambda = 1.039$ and the male model yields $\lambda = 1.036$; therefore, the population is growing. The two models produce remarkably close estimates of λ given the challenges associated with estimating male fertility and survival.

Stable Age Distribution

A striking property of most populations is that if age- or stage-specific mortality and fertility schedules are constant from one generation to the next, each population will converge on its own characteristic and unchanging age

Table 4.4 Values of λ for the models presented
in this chapter

Model[a]	λ
Hypothetical birth-pulse monkeys	0.914
Amboseli baboon females, birth-flow	1.039
Amboseli baboon females, birth-pulse	1.034
Amboseli baboon males, birth-flow	1.036

[a]Amboseli baboon models are for wild-foraging animals only.

(or stage) distribution (Stearns 1992, chap. 2; Charlesworth 1994, chap. 1). This stable age distribution is described by the right dominant eigenvector of the matrix (Caswell 2001, appendix A) and is a direct consequence of the population's vital rates. The population will converge on a stable age distribution regardless of its initial age distribution and regardless of its value of λ. The meaning of a stable age distribution is that the proportional representation of age classes remains the same as the population grows or shrinks. In most natural populations, both stochastic and deterministic processes result in vital rates changing over time, so that populations rarely if ever reach stable age distributions. Nonetheless, populations that experience relative stability will converge on a stable age distribution.

The Amboseli baboon population exhibits a sex difference in its stable age distributions (fig. 4.7), such that a larger proportion of males are in the younger age classes, and a smaller proportion in the older ones, than is the case for females. This difference reflects the generally higher mortality rates of males as compared with females, a typical pattern for many primates and, indeed, for many animals, including humans (Shapiro, Schlesinger, and Nesbitt 1968), red deer (Clutton-Brock, Albon, and Guinness 1988), lions (Packer et al. 1988), and Belding's ground squirrels (Sherman and Morton 1984; see also discussions in Clutton-Brock, Albon, and Guinness 1985; Trivers 1985, chap. 12).

The sex difference in age distributions for the Amboseli baboons has some potentially interesting behavioral implications. At any one time, juvenile and infant males will have fewer adult role models than will juvenile females (35% of males are adults, 53% of females are adults). However, because secondary dispersal results in frequent changes in the identities of the adult males in a group, young males may over time have a larger set of role models, as well as ones that come from more diverse backgrounds, than females do (see Pereira 1988b for a discussion of the importance of same-sex role models in the development of sex-typical behavior in nonhuman primates). Differences in age cohort sizes within each sex also have implica-

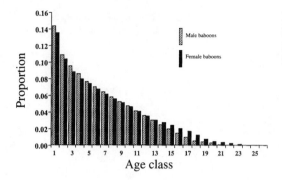

Fig. 4.7. Stable age distributions for wild-foraging female and male baboons in Amboseli.

tions for behavior. For instance, the fact that each cohort shrinks as it ages means that in species such as baboons, in which age cohorts are likely to be paternal sibships (Altmann 1979; Altmann et al. 1996; Smith 2000), older females will have fewer paternal relatives with which to interact than younger females. The consequences of this difference for decisions involving agonistic support, grooming, and patterns of social group fission are just beginning to be explored (Smith 2000).

Reproductive Value

At any given time, the individuals in a population differ in the extent to which they will contribute to future population growth. Their expected contribution depends on the age or stage class they currently occupy, their expected changes in fertility as they move between classes, and on the probability that they will survive to reproduce again. This class-dependent expected contribution to future generations is known as reproductive value, a concept first developed by R. A. Fisher (1930; see also Stearns 1992, chap. 2; Charlesworth 1994, chap. 1; Caswell 2001, chap. 4), and is defined by the left dominant eigenvector of the demographic matrix (see Caswell 2001, appendix A). Reproductive value is usually scaled to the value of the first class (so that the reproductive value of the first class is 1). With each successive interval that a young animal survives, its likelihood of reproducing increases; hence its reproductive value typically increases steadily from birth until near the age of first reproduction. Its reproductive value then drops because the expected number of future young declines as the animal ages. The rate at which reproductive value declines reflects the rates of adult mortality and reproductive senescence, a finding that helps to develop our intuition that selection events occurring late in life have relatively little effect on overall fitness (see fig. 4.8 for reproductive values for Amboseli baboons).

In primate studies, reproductive value has figured prominently in mod-

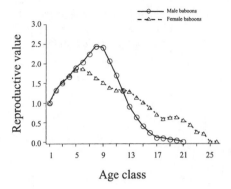

FIG. 4.8. Reproductive values for wild-foraging female and male baboons in Amboseli.

els of, and debates about, the evolution of dominance patterns within matrilines (e.g., Hrdy and Hrdy 1976; Chapais and Schulman 1980; Schulman and Chapais 1980; Horrocks and Hunte 1983). The general argument has been that the reproductive value of each female in the family will have consequences for the relative rank that her kin "allow" her to occupy, because of her effects of their inclusive fitness. Thus, for instance, Schulman and Chapais (1980) propose that when rank is contested between sisters, mothers will support the daughter with the highest reproductive value (see also the critique of this model by Horrocks and Hunte 1983). Recent work with the Amboseli baboons (Combes and Altmann 2001) supports a reproductive value model for the pattern of rank reversal between mothers and daughters. Reproductive value has other potential implications for the evolution of behavior (see, e.g., discussion and cautions in Charlesworth 1994, chap. 5, pp. 237–239).

Perturbation Analysis: Sensitivities, Elasticities, and Other Simulated Life History Changes

One of the most powerful and useful applications of the matrix model is perturbation analysis. Each vital rate in a matrix model has a characteristic sensitivity, which is an estimate of the extent to which λ changes as that vital rate undergoes small changes (and while other vital rates are held constant) (fig. 4.9).

The sensitivity of a matrix element, s_{ij}, is defined as

$$s_{ij} = \frac{\partial \lambda}{\partial a_{ij}}.$$

Thus, a vital rate with a high fitness sensitivity is one for which changes result in a relatively large change in λ. The changes are measured from an initial starting value for that vital rate, which is the specific mean value a_{ij} in the

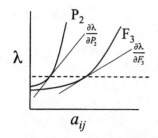

FIG. 4.9. Two vital rates from Figure 4.1 and their hypothetical effect on λ. λ_0 (denoted by the dotted line) is the observed λ in the population. Sensitivities are defined by the slope of the lines tangent to the curves at λ_0. Note that in neither case is the relationship between a and λ linear; in general, there is no reason to expect it to be so.

matrix; a different mean value, derived from a different population or from different time periods for the same population, could have a different sensitivity if the effects on λ of one or more vital rates are not constant throughout the range of that vital rate.

Sensitivity can be calculated as

$$s_{ij} = \frac{v_i w_j}{\sum\limits_{i=1}^{\infty} v_i w_i},$$

where v_i is the reproductive value of class i and w_i is the proportional representation of class i in the population. Note that the calculation of sensitivity is based on changes in fitness in response to infinitesimally small, absolute changes in a_{ij} that are unspecified in magnitude. Sensitivities answer the question, if we perturb each matrix element by some small amount, what is the consequent change in λ? In other words, what is the slope of each relationship between λ and a_{ij}? We cannot extrapolate from the sensitivity analysis to predict the effect of large changes in a_{ij} unless we assume that the relationship between λ and a_{ij} is (at least locally) linear.

A related, alternative measure of the effect on λ of matrix perturbations is elasticity. Elasticities are standardized sensitivities that measure the effects of proportional changes in vital rates. That is, elasticities report the effect of perturbations that are all of the same relative (and not absolute) magnitude. Elasticities answer the question, if we perturb each matrix element by the same relative amount (e.g., 0.05%), what are the consequent changes in λ? Elasticities, e_{ij}, are defined and calculated as

$$e_{ij} = \frac{\partial \ln(\lambda)}{\partial \ln(a_{ij})} = \frac{a_{ij} s_{ij}}{\lambda}.$$

The elasticities of a projection matrix sum to 1 over the entire matrix and can be thought of as the relative contributions of each element a_{ij} to λ. An important difference between sensitivities and elasticities is that matrix elements for which $a_{ij} = 0$ (such as the fertility of the newborn age class) can have nonzero sensitivities, but all matrix elements for which $a_{ij} = 0$ will have $e_{ij} = 0$.

Perturbation analyses for wild-foraging female and male baboons in Amboseli indicated that λ is much more sensitive to changes in survival than to changes in fertility; fertility represents just 9% of the total elasticity in the life history for both males and females (fig. 4.10). Prereproductive survival, in contrast, represents 37% of the total elasticity in the female matrix and 62% of the total elasticity in the male matrix. This difference between males and females results entirely from the difference in age at first reproduction; elasticity values for survival in each of the first four age classes are nearly identical for the two sexes. In comparing a range of social vertebrates, McDonald (1993) showed that survival commonly shows higher elasticity than does fertility, particularly in long-lived species. McDonald also discussed the significance of sex differences in elasticity patterns, noting that such differences may be the consequence of sexual selection and thus may amount to sexually selected characteristics.

Perturbation analysis may also involve direct manipulation of matrix entries. The advantage of direct manipulation is that one is not confined to the infinitesimal changes described by the partial derivatives. For instance,

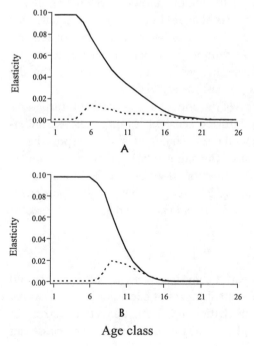

FIG. 4.10. Elasticity in survival (solid line) and in fertility (dashed line) as a function of age class for *(A)* female baboons and *(B)* male baboons in Amboseli.

if a species such as the savanna baboon experiences infant survival values in natural habitats that range from 0.5 to 0.9, elasticity analysis alone will not capture the consequences of this range of values. Instead, matrix entries for the infant age class can be directly manipulated so that the effect of the whole natural range of infant survival on λ can be examined. Mills, Doak, and Wisdom (1999) used this method of manual changes in matrix entries in a very useful discussion of the assumptions and limitations of elasticity analysis.

Perturbation analysis is increasingly being used to assess management strategies for threatened populations. Heppell, Crowder, and Crouse (1996) evaluated the practice of "head-starting" young turtles, in which hatchlings are reared in captivity and then released after the age of high mortality in the wild. The authors concluded that survival elasticity is much lower for hatchlings than for adults, so that headstarting is of little or no value without efforts to increase adult survival as well. Crooks, Sanjayan, and Doak (1998) reached a similar conclusion for cheetahs, in which high cub mortality has been cited as a major factor limiting wild populations. In contrast to the analyses for these threatened populations, our elasticity analyses of the Amboseli baboon models indicated that throughout the range of survival values seen in natural populations, juvenile survival has a much greater effect on λ than either adult survival or fertility (fig. 4.10) (S. C. Alberts and J. Altmann, unpub.). Moreover, this is the case whether we consider starting rates that would result in a declining (λ < 1.00), stationary (λ = 1.00), or increasing (λ > 1.00) population.

Whether a vital rate changes in response to an ecological change, or in response to selection, will depend on the potential plasticity of that vital rate for a given species. From an evolutionary point of view, the great utility and appeal of elasticity is that it provides a measure of the relative strength of selection on vital rates (McDonald and Caswell 1993; Benton and Grant 1999). Whether a given vital rate responds to selection will depend, of course, on the presence of additive genetic variance for the vital rate and on genetic covariance between rates. Elasticities alone will not predict the response to selection, but will estimate the strength of selection on various rates.

Limitations of the Model

In spite of their power and utility, matrix models contain several important assumptions and limitations. Two key assumptions are density independence and time invariance of vital rates. Many populations violate these assumptions. However, methods are available for modifying the models to relax these assumptions or to test the robustness of population projections when these assumptions are violated (e.g., van Groenendael, de Kroon, and Caswell 1988; Benton, Grant, and Clutton-Brock 1995; Benton and Grant 1996, 1999; Grant and Benton 2000; Wisdom, Mills, and Doak 2000). For instance, Benton and Grant (1996, 1999) report that while elasticity analysis is fairly robust to the assumption of time invariance in some circumstances, this is not usually true for short-lived organisms or when variance in vital rates is very high. Wisdom, Mills, and Doak (2000) have developed a probability-based resampling method that addresses this problem by incorporating variance in vital rates into estimates of elasticities (Benton and Grant 1999; Wisdom, Mills, and Doak 2000).

An increasingly common method of exploring the effect of variance in vital rates is the use of stochastic models, which investigate the effects of environmental variance, demographic variance, or both (Armbruster and Lande 1993; Gross et al. 1998; Kendall 1998). Stochastic models may reveal rather different population dynamics than deterministic models of the sort explored here. They are important tools where variable environments result in widely varying vital rates or where populations are small, so that demographic stochasticity becomes an issue in population dynamics (Caswell 2001).

Another important limitation of matrix models is that they do not generally incorporate covariance among vital rates; sensitivities and elasticities are explicitly calculated as partial derivatives of single matrix elements, holding all other elements constant. Empirical data have shown, however, that covariance among life history traits exists (e.g., Stearns 1989b; Benton, Grant, and Clutton-Brock 1995). In primates, for example, infant survival sometimes affects the mother's future reproduction (e.g., Altmann, Altmann, and Hausfater 1978 for baboons; Tanaka, Tokuda, and Kotera 1970 for macaques), and reproduction may increase maternal mortality (e.g., Altmann 1980 for baboons; Westendorp and Kirkwood 1998 for humans). Van Tienderen (1995) provides an excellent introduction to the use of integrated sensitivities, which incorporate covariance among vital rates in the calculation of sensitivities. They are calculated as the ordinary (not partial) derivatives of λ on a matrix entry, as follows:

$$\frac{d\lambda}{da_i} = \frac{\partial\lambda}{\partial a_i} + \sum_{j \neq i} \frac{\partial\lambda}{\partial a_j}\frac{\partial a_j}{\partial a_i}$$

Van Tienderen discusses the problems of estimating covariances among matrix entries and provides examples of such covariances and of integrated sensitivities using both animal and plant taxa.

An additional caution in interpreting elasticities is that, because they are measures of how λ changes with infinitesimal changes in a vital rate, they may not accurately predict the consequences of large perturbations in vital rates (Benton and Grant 1999; Mills, Doak, and Wisdom 1999), such as might occur with a major environmental change or the introduction of human-associated food enhancement. This issue is of particular importance when environments are highly variable, so that vital rates change greatly from one generation to the next, or when management techniques for threatened populations are likely to result in large changes in vital rates. Mills, Doak, and Wisdom (1999) show that large perturbations may actually cause changes in λ that are of the opposite sign from those predicted by elasticities, particularly when several rates are simultaneously changed by different amounts. They caution that manual perturbations of the matrix (in which matrix elements are modified directly and λ sensitivities are recalculated) are an important supplement to standard sensitivity analysis.

Summary and Conclusions: Why Employ a Matrix Model?

There is wide general consensus that matrix models are the best approach to understanding population dynamics, and this makes them the method of choice for researchers working on endangered, threatened, or rare species or species that require management of any sort. However, other methods are available for examining selection on life history components (Arnold and Wade 1984a,b; Brown 1988). Matrix models have some advantages over these methods, as well as some limitations. (The methods of Arnold and Wade and Brown also have advantages and disadvantages relative to each other that will not be discussed here; the reader is referred to Brown 1988 for this discussion.)

First, for researchers working on primates or any other long-lived species, the multivariate methods of Arnold and Wade and Brown for estimating fitness and the strength of selection require extensive longitudinal data on individuals. Matrix methods can certainly utilize such detailed long-term data, but they require at a minimum careful, repeated yearly census data over some period of time, with individuals in the census accurately assigned to age classes. Indeed, in some circumstances, matrix methods do not require identification and tracking of individuals from year to year. If individuals can be accurately assigned to age classes so that age structure and age-specific fertility can be estimated, then cross-sectional data without individual identification is sufficient to construct a matrix model. The difficulty here is that without some information on known individuals, mortality can be estimated only by examining the age distribution, which may over- or underestimate survival and thus inflate or deflate λ respectively. Moreover, one will not generally be able to analyze covariance among traits in the absence of information on individuals.

Second, the formal structure of matrix models makes it difficult to overlook any parameters; the survival and fertility of each recognized age or stage must be included. This is not true of the multivariate methods, in which the researcher chooses a set of fitness components presumed to encompass total fitness. For instance, in the male bullfrog example developed by Arnold and Wade (1984b), total fitness is calculated as the product of number of mates, zygotes per mate, and hatchlings per zygote. Survival from hatchling to adult and adult survival are missing as components of fitness. Thus, the structure of the model is determined by the researcher's a priori beliefs about which phases of the life cycle are important. Such assumptions will limit the capacity of the model to capture important life history variance and may result in misestimations of the strength of selection. For instance, in their male bullfrog example, Arnold and Wade demonstrate strong selection for large body size, but this finding is difficult to interpret without knowing whether large size also affects adult survival, which is excluded from the model.

Age-specific changes in fertility and survival are also largely ignored by the multivariate methods. Brown's approach, in particular, designed to overcome the limitations of the Arnold and Wade approach (Brown 1988), explicitly averages fitness components over the entire life span and disregards changes with age. The Arnold and Wade approach can incorporate age-specific changes by separating the fertility component of fitness into several age-specific fertilities, but this is not built into the approach, and again depends on whether the researcher decides a priori that such changes may be important. Matrix models, by forcing us to examine the entire life cycle and to explicate age-specific changes in survival and fertility, draw our attention to age-specific changes in fitness, which theory indicates are of great importance in the evolution of life histories (Roff 1992; Stearns 1992; Charlesworth 1994).

Matrix models also make it relatively easy to examine the effect of variation that is not directly observed in the study population. The multivariate methods are confined to the variation directly observed in the study population, while matrix methods allow researchers to explore the entire range of variation in life history parameters seen in the species through perturbation analysis. Thus, while both methods identify selection pressures, matrix methods allow more explicit analysis of the effect on fitness of varying life history parameters.

Matrix models also have significant limitations, some of which are described above. An important limitation not yet mentioned is that matrix models in themselves provide no clues about which phenotypic traits influence vital rates. Thus, a matrix model that paralleled the analysis of male bullfrogs developed by Arnold and Wade (1984b) would identify the phases of the life cycle under the strongest selection pressure (for example, adult fertility or zygote survival), but would not estimate the effects of male body size on these critical life cycle phases, or even identify male body size as an important trait. Similarly, perturbation analysis of the baboon matrix model indicates that infant and juvenile survival are under strong selection, but does not identify factors that contribute to variance in infant and juvenile survival.

However, matrix methods are excellent guides for subsequent analyses of sources of variance in fitness. Analyses of the sort exemplified in Pereira and Leigh (chap. 7), Godfrey et al. (chap. 8), and Ganzhorn et al. (chap. 6; all this volume), which describe detailed examinations of the causes and consequences of variation in particular life history stages, would be especially powerful if informed by a matrix analysis that identified critical stages of the life history. Another approach would be to follow a matrix analysis with a path analysis or multiple regression. Van Tienderen (2000) proposes a hierarchical method that incorporates matrix models and elasticity analysis with multivariate selection analysis of important phenotypic traits. He also provides an excellent discussion and comparison of the parameters used in multivariate models versus those used in matrix models (van Tienderen 2000). The two approaches are complementary and together offer a richness and completeness that is not possible with either alone.

Matrix models have multiple applications that make them flexible and useful tools for asking both evolutionary and ecological questions. They can provide insights into management strategies for threatened populations, identify life history parameters that are under selection within a given species, or even facilitate interspecific comparisons that may shed light on long-standing problems in life history theory. Their formal structure ensures that all life history stages of the animal are included in the analysis, so that they help to develop and correct our intuition about the biology of our study animals. The limitations of matrix models are increasingly being resolved, so that density dependence, variability over time, and covariance among life history parameters can be incorporated into the models. Especially in combination with other, more traditional techniques for identifying the importance of phenotypic variance, they provide an outstanding tool for primate life history analysis.

Acknowledgments

We gratefully acknowledge primary financial support from the Chicago Zoological Society and from grants NSF IBN-9223335, IBN-9422013, and IBN-9729586 to J. A. The Office of the President of Kenya and the Kenya Wildlife Service provided permission to work in Amboseli; R. Leakey, J. Else, N. Kio, D. Western, and the staff of Amboseli National Park provided cooperation and assistance. M. Isahakia, C. S. Bambara, J. Mwenda, O. Mushi, C. Mlay, the members of the pastoralist communities of Amboseli and Longido, and the Institute for Primate Research in Nairobi provided assistance and local sponsorship. We thank Juliet Pulliam for writing the Matlab program shown in appendix 4.2, and for her helpful contributions in thinking about birth-flow survival and fertility estimates and how best to calculate them using the Amboseli data. W. F. Morris also kindly provided a Mathematica program for calculating eigenvalues and eigenvectors, which we used in our initial analysis of the Amboseli data. We thank W. F. Morris, D. B. McDonald, J. H. Jones, L. F. Keller, E. Simms, S. Levin, T. Wooton, H. Caswell, P. Grant, and D. Rubenstein for helpful discussions of our model. J. Hollister-Smith and J. H. Jones provided helpful comments on the manuscript.

Many people have contributed to the Amboseli long-term field data; particularly important contributions have been made by R. S. Mututua, S. Sayialel, A. Samuels, P. Muruthi, S. A. Altmann, and J. K. Warutere. We also thank S. L. Combes, B. King, J. Mann, S. McCuskey, R. Noë, B. Noë, M. E. Pereira, D. Post, T. Reed, C. Saunders, J. Scott, D. Shimizu, J. Shopland, J. Silk, S. Sloane, K. Smith, J. Stelzner, K. Strier, J. Walters, and D. Takacs for their contributions to the long-term data. M. Else and especially V. Somen have helped facilitate the fieldwork from Nairobi for many years.

Appendix 4.1

Reproductive Rate Terminology

The terms "fecundity" and "fertility" are used in a variety of ways in the literature. When developing a model and researching existing models, it is important to identify the meaning that each author attributes to these terms, rather than simply assuming a standard usage. For instance, the term "fecundity" is used to describe the top row matrix elements (F_i) by some authors (e.g., van Groenendael, de Kroon, and Caswell 1988; Heppell, Walters, and Crowder 1994; Pfister 1998), while others use the term "fertility" for F_i (e.g., McDonald 1993; Crooks, Sanjayan, and Doak 1998). Following Jenkins (1988), McDonald and Caswell (1993), and Caswell (2001), we have used the term "fertility" to indicate the values (F_i) in the top row of the projection matrix: the expected number of newborns that will have been produced by time ($t + 1$) per individual in age class i at time t. In this terminology, "fertility" refers to the realized reproductive rate ($m_i P_i$), while "fecundity" refers either to the unrealized maximum potential reproductive output (Caswell 2001, chap. 2) or to the uncorrected birth function, m_i (Jenkins 1988).

Similarly, various terms have been used to denote m_i. We have used the term "birth rate," but others may use the terms "natality" (e.g., Sade et al. 1976), "fertility rate" (e.g., McDonald and Caswell 1993), or "fecundity" (e.g., Jenkins 1988). Again, rather than assuming a standard usage, the meaning implied by the author must be identified.

Appendix 4.2

Matlab Program for Calculating Eigenvectors and Eigenvalues

This program employs a birth-pulse model to find the stable population growth rate (λ), stable age distribution, age-specific reproductive values, and various elasticity and sensitivity values for an age-structured population based on survival and fecundity values. This program was written in Matlab version 5.3.

STEP 1: Create a Leslie matrix with survival and fecundity values. First we create two vectors, P (which contains survival values) and F (which contains fecundity values). To do so, age-specific survival and birth rate values should be plugged into the vectors below, with values separated by commas.

Insert age-specific survival values, separated by commas, between brackets in command.

```
P=[ ];
```

Insert age-specific birth rates, separated by commas, between brackets in command.

```
B=[ ];
F=B.*P;
```

The survival and fecundity values are then plugged into the appropriate slots in an ($n*n$) zero matrix to create the Leslie matrix, **M.** That is, the $n - 1$ entries in the survival vector are plugged into the subdiagonal, and the n entries in the fecundity vector are plugged into the first row.

```
dimensions=size(F);
n=dimensions(2);
M(n,n)=0;
for r=[2:n]
    M(r,r–1)=P(r–1);
end
for c=[1:n]
    M(1,c)=F(c);
end
```

The Leslie matrix is then displayed.

```
fprintf('Leslie Matrix:') M
```

STEP 2: Find the dominant and largest subdominant eigenvalues and calculate the damping ratio.

First we create a matrix, **Right_Eigenvectors,** of the right eigenvectors of **M** and a diagonal matrix, **Values,** of the eigenvalues of **M.** From **Values** we then create a vector **Eigenvalues,** which lists the eigenvalues of **M.**

```
[Right_Eigenvectors,Values] = eig(M);
```

```
Eigenvalues=eig(M);
```

The dominant eigenvalue, lambda, is then found and printed by selecting the maximum value from **Eigenvalues.** Note that the dominant eigenvalue of a Leslie matrix will always be both positive and real.

```
fprintf('Dominant Eigenvalue:')
```

```
lambda=max(Eigenvalues)
```

Next, a vector **Magnitudes** is created that lists the absolute values of each of the eigenvalues. The entry occupied by the dominant eigenvalue is then set to zero.

```
Magnitudes = abs(Eigenvalues);
for j=[1:n]
    if Magnitudes(j)= =lambda
        dominant=j;
        Magnitudes(j)=0;
    end
end
```

The magnitude of the largest subdominant eigenvalue, abs_lambda2, is then extracted from **Magnitudes.**

```
fprintf('Magnitude of Largest Subdominant Eigenvalue:')
abs_lambda2=max(Magnitudes)
```

STEP 3: Find the stable age distribution. First the right dominant eigenvector, **RDE,** is extracted from **Right_Eigenvectors,** and the sum of its entries, sadnormalizer, is calculated in order to normalize the stable age distribution.

```
for j=[1:n]
    RDE(j)=Right_Eigenvectors(j,dominant);
end
sadnormalizer=sum(RDE);
```

Then the normalized stable age distribution is calculated from **RDE** and sadnormalizer.

```
Stable_Age_Distribution=RDE/sadnormalizer
```

STEP 4: Find the age-specific reproductive values. First, we create a matrix, **Left_ Eigenvectors,** of left eigenvectors. (For an explanation of the methodology used, see Caswell 2001, pp. 92–94.)

```
Left_Eigenvectors=conj(inv(Right_Eigenvectors)) ';
```

We then extract the left dominant eigenvector, **LDE,** from **Left_Eigenvectors.** We also extract the reproductive value of the first age class, rv1, from **LDE** in order to scale the reproductive values.

```
for j=[1:n]
    LDE(j)=Left_Eigenvectors(j,dominant);
end
rv1=LDE(1);
```

The reproductive values are then scaled to the reproductive value of the first age class, yielding the vector **Reproductive_Values,** which lists the relative reproductive values for each age class.

```
Reproductive_Values=real(LDE/rv1)
```

STEP 5: Calculate sensitivities for all entries of **M.** The sensitivity of a Leslie matrix entry a_{ij} is the partial derivative of the population growth rate, λ, with respect to a_{ij}. (For further details, see Caswell 2001, pp. 206–211.)

```
Sensitivities(n,n)=0;
for i=[1:n]
    for j=[1:n]
Sensitivities(i,j)=((Reproductive_Values(i)*Stable_Age_Distribution(j))/
sum(Stable_Age_Distribution.*Reproductive_Values));
    end
end
```

```
fprintf('Dominant Eigenvalue')
Sensitivities
```

STEP 6: Calculate elasticities for all entries of **M**. The elasticity of a Leslie matrix entry a_{ij} is a scaled sensitivity value, giving the proportional change in the population growth rate, λ, that results from a proportional change in a_{ij}. (For further details, see Caswell 2001, p. 132.)

```
Elasticities(n,n)=0;
for i=[1:n]
    for j=[1:n]
        Elasticities(i,j)=((M(i,j)*Sensitivities(i,j))/lambda);
    end
end
fprintf('Dominant Eigenvalue')
Elasticities
```

5 Puzzles, Predation, and Primates: Using Life History to Understand Selection Pressures

CHARLES H. JANSON

Primatologists who maintain that predation is a major factor favoring sociality in primates face a serious quandary. Primate sociality should evolve in response to predation risk—the likelihood that an individual will be killed by a predator under a specified set of conditions (Janson 1998). All other things being equal, populations of prey exposed to higher predation risk should evolve higher levels of antipredator defenses, which include traits of interest to primatologists such as group size and cohesion (Dunbar 1988). The problem is that these antipredator traits may alter predation risk to the point that present-day predation risk may not reflect the original selection pressures (Abrams 1993). Thus, testing the adaptive value of sociality by looking for correlations across species between existing social traits and current levels of individual predation risk (as assessed by a population's predation rate) may be meaningless (Hill and Dunbar 1998).

Before exploring this problem further, it is necessary to clarify the distinction between predation risk and predation rate. Despite much debate about these terms (see Janson 1998), common usage makes the distinction clear: "risk" is the probability of death or severe injury to an individual, whereas "rate" is the frequency per unit time of such events in a population of individuals. In humans, we speak of the risk that someone will be killed in a car accident, but of the rate of traffic deaths in a country per unit time. It would make little sense to speak of an individual's rate of death, as such an event can occur only once in the individual's life. Because a population is simply the sum of its constituent individuals, individual risk and population rate must be equatable. An average individual's current predation risk should be identical to its population's current predation rate. This equality is important because predation rate is what can be estimated in real animals (despite many difficulties: e.g., Cheney and Wrangham 1987; Boinski and

Chapman 1995), but it is individual predation risk that is related to the selection pressure imposed on prey by predators. Luckily, individual predation risk can be assessed by the population predation rate. Thus, I use the term "current predation risk" instead of "current predation rate" in the rest of this chapter, except when referring specifically to empirical studies that estimate predation rates. Other events, such as attacks by predators, that can occur many times during an individual's lifetime are still quantified in the text as rates.

One problem in testing the adaptive value of social traits as a response to predation is that current predation risk, after the evolution of antipredator traits, may not reflect the underlying differences in predation risk between populations (Hill and Dunbar 1998). This problem could be overcome if primates were variable enough so that each population included some individuals living under conditions similar or identical to those of some individuals in all other populations. An obvious case would be truly solitary individuals (not just individuals temporarily separated from their social group). In this case, the current predation risk of these comparable individuals in different populations could be used as a direct measure of the real differences in underlying predation risk between the populations. Unfortunately, no such uniformity exists: primate populations of different species typically contain quite distinct sets of antipredator traits, showing little or no overlap with most other populations, thereby complicating any attempt to compare existing levels of predation risk. As a result, it is difficult to test the proposition that primate antipredator traits are responses to different levels of current predation risk.

An alternative strategy is to estimate what I have named the "intrinsic" predation risk (Janson 1998), which should reflect the likelihood of prey death from predation in a given population if the prey did not use any antipredator defenses (Hill and Dunbar 1998). Intrinsic predation risk can be assessed using estimates of predator density, attack frequency per predator, predator success per attack, and individual vulnerability per successful attack (on a group of prey). Some studies have simply used the diversity of predators in a habitat as an index of predation risk (Anderson 1986), while others have used rough gauges of attack rate (Hill and Lee 1998). A few have employed more refined modeling of predator-prey interactions in combination with habitat structure to estimate the probability that a prey individual (or group) will escape a predator's attack (e.g., Lima 1987; Lima and Dill 1990; Cowlishaw 1997).

Let us assume that we can accurately gauge differences in intrinsic predation risk between populations. What should be the expected evolutionary response of the prey? Elsewhere I have argued (Janson 1998) that natural selection on antipredator traits should act to reduce any differences in intrinsic predation risk between similar prey populations, but should not eliminate them. The basic reason is simple: increasing antipredator traits require increasing fecundity costs (e.g., van Schaik 1983), which should act as a brake on the evolution of antipredator defenses. For instance, if population X of a primate species has twice the intrinsic predation risk as population Y, X should evolve higher levels of antipredator defenses than Y, but not so high that it ends up with a lower current individual predation risk (= population predation rate) than Y. For instance, guppies living in high-predation streams have higher predation rates than their cousins that live in low-predation streams, despite the evolution of extensive differences between the populations (Reznick 1996). Similarly, meerkats living in a high-predation environment have significantly higher predation rates than those in a low-predation environment (Clutton-Brock et al. 1999). This argument suggests that we should still be able to detect differences among existing primate populations in their current predation risks that parallel the differences in their intrinsic predation risks, despite their having evolved differing antipredator traits. In this case, the degree of development of antipredator traits should correlate positively with both intrinsic and current predation risks. However, this expectation appears to fail, judging from a series of puzzles that challenge the interpretation of primate sociality as a response to predation risk (see also Hill and Dunbar 1998).

Puzzle 1: Larger-bodied primates typically live on the ground or in open habitats with more or larger predators (Anderson 1986). If increased body size is in part an adaptation to this higher intrinsic predation risk, the above argument suggests that we should still find a positive relationship between primate body size and current predation risk. However, current population predation rates actually decrease with increasing body size (Cheney and Wrangham 1987; Isbell 1994). Thus, current predation risks actually vary in a direction opposite to variation in apparent antipredator traits across species.

Puzzle 2: Primate group sizes increase in areas with more predators or higher attack rates (Anderson 1986; Hill and Lee 1998), yet measured predation rates decrease with increasing mean group size across species and habitats (Isbell 1994; Hill and Dunbar 1998). If increasing group size is an adaptation to increasing intrinsic predation risk, then current predation risk ought to increase with current group size, even if only weakly. Again, current predation risks appear to vary in a direction opposite to the degree of expression of this putative antipredator trait.

Puzzle 3: Current predation risks decrease with increasing body size (e.g., Isbell 1994), and so larger species would appear to be safer from predators, yet primate group sizes increase with increasing body size (e.g., Clutton-Brock and Harvey 1977). If increased group size is an adaptation against predation (e.g., van Schaik 1983; Terborgh and Janson 1986), then group size ought to decline in larger species that are exposed to fewer predators. In fact, the latter trend does occur among the very largest primate species in both terrestrial and arboreal settings, but not across the remaining 50–90% of the range of body masses (Janson and Goldsmith 1995).

In all of these puzzles, the problem is that evolution seems to have overcompensated. Species exposed to higher intrinsic predation risk appear to have evolved putative antipredator traits (large body size, large group size) to the point that they have not only reduced the difference in current predation risks relative to species in lower-risk situations, but have actually reversed it. It is hard to imagine an evolutionary explanation that would lead to such an outcome.

There are at least three possible resolutions of these paradoxes. First, as predation risk increases, larger group size and body size may present net costs instead of benefits, so that our expectations were wrong to start with. This possibility is not likely to be a general explanation, however. There are conditions that might favor smaller group size under higher intrinsic predation risk (see Janson 1998), but the general pattern of larger group size reducing predation risk is supported by a considerable body of theory and data on many kinds of animals. Similarly, it is widely accepted that very large-bodied prey are effectively free from predation risk (Colinvaux 1978).

Second, the relationship between intrinsic predation risk and putative antipredator traits may be confounded by one or more intervening variables that are correlated both with predation risk and with these traits (body size, group size). One possibility is food competition. For instance, Cheney and Wrangham (1987) suggested that the large group sizes of large-bodied savanna primates might reflect the low patchiness of their food rather than high predation risk in that habitat. However, when food competition is controlled for statistically, the pattern of puzzle 3 (increasing group size with body size) does not change appreciably (Janson and Goldsmith 1995). Thus, it seems unlikely that differences in food competition are the cause of the puzzling relationships listed above.

The third possible resolution, and the one developed in this chapter, is that life history differences correlated with body size somehow influence the relationship between intrinsic predation risk and the *strength of selection* on antipredator traits (cf. Boinski and Chapman 1995). In other words, a given predation risk of $x\%$ may have quite different consequences for fitness in a small versus a large animal. Hill and Dunbar (1998) postulated that the vital correlate was fecundity: "animals may accept higher levels of predation rate if their potential reproductive rate is high enough to compensate for the losses incurred, especially when these losses are confined to immature animals." This argument works well for juvenile mortality, in which case the risk of infant loss to predation is one of several variables that parents have to factor into parental care decisions.

However, the fecundity argument is difficult to justify when predation affects adults. Predation on adults should have no direct effect on adult fecundity, but only on adult life span. Thus, whatever reduction in life span is caused by increased predation, its fitness effect should be independent of potential fecundity. Thus, regardless of potential high or low fecundity, individuals should be under equal selection pressure to reduce adult predation risk and increase their reproductive life span, all other things being equal. Of course, all other things are not equal: fecundity is highly correlated with many other life history variables, which may be as or more important in shaping how antipredator traits affect fitness under different densities or types of predators. Hill and Dunbar (1998) do not explore other life history variables, so it is not clear whether fecundity is the primary variable affecting the expression of antipredator traits or is merely correlated with some other life history trait that is more important.

How can we decide which life history trait or traits might resolve our puzzles? Trying to use measured values of many distinct traits in correlational analyses is not likely to be convincing, as many of the traits are so tightly correlated that separating their causal effects is difficult or impossible statistically. Instead, I adopt a theoretical approach that permits us to vary major life history components independently of one another and body size, or allows them to covary in ways supported by allometric relationships among primate species (e.g., Harvey, Martin, and Clutton-Brock 1987). In either case, we can examine the effects of variation in one or more life history traits on the evolution of current predation risks and antipredator traits. My model explicitly includes a trade-off between avoiding predation and acquiring food (and thus increasing fecundity), a conflict for which there is ample evidence (e.g., Janson 1990; Cowlishaw 1997).

To give away the answers early, the conclusions of the model can be stated as follows, keyed to the three puzzles presented above. First, the intrinsically longer life span of individuals in larger-bodied species makes a given predation risk more costly in lost potential fitness than it would be to members of a smaller species with a shorter life span. Thus, all other things being equal, species with longer potential life spans should be under stronger selection pressure to reduce current predation risk than species with shorter potential life spans. Note that this result does not rely on the notion that larger species have a lower intrinsic risk of predation than smaller ones. Differences in fecundity have little effect in this model, in contrast to the postulate of Hill and Dunbar (1998). Second, given the first result, individuals in large, long-lived species should exhibit greater levels of antipredator behavior (including sociality) than small, short-lived species under otherwise identical conditions of food abundance and predator density. Combined with the first result, this means that current predation risk should decline with increasing group size, although this effect should vanish when potential life span is held constant. Third, if intrinsic predation risk declines only slightly with increasing body size, then larger species should be more social because individual fitness increases via increased longevity, despite some fecundity costs. In contrast, if intrinsic predation risk drops rapidly with increasing body size, then larger species should be less social because predation risk is too weak a force to counteract the fecundity costs of increased sociality. Thus the general observed increase in sociality in larger species is consistent with a weak effect of body size on intrinsic predation risk, coupled with a real benefit of sociality in reducing predation risk.

The Model
Defining Fitness

I define "fitness" as the expected lifetime production of offspring per individual born of a given phenotype. Strictly speaking, this definition holds true only if the population growth rate is zero, but over sufficiently long evolutionary time spans, the growth of any population is likely to be indistinguishable from zero (or the world would be overrun with your research species, or it would be extinct!). In this case, if annual adult fecundity, f, and mortality, d, are constant, lifetime reproductive output per individual reaching adulthood can be shown to be simply f/d, based on the widely used Euler's equation (Janson 1998). Adult mortality d is the sum of predation-independent sources of mortality with a constant risk, D, per year and a predation risk, μ, per predator per prey per year; thus, $d = D + \mu P$, where P is the density of predators. (For definitions of all parameters used in the model, see table 5.1.)

To obtain fitness as defined above, lifetime reproduction per adult needs to be multiplied by the likelihood that an individual born will survive

Table 5.1 Symbols used in the model and their definitions

Symbol	Definition
C	Riskiness in foraging; increases both feeding success and predation risk
S	Body size or mass
$f, f(C)$	Fecundity per female as a function of riskiness in foraging (see fig. 5.1)
f_{max}	Maximum fecundity per female regardless of food intake
b	Fecundity saturation constant, which determines how quickly fecundity increases with increased food intake (see fig. 5.1)
d	Death rate per female, the sum of predation rate and other nonpredation mortality $= P\mu + D$
$D, D(S)$	Minimum mortality rate, the inverse of maximum life span, a function of size
$j, j(S)$	Juvenile period or age at maturity, a function of body size
P	Density of predators (arbitrary scale)
$\mu(C)$	Per-predator predation risk, which increases with riskiness in foraging
$P\mu(C)$	Total predation risk per prey individual as a function of riskiness in foraging
L_T	Expected lifetime mortality exposure, in years; approximately equal to total life span if juvenile and adult mortality risks are similar.
FPR	Field predation rate, determined by fitness-maximizing level of riskiness

to adulthood. This juvenile survival can be modeled in various ways. If annual mortality is constant, then survival across any given period is just e^{-dt}, where d is the death risk per time and t is the total time period considered. Because survival depends only on the product dt, this term can be considered the total force of mortality during the period considered—in this case, the time between birth and adulthood. In this model, the total force of mortality during the juvenile period is considered proportional to the adult predation risk, and thus can be written as $j\mu P$. Here, j is called the juvenile risk factor and is measured as the length of the juvenile period times the factorial predation risk of juveniles relative to adults. For instance, if juveniles suffer twice the risk of predation as adults in the same groups, and the juvenile period lasts four years, then $j = 2 * 4 = 8$. In fact, an empirical analysis based on data in Ross and Jones (1999) shows that juvenile average annual mortality is correlated with adult average annual mortality across species, and is roughly 1.5 times as high (C. H. Janson, unpub.). Thus, for a rough approximation, j can be taken as 1.5 times the time to maturity. I assume that juvenile risk is constant across ages; under these conditions, juvenile survival is written as $e^{-j\mu P}$. A more general (but more complex) expression can be derived for risk that changes with juvenile age (see Charnov 1993). Altogether, fitness is written as

$$W_1 = \frac{fe^{-jP\mu}}{(D + P\mu)} \tag{5.1}$$

Modeling Trade-offs between Reproduction
and Predation Avoidance

Following the approach of Abrams (1993), the essence of the trade-off between food acquisition and predation avoidance is encapsulated in a single parameter, C, which represents foraging effort—or, more precisely, the willingness to employ risky foraging behaviors that enhance food intake at the expense of increased predation risk. Risky foraging behaviors include longer daily foraging activity, smaller group sizes, foraging farther apart from group members in existing groups, and reduced vigilance against predators (Lima 1988; Janson 1992). Thus, "risky" behaviors include many aspects of social structure that socioecological studies have attempted to explain. Foraging behaviors that increase food intake with no effect on intrinsic predation risk are assumed to have evolved to fixation and thus are not modeled as part of current natural selection. Both fecundity (f) and per-predator, per-prey mortality risk (μ) increase with risky behavior (C). However, f is a saturating function of C (fig. 5.1), whereas μ should be an accelerating or possibly linear increasing function of C (fig. 5.2).

Fecundity should also depend on resource density, R, but I assume that R is not correlated with foraging behavior or body size, and thus hold it constant for the rest of this chapter. In this model, I also include body size, S, which affects many of the life history variables. Thus, fecundity as a function of foraging riskiness, $f(C)$, is expanded to a function of riskiness, resource abundance, and body size, $f(C,R,S)$. Similarly, the predation rate $\mu(C)$ becomes $\mu(C,S)$. In addition, the length of the juvenile period, j, and the mortality risk that is not related to predation, D, can be considered functions of body size, $j(S)$ and $D(S)$. Altogether, the expression for fitness is

$$W_1 = \frac{f(C,R,S)e^{-j(S)P\mu(C,S)}}{[D(S) + P\mu(C,S)]} \tag{5.2}$$

Working with this expression is difficult, so it is more convenient to analyze the natural logarithm of fitness, $W_2 = \ln(W_1)$. W_2 is monotonically related to W_1, so any maximum of the latter is also a maximum of the former. Also, the slopes of W_2 with respect to any variable will have the same direction (positive or negative), although not the same magnitude, as the slopes of W_1 with respect to the same variable.

$$W_2 = \ln[f(C,R,S)] - j(S)P\mu(C,S) - \ln[D(S) + P\mu(C,S)] \tag{5.3}$$

To understand the evolution of social traits that are included in the risky behavior parameter C, we need to solve for the optimal (fitness-maximizing)

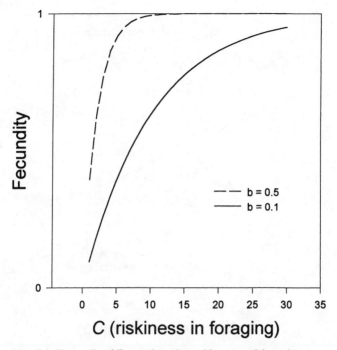

FIG. 5.1. Fecundity (f) as a function of increased foraging rate, which depends on riskiness (C). The equation used is $f = f_{max}(1 - e^{-bC})$, where maximum fecundity $f_{max} = 1$ (reflecting the fact that many primates typically produce only one offspring at a time) and $b = 0.1$ or 0.5. The parameter b determines how rapidly a given increase in foraging rate translates into higher fecundity. The particular values chosen to model b here are for illustration only, and the qualitative predictions in table 5.2 do not depend on the choice of b. Relative to smaller species, primates of larger body size should have lower values of b and thus require greater increases in foraging rate to achieve a given absolute increase in fecundity.

level of C. To do this, we set the derivative of W_2 with respect to C equal to 0 and solve for C. The derivative of W_2 with respect to C, leaving out the variables in parentheses, is

$$\frac{dW_2}{dC} = \frac{\partial W_2}{\partial C} = \frac{1}{f}\frac{\partial f}{\partial C} - jP\frac{\partial \mu}{\partial C} - \frac{P\dfrac{\partial \mu}{\partial C}}{D + \mu P} \tag{5.4}$$

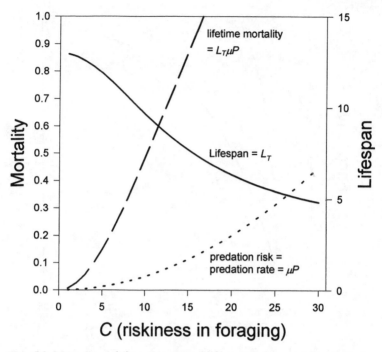

FIG. 5.2. Mortality, $\mu(C)$, and expected life span values as a function of riskiness, C. The curve for $\mu(C)$ in this example is $g^2 C^2$, with the constant $g = 0.1$. Expected life span equals $j + [1/(D + \mu P)]$, with $j = 3$, $D = 0.05$, and $P = 60$. The particular parameters (g,j,D,P) used here are arbitrary and do not affect the qualitative predictions in table 5.2. Total lifetime mortality rate is the product of expected life span and mortality rate. Note that the individual's predation risk, μP, is the same as the population's predation rate.

Setting equation (5.4) equal to zero, we obtain the relatively simple expression for the condition that yields the value of C that maximizes fitness:

$$\frac{1}{f}\frac{\partial f}{\partial C} = L_T P \frac{\partial \mu}{\partial C} \tag{5.5}$$

Here, $L_T = j + 1/(D + \mu P)$, which is about equal to the total expected life span if j is simply the length of the juvenile period in years. More generally, if j represents the total juvenile risk factor defined above, then L_T equals the total number of expected years of exposure to predation risk at levels equal to adult risk. I refer to μ alone as the "basic predation risk" function, and to μP as the "total predation risk" function. Both these functions represent intrinsic risks, in the sense that they could be compared across populations for any standard value of risky behavior C. The current annual predation risk is

μP evaluated at an individual's or population's *actual* value of risky behavior
C. The conclusion from equation (5.5) can be stated thus: "The optimal level
of risky behavior is attained when the proportional change in annual fecundity due to increased riskiness is exactly balanced by the total increase in
lifetime mortality caused by increased riskiness."

To solve for C^*, the optimal level of risky behavior, without postulating
specific functional forms for f and μ, we can use the general curve shapes
given in figures 5.1 and 5.2. Both f and μ increase with C, but f is a saturating function of C, whereas μ is an accelerating function. Evaluating the appropriate expressions on the two sides of equation (5.5) and graphing them
as functions of C yields figure 5.3. Here, the change in relative fecundity,
$(1/f)\partial f/\partial C$, is positive, starting at infinity when $C = f = 0$ and approaching 0
as C becomes large. Conversely, the change in lifetime mortality, $L_T P \partial \mu/\partial C$,
starts at or near zero when C is small and increases as C increases, perhaps

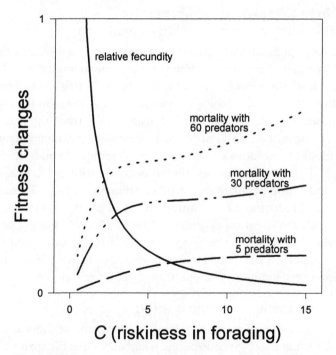

FIG. 5.3. The fitness components due to changing fecundity
and mortality as a function of increased riskiness, C, at three
predator densities. Where the fecundity and mortality curves
cross is the optimal value of riskiness. Higher densities of
predators always favor reduced riskiness (increased sociality), all other things being equal.

in a complex way, as shown here. Where these two curves cross, equation (5.5) is satisfied, and the corresponding value of C is the optimal one.

This graphical solution allows us to explore the interaction between predator density, life history traits, and optimal social structure (as measured by risky behavior). I present them in two parts, comparisons within species and comparisons between species, because the results differ strikingly between the two. For within-species comparisons, I hold the fecundity function f, juvenile period j, and minimum mortality risk D constant, allowing variation only in total predation risk μP. For between-species comparisons, I generally keep predator density P constant, but allow variation in the other parameters or functions. To mimic the results expected of real primates, I constrain these parameters and functions to vary according to allometric relationships with body mass (S) similar to those observed among primates (e.g., Harvey, Martin, and Clutton-Brock 1987). A summary of the results is presented in table 5.2.

Results from Comparisons within Species
Comparisons among Groups within a Population

Within a single population, all functions and parameters should be the same for all individuals, except for random variation due to mutation and recombination. Thus, any variation in fecundity, predation risk, and group size should directly reflect the ecological mechanisms assumed in the model, as no evolutionary change is allowed. For instance, the trade-off assumed between foraging success and antipredator defenses (including larger social groups) predicts a negative effect of increasing group size on fecundity (see van Schaik 1983). Predation risk should decrease with group size, a trend documented within populations for other animal taxa (e.g., Treherne and Foster 1982; Lindstrom 1989; Clutton-Brock et al. 1999), but not yet for within-population variation in primates (see Isbell 1994). Finally, predation risk should increase with increasing fecundity because both are positively related to increased riskiness, by assumption. No analysis of this relationship within populations has been published yet for primates.

Comparisons among Populations within Species

Populations of the same species are likely to share similar evolved values of the fecundity function f, juvenile period j, and maximum life span $1/D$. Some adaptive variation in these values may exist, but it is likely to be small compared with variation among species. Thus, I examine variation in the only remaining function, total predation risk, $P\mu(C)$.

The first simple prediction from figure 5.3 is that if all other factors are

Table 5.2 Summary of predictions of the theoretical model at different levels of comparison, and with distinct parameters allowed to vary

Level of analysis	Parameters allowed to vary	Predicted correlation between					
		f, G	FPR, G	FPR, f	G, S	FPR, S	P, G
Within population, across groups	None (random deviations only)	Negative	Negative	Positive			
Within species, across populations	μ, P, and their product	Negative	Positive	Negative			Positive
Across species	Maximum fecundity f_{max}	None	None	None	None	None	
Across species	Fecundity saturation constant b	(Positive)	(Negative)	(Negative)	(Negative)	(Positive)	
Across species	D or j only, or both D and j	Negative	Negative	Positive	Positive	Negative	
Across species	Predator density P	Negative	Positive	Negative	Negative	Positive	Positive
Across species	D, j, f vary allometrically with S	Negative	Negative	Positive	Positive	Negative	
Across species	D, j, f, and μ vary allometrically with S; μ–S allometry weak	Negative	Negative	Positive	Positive	Negative	
Across species	D, j, f, and μ vary allometrically with S; μ–S allometry strong	Positive	Positive	Positive	Negative	Negative	
Across species (mostly)	Observed trends[a]	Negative	Negative	Positive	Mostly positive	Negative	Unknown

Note: G = group size. For remaining symbols, see table 5.1. By the definition of risky foraging, increasing G is equivalent to decreasing riskiness C in the model. Predictions noted in parentheses are weak effects.

[a] Observed trends introduced as puzzles are provided in this row.

held constant, a higher density of predators P or a higher predation risk function $\mu(C)$ should always favor lower levels of risky behavior—that is, larger, more cohesive groups with more vigilance and reduced foraging effort. This result depends on the proposition that increasing risky behavior (e.g., solitary foraging) will increase predation risk, but this assumption does not hold for cryptic species. For these species, increasing predator density should favor smaller groups, as may be the case for primate species faced with intense predation by humans (Cheney and Wrangham 1987; Janson 1992).

Because increased group size entails foraging costs, individuals in larger groups have reduced fecundity. These foraging costs also ensure that individuals that live under higher intrinsic predation risk end up with higher current predation risk at the evolutionarily optimal level of risk-taking (see Janson 1998). Thus, comparisons of fecundity, predation risk, and group size among populations of one species should show fecundity declining with group size and population-level current predation risk increasing with mean population group size, but declining with increased fecundity. The latter two predictions are opposite to the trends predicted from the previous comparisons among groups within a population, unless the groups occupy areas of different intrinsic predation risk.

Results from Comparisons among Species

The major question addressed with this graphical model is that posed by Boinski and Chapman (1995) and Hill and Dunbar (1998): does the fitness effect of predation risk depend on life history traits? Life history traits are so generally correlated with body size that it is difficult, if not impossible, to tease out which of many intercorrelated variables is the one causing a given pattern in real data. The virtue of the analysis presented here is that it is possible to vary distinct life history components separately to analyze their relative importance. Once we understand their relative contributions, we can let them covary in patterns similar to those observed in real animals to predict what patterns we should actually observe in nature. The rest of this chapter follows this approach by presenting first the effects of varying only fecundity, then longevity and mortality, and ending with combinations of these components that covary according to the body size allometries observed across primate species. To illustrate the results, I use the particular curves graphed in figures 5.1 and 5.2 for fecundity and mortality, but the results can be shown to be independent of these particular functional forms, as long as the shape of the curve is preserved (C. H. Janson, unpub.). In each case, I calculate the optimal level of riskiness C^* and the associated values of current total predation risk $P\mu(C)$, fecundity $f(C)$, and body mass S where appropriate. For ease of interpretation in the graphs, I translate riskiness into "group size" by the simple formula G(group size) $= 50/C$.

Effects of Variation in Fecundity Only

The effect of larger body size on f will be to reduce the maximum fecundity reached regardless of food intake, and probably to reduce the rate at which fecundity saturates as food intake increases (lower b parameter; see fig. 5.1). Because the fecundity curve in figure 5.3 depends on *proportional* changes in fecundity, it turns out that changes in the maximum fecundity do not change the predicted optimal level of risky foraging behavior at all. Changing how quickly fecundity increases with food intake (by varying b) does affect the optimal level of risky foraging behavior. However, the effect of changing b is small: animals with lower b have slightly higher $(1/f)\partial f/\partial C$ curves, leading to slightly higher optimal levels of riskiness (fig. 5.4). Larger animals should have lower b values, as higher absolute food intakes should

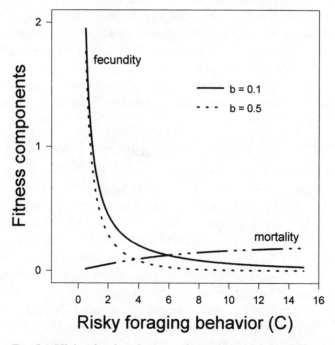

FIG. 5.4. Higher b values increase the rate at which fecundity approaches its maximum as risk C increases (see fig. 5.1), and favor slightly lower optimal risk levels (increased sociality).

be required to make larger offspring. Given this trend, larger species should show reduced sociality and higher current predation risks than smaller ones when all other parameters are kept constant. Thus, the effect of variation in only the fecundity function does not help to explain two of the three puzzles above; only the pattern of declining current predation risk in species with larger groups is predicted.

Effects of Variation in Longevity Components Only

Life span $L_T = j + 1/(D + \mu P)$ should be a decreasing function of increased risk-taking in foraging behavior C (see fig. 5.3) because high risk implies a high predation rate, and thus a reduction in annual survival. Variation in expected life span for a given value of risk-taking can occur because of changes in age at maturity (reflected in j), minimum adult mortality risk (D, which should be inversely proportional to maximum life span), or the basic predation risk function $\mu(C,S)$. Increasing j or decreasing D has the same effect of increasing L_T at a given C value, thus raising the lifetime mortality curve and favoring less risky, more social but (by assumption) less rewarding foraging behavior. If the predation risk function μ is kept the same, these changes will also produce lower current predation risks (changing μ is handled in the next section). Because larger primates are likely to have a greater age at maturity j and a smaller minimum death risk D, these predicted effects of variation in life span are exactly those seen in the three puzzles listed above. Larger species will be more social and have lower current predation risks, and thus measured predation rates will decrease in more social species (fig. 5.5).

Effects of Variation in the Total Predation Risk Function

Changes in the shape of the predation risk function $P\mu(C)$ will affect both the expected life span L_T and the derivative $P\partial\mu/\partial C$, which I call the "marginal predation risk," but in opposite ways. For instance, increasing the density of predators P will shorten the life span L_T, but will increase the marginal predation risk. Which change is larger? The effect of changes in $P\mu$ on expected life span L_T is diluted by the presence of the other variables (j, D) that comprise L_T, whereas no such dilution affects the marginal predation risk. If total predation risk varies only because of a change in predator density P, then the change in marginal predation risk outweighs that in expected life span. More complex results can occur if μ changes shape, but we will not explore these here, as it is difficult (to put it mildly) to assess the shape of μ (but see Cowlishaw 1997 for a start). Thus, in table 5.2, I present the predicted correlations among group size, current predation risk, fecundity, and

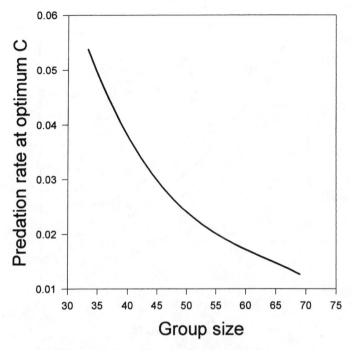

FIG. 5.5. When only longevity parameters j and D vary, predation rate at the optimal level of riskiness decreases with group size.

body mass when the marginal predation risk $P\partial\mu/\partial C$ varies only as a result of changes in predator density P. The predicted trends generally run counter to those of the three puzzles described at the beginning of this chapter.

Effects of Allometric Covariation in Life History Components with the Intrinsic Predation Risk Kept Constant

We can now consider the effects of combined changes in fecundity and life span components, keeping the basic predation risk function $\mu(C)$ constant. For this analysis, I calculated raw allometric coefficients between log-transformed variables (age at maturity, maximum life span, and fecundity) on the log of female body mass, using species data from Harvey, Martin, and Clutton-Brock (1987) updated with values from Ross and Jones (1999). I did not perform analyses on phylogenetic contrasts, as my purpose was only to estimate roughly realistic values for entering into the model, not to test precise hypotheses about correlated variables. The resulting allometric equations were log(age at maturity) = 0.3853 + 0.366 * log(female mass), log(fecundity) = 0.043578 − 0.367 * log(female mass), and log(maximum

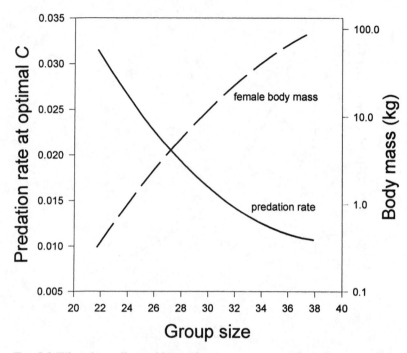

FIG. 5.6. When fecundity and longevity parameters are allowed to vary allometrically with body mass and total predation risk μP declines only slightly
with increasing body mass, current predation rates are predicted to decline
with group size, and group size should increase with body mass. As a consequence, predation rates will decline with increasing body mass, even if total
predation risk μP is independent of body mass.

life span) $= -0.245 - 0.23138 * \log(\text{female mass})$. Equating age at maturity
to j, fecundity to maximum fecundity f_{max}, and maximum life span to $1/D$,
I then calculated the corresponding values of f, j, and D for female masses
of 0.3, 1, 3, 10, 30, and 100 kg. The six resulting sets of values were combined
with a single basic predation risk function ($0.0001 * C^2$), one predator density (60), and a fecundity function (see fig. 5.1, with $b = 0.1$) to calculate the
optimal riskiness level for each body size. The calculations show that the
results of combining allometrically linked variation in fecundity and life
span (fig. 5.6) mirror those of variation in life span components alone (see
fig. 5.5). This outcome should not be surprising, given that variation in fecundity had weak effects on optimal C levels. Thus, the three puzzles introduced earlier can be explained using allometric variation in life history components if intrinsic predation risk is held constant independent of body size.
It is important to note that the match between observed and predicted
trends is due to variation in the longevity component, not to variation in fe-

cundity, even though the two components are perfectly correlated in this analysis.

Effects of Allometric Covariation in Fecundity, Longevity, and Intrinsic Predation Risk

In the previous section, keeping the per capita mortality function $P\mu$ independent of body mass was equivalent to stating that larger animals do not benefit from a reduced likelihood of predation per unit time at a given level of risk-taking behavior. Although counterintuitive, this assertion is not crazy: the size of predators and their prey consumption rate may well parallel increases in prey body mass across small to moderate sizes. For instance, in the New World, tamarins may fall prey to a variety of small understory hawks and hawk-eagles (Robinson 1994), but are not found in the diet of harpy eagles (Rettig 1978). In contrast, capuchin monkeys, one order of magnitude larger than tamarins, are relatively immune to the smaller understory hawks, but are still vulnerable to hawk-eagles and are a favored prey of harpy eagles (Rettig 1978). Observed rates of attack by any kind of raptor on tamarin groups and capuchin groups in the Manu National Park did not differ appreciably (C. H. Janson, unpub.). Cheney and Wrangham (1987) note that larger primate species are attacked by fewer kinds of raptors, but by more species of mammalian carnivores, than are smaller species. In any case, even if the basic predation risk declines with increasing prey mass, how strongly it does so has a marked effect on the predicted patterns of covariation between antipredator traits and observed predation risks.

Above some body size, prey must benefit from reduced intrinsic predation risk. In most ecosystems, predators are much less abundant than their prey (e.g., Colinvaux 1978). Because prey density also decreases with increasing body mass (e.g., for mammals: Robinson and Redford 1986), there must be a prey adult body size above which a predator population cannot be supported, although juveniles may still be at risk. For arboreal primates in both the New World and Africa, this size appears to be about 7–8 kg (Struhsaker and Leakey 1990), while for terrestrial species it may be about 30 kg, about the size of a chimpanzee. To model the decreasing intrinsic predation risk of larger species, I tried several allometric coefficients to relate the constant g of the basic predation risk function (see fig. 5.2) to body mass, including 0 (the case covered in the previous section), -0.125, -0.25, and -0.5.

When intrinsic predation risk declines only slightly with body mass (allometry of -0.125), the effect of increased j and reduced D at larger body sizes overwhelms the effect of reduced predation risk, and the results look qualitatively like figure 5.6, although the slopes are shallower. Thus, the

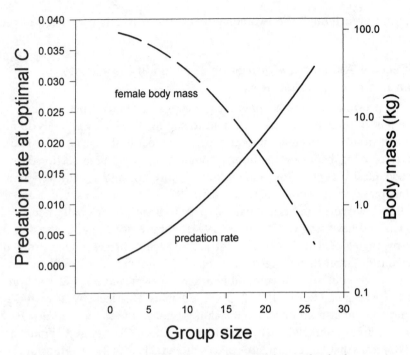

Fig. 5.7. When predation risk declines rapidly with increasing body mass, the trends in figure 5.6 are reversed: predation rate increases with group size and decreases with body mass. Predation rates will still decrease in larger species because of both increased minimum longevity and strongly reduced predation risk.

three predation risk puzzles can still be explained when life history components and intrinsic predation risk vary allometrically with body mass, as long as the effect of mass on predation risk is weak. When intrinsic predation risk declines more steeply with body mass (allometries of -0.25 and -0.5), the reduced predation risk of larger species dominates, and the results change radically. Current predation risks now decline steeply with body mass, larger species live in smaller groups, and current predation risks increase with group size at the optimal risk level (fig. 5.7), reflecting the influence of body mass on the basic predation risk function.

Testing the Model

The predictions of the model are summarized in table 5.2. The key aspect of these findings is the importance of longevity in the model. In many cases, relationships between variables would disappear, or would even be reversed, if longevity differences between species were held constant or controlled statistically. I attempted to test some of the predictions with data, but it was hard to find reliable published estimates of field longevity (related to μP) or maximum life span (inversely related to D), although good estimates of age at maturity (j) are available. Existing data on field longevity are summarized in Ross and Jones (1999), but are available for only twenty-five primate species, of which published estimates of predation rate exist for only nine. Estimates of maximum longevity are prone to considerable error. For instance, the maximum life span for geladas is reported as 19.3 years (Barton 1999; Ross and Jones 1999), yet roughly a fifth of the geladas at the Bronx Zoo, New York, are currently over 20 years old, and the oldest individual there died recently at 36 (Colleen McCann, pers. comm.), nearly twice the currently published value.

A general inference from the model is that the known patterns relating current predation risk, group size, and body mass (the three puzzles) seem to be consistent with relatively small declines in total predation risk (μP) with increasing body mass across most of the mass range of primates within either forest or savanna habitats. Only at the upper end of the body mass range should total predation risk be found to decline rapidly. The effect of prey body size on intrinsic predation risk will not be easy to measure, but a start could be made with relatively simple density estimates of predators and prey, combined with knowledge of the predators' diets. Assuming that predators are largely opportunists (e.g., Emmons 1987), a reasonable measure of predation risk that is independent of prey behavior would be

$$\frac{\sum_k \frac{A_{ik}}{\sum_i A_{ik}} E_k}{A_{ik}} \tag{5.6}$$

Here, A_{ik} is the abundance of species i that occurs in the diet of predator k, and E_k is the consumption rate of all prey consumed by the population of predator k. The numerator of equation (5.6) gives the number of individuals of prey species i that would be eaten by all the predators that consume it if species i were taken strictly in proportion to its abundance relative to other prey used by its predators. The denominator is the total population size of species i, which converts the predators' consumption rate into a prey mortality risk. Because this index treats all potential prey as equally available, it should be little influenced by the particular antipredator behaviors of a given primate species. If my model correctly accounts for the puzzling patterns listed above, then the predation risk index should decline only weakly with primate body size for diurnal arboreal primates from 200 to 6000 g and for terrestrial primates from 2 to 25 kg. Above these ranges, predation risk for primates should drop rapidly with body mass.

   In testing the predictions in table 5.2, it is important to note the level of comparative analysis, as different predicted trends emerge when comparing groups within populations, between populations of the same species, and between species. Essentially, the lower the taxonomic level of comparison, the less evolutionary change will have occurred in antipredator behavior among the units being compared. Thus, at the lowest level, comparisons should directly reflect the effects of the mechanisms assumed in the model, while these mechanisms may be obscured when extensive evolutionary change in antipredator behavior or life history variables intervenes.

   Hill and Dunbar (1998) are the only authors to have analyzed variation in predation risks and group sizes among primate species in the context of a life history variable. They argued that populations or species with higher fecundities can tolerate higher current predation risks (as measured by predation rate data), and their data analysis supported this positive trend. However, the theory outlined above predicts that variation in fecundity (independent of other life history traits) should have little direct effect on field predation rates, and that what effect it has should run opposite to Hill and Dunbar's (1998) results (see table 5.2, variation in fecundity saturation constant b).

   Can these seemingly incompatible predictions be reconciled? Yes, in part. In the model presented here, current predation risks will in fact be positively correlated with fecundity values when longevity components vary across species and intrinsic predation risk is relatively constant or is related only weakly to body size (fig. 5.8). If these conditions apply to the data of Hill and Dunbar (1998), then their results can be reconciled with some of the predictions in table 5.2. However, their rationale cannot: the correlation of fecundity with current predation risk is not a causal effect, but occurs because of the influence of longevity on both parameters.

   There remains the fact that Hill and Dunbar's (1998) empirical analysis finds that field predation rates are better correlated with fecundity than with body size. According to the theory developed here, a different correlate of body size, longevity, is the likely cause of the interconnected trends linking current predation risk, body mass, fecundity, and group size. If the correlation of current predation risk with fecundity arises only as a consequence of the trade-off between foraging success and predation avoidance, why might fecundity be more highly correlated with observed predation rate than is body size? One possible reason is that fecundity might be more tightly correlated with longevity than is body size. However, a correlation analysis of the data in Harvey, Martin, and Clutton-Brock (1987), without adjusting for phylogeny, does not show such a pattern across primates in general. Another possible reason is that field predation rates and fecundity are indeed causally linked, but in a different way than postulated by Hill and Dunbar (1998). It

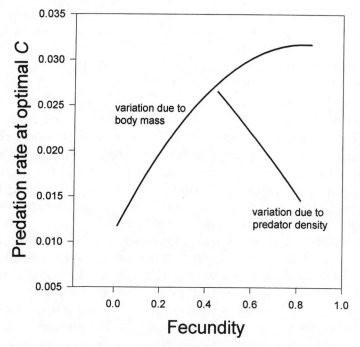

FIG. 5.8. The correlation of field predation rates with fecundity across populations will depend on the source of variation in optimal riskiness. If the source is changes in longevity components, as would be expected for comparisons among species in a single habitat, then field predation rates should increase with field fecundities. However, if the source is only changes in predator density, as would be expected in comparisons of the same species in different populations or habitats, then field predation rates should decline with increasing field fecundities.

is well known that the loss of an infant shortens the interbirth interval in many primate species (e.g., van Schaik 2000a). Thus, among groups or populations of a species that differ in rates of predation on infants, higher predation rates should cause higher fecundity rates. This mechanistic link may be sufficient to make the correlation of observed field predation rates with fecundity somewhat higher than that with body size. This trend would hold even if body size–related changes in longevity were the real cause of evolutionary changes in optimal levels of risky foraging behavior, which then translate into current predation risks. Although this line of reasoning applies best to comparisons within species, it is still valid among species, especially when the species are in the same genus and share similar basic fecundities.

Nearly half of Hill and Dunbar's (1998) data for diurnal primates consists of congeneric or conspecific replicates.

Discussion
Resolving the Puzzles of Primate Predation Rate and Predation Risk

The theory presented here makes clear the distinction between the concepts of predation risk and predation rate. Predation risk is an individual's chance of being killed by a predator per unit time; the average of these risks across the individuals in a population yields the rate of predation as a fraction of the population per year. The predation risk function in this model is the curve $P\mu(C)$ relating per capita predation risk to risky foraging traits (see fig. 5.2). What I call intrinsic predation risk is the value of the $P\mu(C)$ curve at a particular value of C that corresponds to a lack of social antipredator traits (see also Hill and Dunbar 1998). If the full $P\mu(C)$ curve were known for a given primate population, it could be used to test the model presented here directly, using the predictions presented above. Primate populations rarely demonstrate the full range of C values, however, and so the full curve could not be evaluated even if predation risks could be measured for individuals or populations employing a given C value. However, indirect methods can be used to assess a population's intrinsic predation risk (see above and Janson 1998). Differences in intrinsic predation risk can then be compared with social or other antipredator traits (e.g., Cowlishaw 1997; Hill and Lee 1998).

Correlations between primate social structure and measures of intrinsic predation risk typically support the general theory that sociality evolves in part as a defense against predators (Anderson 1986; Hill and Lee 1998). Yet, when similar correlations are attempted with current predation risk as estimated from field predation *rates,* the relationships are of the *opposite* sign to those found using measures of intrinsic predation risk (Hill and Dunbar 1998). This paradox is not just a result of the prey's evolutionary response to intrinsic predation risk, as suggested by Hill and Dunbar (1998). Because of the fitness costs of antipredator defenses, predation rates should be positively correlated with the underlying changes in predation risk (Janson 1998). Such a positive correlation is clearly observed within a species when predators are added or removed, as shown in many experimental ecological studies (e.g., Schmitz, Hamback, and Beckerman 2000), even when the prey change their antipredator behavior, but not their life history traits. Therefore, the paradoxical correlations between sociality and predation rates across primate species must be confounded by some other variable.

Introducing life history traits into the relationship between intrinsic predation risk and sociality provides a single explanation for all three of the predation rate puzzles of primate socioecology. Because larger species have longer potential life spans and thus gain greater lifetime fitness benefits from reducing their mortality, predation rates should decline with body size across species. This trend holds even if the predation risk curve does not change with body size. The cause of reduced predation rates in this case is a reduction in risky behavior (increased sociality), so predation rates are predicted to decline with increased sociality. These two trends combine to predict that larger species should live in larger social groups. All three of these correlations persist as long as the total intrinsic predation risk does not decrease markedly with body size. In contrast, when predation risk drops rapidly with body mass, as may occur at the upper limit of primate body sizes, increased sociality should occur in smaller species and be associated with higher predation rates.

Why Does Primate Body Size Increase under Higher Predation Risk?

I have taken as given the observation that primates are larger in high-risk habitats and have used this trend to resolve the resulting puzzles linking social antipredator traits and predation risk. However, I have skirted the issue of whether large body size itself is an antipredator adaptation. Life history theory predicts that the higher mortality rate imposed on a species by increased predation should favor earlier maturation and greater reproductive effort, both of which are typically associated with smaller, not larger, body size. Much comparative life history analysis supports these predictions (e.g., Promislow and Harvey 1990), and the evolution of such differences has even been observed within a few generations in guppies when predators were introduced into low-predation streams (Reznick 1996). Thus, it is not easy to argue that increased body size in primates living in high-risk savanna environments is a natural evolutionary response to higher intrinsic predation risk.

Two possible explanations can be considered. First, increased body size in primates may offer a reduction in intrinsic predation risk greater than the associated cost in reduced fecundity. Because the cost in reduced fecundity is roughly known from correlational analysis to have an allometry of 0.36, the reduction in intrinsic predation risk would have to have a larger allometric slope than 0.36. Such a slope, however, would conflict with the requirement that intrinsic predation risk decline only weakly with body mass. Thus, if body mass is in fact an adaptation to high predation risk, it negates the theory presented here to resolve the socioecological puzzles given above. Second, the larger body size of primates in savannas may have nothing to do with predation risk, and instead may be a response to the more efficient energetics of terrestrial versus arboreal locomotion, or possibly a necessity imposed by sparser food distributions on the ground than in trees. Other authors have argued that increased sexual dimorphism in terrestrial primates may be permitted by relaxed energetic or movement constraints on male body size compared with those in the trees (Leutenegger and Kelly 1977; Clutton-Brock, Harvey, and Rudder 1977). If locomotion is also easier for females on the ground than in trees, then small body sizes in arboreal species may be an adaptation for locomotor efficiency rather than an adaptation to low predation risk.

Robustness of the Model

The model presented here offers a general approach to modeling the trade-offs assumed in socioecology. The use of Euler's equation allows a definition of fitness that is congruent with life table and other methods of exploring variation in adaptive tactics (see Alberts and Altmann, chap. 4, this volume). However, to make the analysis tractable, a number of simplifications were necessary, including the assumption of stable population size and constant adult fecundity and death rates. How realistic are these assumptions? Although primate populations do fluctuate, primates as a group have far more stable populations than described for many other mammals. Given that the majority of primates exist in mature forest or relatively stable savanna ecosystems, it is unlikely that their populations will experience marked long-term growth or decline over the periods of many generations needed to evolve the traits being modeled here. The assumed constancy of adult fecundity and death rates is not strictly correct, but it is also not too far off, based on existing data (Dunbar 1987a). The most obvious deviation occurs because of aging, with fecundity dropping and mortality increasing after a period of relative constancy following the age of maturity (Dunbar 1987a). However, the evolutionary effect of senescence is likely to be small, as the relatively few individuals that survive to older ages constitute a small fraction of the reproductive gene pool at any given time.

Encapsulating the trade-off between foraging success and intrinsic predation risk in a single parameter, foraging riskiness (C), simplifies the mathematical analysis, but hides much detail. First, all the behaviors that contribute to riskiness (small group size, low cohesion, long activity periods, low levels of vigilance, etc.) are confounded. However, specific functional forms could be proposed for these effects (as was done for bird flocks by Lima 1987, 1988; Lima and Zollner 1996), albeit at a cost of reduced generality. It is not a problem if some behaviors reduce predation risk without a loss of foraging success; these behaviors are assumed to evolve to fixation and are not the subject of this analysis. Second, the relationship of fecundity to foraging success, and thus to foraging riskiness, will depend on the distribution of food patch productivities (e.g., Janson 1992) and the relationship between foraging success and intrinsic predation risk among habitats (e.g., Cowlishaw 1997). If particular species and study sites are of interest, these relationships can be modeled based on theoretical and empirical findings (e.g., Janson 1988).

Finally, the assumption that optimality is the appropriate criterion for predicting actual behaviors can be questioned (e.g., Pulliam and Caraco 1984). In particular, group size may be determined by individual decisions that pose a conflict of interest between group members and solitaries—solitaries may gain more by joining a group than individual group members lose by admitting an extra individual, even if their doing so increases the group size beyond its optimal point (Clark and Mangel 1984). The resulting ESS group size may produce a mean fitness far below the optimal. These conflicts can be resolved to some extent by increasing the mean degree of relatedness among group members (Rannala and Brown 1994). In any case, the predicted trends in the model presented here remain qualitatively similar if the ESS group size is used instead of the optimal group size (C. H. Janson, unpub.). Conflicts of interest may also exist among members of a single group because the fitness benefit of risk-reducing behaviors declines with the number of individuals performing the behavior (e.g., for vigilance, see Pulliam, Pyke, and Caraco 1982). Such frequency dependence will generally lower the predicted equilibrium level of antipredator behaviors (Pulliam, Pyke, and Caraco 1982; Packer and Abrams 1990). However, because the predictions of this model depend only on the general shapes of the fecundity and mortality curves as functions of riskiness, it is unlikely that these complications will change the general conclusions produced here.

Summary and Conclusions

This chapter was inspired by several puzzling relationships between primate sociality and ecological traits that should affect predation risk. In each case, primate species possessing traits that are putative adaptations to high intrinsic predation risk (large group size, large body size) have observed rates of predation that tend to be lower than those of other species. This outcome is not what is expected, so it is difficult to explain these across-species patterns. I take up the suggestion by Boinski and Chapman (1995) and Hill and Dunbar (1998) that life history variables may help to explain these puzzles.

To do so, I use a theoretical model in which I can allow different life history components and selection pressures to vary independently of one another, unlike the highly correlated allometries observed in nature. The model indeed produces one set of predictions that perfectly matches the puzzling patterns observed among primates. This set occurs when primate longevity varies among species but intrinsic predation risk does not, or decreases only weakly with increasing body size.

The results of this model suggest that previous interpretations of the empirical relationship between predation and sociality (including my own: Janson and Goldsmith 1995) have been flawed. The degree of sociality of a primate species has long been interpreted as an optimal compromise between foraging costs and antipredation benefits (e.g., Alexander 1974; van Schaik 1983; Terborgh and Janson 1986; Janson 1990). The benefits of antipredator behaviors are assumed to increase with intrinsic predation risk (e.g., Dunbar 1988), which increases with predator diversity and density (Anderson 1986; Hill and Lee 1998) as well as distance to potential escape routes or refuges (Lima and Dill 1990; Cowlishaw 1997). Under this scenario, larger primates should generally be exposed to lower intrinsic predation risk than smaller ones in a given environment, as large predators can consume smaller prey than is optimal, but small predators simply cannot kill overly large prey. However, this reasoning leads to the prediction that larger species should be less social, a trend that runs opposite to that observed among the majority of primate species (Clutton-Brock and Harvey 1977; Janson and Goldsmith 1995).

Previous explanations of this paradox have not been satisfying. The early suggestion that differences in food competition might explain the existence of large social groups in large-bodied primates (Cheney and Wrangham 1987) has not held up to empirical test. The trend toward increased group size in larger primate species is not radically changed when differences in food competition between species are controlled statistically (Janson and Goldsmith 1995). This finding caused us to suggest an alternative explanation for the pattern: a trend in the use of crypsis as an antipredator strategy (Janson and Goldsmith 1995). We argued that the larger groups of medium-sized, usually fruit-eating, primates might be favored because small cryptic groups were simply not an option for these species; they are too big and active to hide effectively even when solitary (Janson 1998). Thus, large group size would be a second-best option when small, cryptic groups are not possible. The difficulty with this argument is that existing theory suggests that natural selection on group size should be disruptive (Taylor 1979). If predation rates decrease in smaller groups because of crypsis, then individuals should be solitary, or nearly so, whereas if predation rates decline in larger groups, then groups should be as large as ecologically possible. There is no theoretical argument based on crypsis to justify a gradual increase in group size with body mass, such as that observed (Clutton-Brock and Harvey 1977; Janson and Goldsmith 1995).

A resolution of the group size–body size paradox requires acceptance of the general results of the model presented here, as well as the premise that intrinsic predation risk declines slightly or not at all across most of the body size range of diurnal primates in a given habitat. If predation risk does not decline while longevity increases with increasing body mass, then larger species will pay the cost of high predation rates over more years than will smaller species, and thus will gain a greater fitness benefit by reducing predation rates to low values. This trend should favor increased sociality in larger species even if intrinsic predation risk declines slightly with body mass (see fig. 5.6). Only when body mass approaches the upper limit of size found among species in a given habitat will predation risk decline to the point that even modest levels of sociality are sufficient to achieve acceptably low levels of predation. Under this scenario, both very small and very large species show reduced sociality because the increase in lifetime predation due to risky foraging behavior is small compared with the benefits of increased fecundity. The difference is that the benefit of reducing predation is limited in small species by their short life spans, whereas in large species it is limited by very low absolute rates of predation.

The life history resolution of the paradox that group size increases with body size has several virtues over previous explanations. The first is that it emerges from a well-defined mathematical theory, so that both its assumptions and its predictions are in principle testable. Second, it is robust; it applies equally well to terrestrial species and arboreal ones, albeit with different ranges of body size. Third, it can explain a gradual change in group size with body size, as the causal variable (longevity) changes gradually with body mass. Clearly, the inclusion of life history casts a new perspective on hitherto paradoxical trends in adaptive traits in primates. It is time to consider what other primate adaptations may be affected indirectly by slow life history characteristics.

Acknowledgments

This chapter benefited from lively discussions with Robin Dunbar and Russ Hill, and from careful critiques by two anonymous reviewers. This chapter is contribution number 1065 from the Graduate Program in Ecology and Evolution at the State University of New York at Stony Brook.

6

Adaptations to Seasonality: Some Primate and Nonprimate Examples

JÖRG U. GANZHORN, SUSANNE KLAUS,
SYLVIA ORTMANN, AND JUTTA SCHMID

Animal populations are influenced by both bottom-up and top-down ecological processes. Within this framework of constraints, communities, especially those of long-lived organisms, can be structured by intraspecific and interspecific competition for limited resources. This notion applies to community ecology in general (Hunter and Price 1992; Leibold 1997) as well as to primate communities and primate societies in particular (Sterck, Watts, and van Schaik 1997; cf. Fleagle, Janson, and Reed 1999; Boinski, Treves, and Chapman 2000; Kappeler 2000). Temporal and spatial characteristics of food abundance, food distribution, and food quality are considered prime factors acting from the bottom up (Ostfeld and Keesing 2000). As a consequence, they feature prominently in many hypotheses concerning the evolution of life history traits, especially in relation to community composition, population density, and the distribution of females in space and time (van Schaik, Terborgh, and Wright 1993; Fleagle 1999).

However, we still lack a unifying concept to explain how food may be limiting and which characteristics of the nutritional base are relevant in evolutionary terms. Possible limiting constraints range from micro- and macronutrients (Oates et al. 1990; Ganzhorn 1992; Yeager, Silver, and Dierenfeld 1997; Altmann 1998; Conklin-Brittain, Wrangham, and Hunt 1998) to adaptations for detoxifying secondary plant chemicals (Wrangham, Conklin-Brittain, and Hunt 1998; Lawler, Foley, and Eschler 2000) to morphological features, such as dentition, gut or body size, associated metabolic limitations (cf. Chivers, Wood, and Bilsborough 1984; Hughes 1993), and food availability. The latter appears to be the most obvious factor, but it turns out to be among the most difficult to measure in a way that reflects reality, both at the level of food abundance (e.g., Chapman, Wrangham, and Chapman 1994) and, especially, with respect to energy (nutrient) intake in relation to energy (nutrient) expenditure (Janson, Stiles, and White 1986; Oftedal 1991; Barton and Whiten 1994; Altmann 1998). In particular, food availability per se does not always need to coincide directly with reproductive success. In rodents, for example, some species respond to pulsed mast fruiting with increased reproduction (Ostfeld and Keesing 2000), while other sympatric rodents of the same size suspend reproduction after mast fruiting (Bieber 1998; Schlund, Scharfe, and Ganzhorn 2002). Thus, the same phenomenon may impose different constraints on seemingly similar organisms. Phrased another way, divergent adaptations can represent different solutions to the same problem, and different species, sexes, and age groups may be limited by different factors (e.g., Pereira and Leigh, chap. 7, this volume). It is likely that there are adaptational "treasures" that we have as yet failed to discover in primates, due partly to our own (hominid) biases in expectation.

In this chapter, we first illustrate the gaps in our understanding of the relationships between food availability and primate densities. In so doing, we reinforce the conclusion of Chapman et al. (1999) and Ganzhorn (2002) that the lean seasons and measurements of food supply that we usually consider limiting may actually not be so for many populations. Second, we outline a few examples of primate adaptations for coping with seasons of low food supply. Finally, we use examples from nonprimate taxa to draw attention to some possible solutions to changing food availabilities that have not yet been explored sufficiently by primatologists.

Biomass and Resource Use

The biomass of consumers at any given place is expected to increase along with plant productivity. This expectation is supported by many case studies showing that primate individuals and species concentrate in areas with the most food, by the assumption that the availability of mature leaves during lean seasons relates to population size for folivorous primates, and by reports of energy deficits or population declines during times of severe or unusually prolonged food shortage (e.g., Hladik 1977; McKey 1978; Foster 1982; Kay et al. 1997; Altmann 1998; Gould, Sussman, and Sauther 1999). Nevertheless, since more data have accumulated and attempts have been made to measure plant productivity, it has become clear that neither annual food production nor seasonal food bottlenecks regulate primate populations in a uniform way, if at all (Chapman et al. 1999). Contrary to the assumption that food availability in forests with distinct dry seasons is lower than that in seemingly less seasonal forests, studies on plant chemistry have shown that leaf quality decreases with increasing leaf life span, which increases with rainfall (Bryant, Chapin, and Klein 1983; Cunningham, Summerhayes, and Westoby 1999). Consequently, the qualities of leaves (protein to fiber ratio) in deciduous dry forests exceed those in evergreen wet forests; this difference is found in comparisons including both tree communities in general and related plant species growing under divergent seasonal conditions (Ganzhorn 1992; Cunningham, Summerhayes, and Westoby 1999). Thus, in most years, deciduous forests, which provide a better nutritional base for a few months per year, allow higher population densities than the always abundant leaves of evergreen forests. This difference is also reflected by the finding that primates from more seasonal habitats have higher intrinsic rates of increase than primates from less seasonal habitats (Ross 1992b), allowing quick recovery from population crashes after unusually long and severe dry seasons.

In areas not influenced severely by human disturbance, leaf quality is linked to the biomass of folivorous primates in a surprisingly uniform way across Africa, Asia, Madagascar (Oates et al. 1990; Ganzhorn 1992), and possibly Amazonia (Peres 1997). Allometric relationships between species body mass and mean population density allow us to decouple the effect of seasonality on primate biomass from site and species effects and thus to examine whether primate population densities differ between dry and evergreen forests. In general, there are negative correlations between body mass and mean population density for lemurs from Madagascar and for howler monkeys from the Americas. Corrected for body mass, however, comparisons of residuals show that lemurs and howler monkeys of the same body mass reach a higher population biomass in seasonal deciduous than in evergreen wet forests (fig. 6.1).

These findings imply that primate populations are limited by the quality and/or abundance of high-quality food during a few weeks of high food abundance and are little influenced by the lean times of the year. This conclusion is consistent with the observation that food competition seems most severe when food is most abundant (Sauther 1993). However, since there is definitely less food available during lean seasons, the questions remain: why is the rich season so important, and what kinds of mechanisms do primates utilize to make it through the lean season (Pereira et al. 1999)? In the following review we feature examples of different strategies that primates and other taxa employ to cope with the challenges of lean seasons.

Strategies for Coping with Seasonally Changing Food Supplies
Behavioral Adaptations

Primates have different options to optimize energy expenditure in relation to energy intake. During the lean season they can change group size, travel

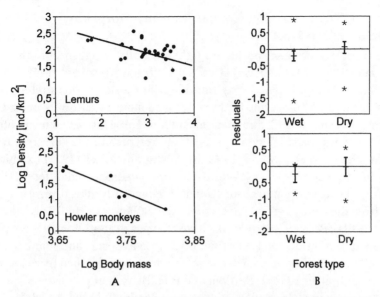

FIG. 6.1. *(A)* Least-squares regression between body mass (g) and mean population density (individuals per km²) of lemurs and howler monkeys. *(B)* Differences between the population densities predicted by the regression (based on mean population densities) and the densities of distinct populations in wet and in dry forests. Sample sizes for lemurs were *n* = 54 in wet forests and *n* = 37 in dry forests; for howler monkeys, *n* = 9 in evergreen forests and 12 in semi-deciduous/deciduous forests. Values are means, 95% confidence limits, minima and maxima. Only data from sites without hunting pressure were used; floodplains, gallery forests, and montane forests were also excluded. (Data from Ganzhorn, Wright, and Ratsimbazafy 1999 for lemurs; Peres 1997, appendix, for howler monkeys.)

longer distances in search of food, switch to low-quality food, change diet composition, use food patches of different sizes, or save energy by simply reducing their activity and waiting until the period of food scarcity is over. Factors such as body size (Kleiber 1932; McNab 1983), climate (Hulbert and Dawson 1974; McNab 1979), and food habits (McNab 1980) affect the energetics of endothermic animals and therefore influence the strategies utilized to cope with periods of energy shortage. A variety of seasonal behavioral adjustments have been reported for prosimians (e.g., Morland 1993; Overdorff 1993; Pereira et al. 1999; Curtis and Zaramody 1999; Wright 1999; Gursky 2000), New and Old World monkeys, and apes (e.g., Terborgh 1983; Wrangham, Conklin-Brittain, and Hunt 1998; Conklin-Brittain, Wrangham, and Hunt 1998; White 1998; Knott 1998).

Seasonal Changes in Diet Composition and Morphology of the Digestive Tract

The majority of primate species are rather catholic with respect to their diets (Oates 1987). Many species classified as "frugivorous" or "granivorous" switch to insects or leaves during times of fruit shortage. This strategy parallels that of many passerine birds, which feed almost exclusively on insects in spring but shift to fruit and seeds in autumn. These birds, and many other animal species, completely and repeatedly restructure the digestive tract within days or weeks (Starck 1999a,b). Starck and Kloss (1995) examined the guts of Japanese quails *(Coturnix coturnix japonica)* fed different diets (fig. 6.2). All experimental groups differed significantly in resorptive surface area from the control group fed a standard diet. For passerine birds, such as bearded tits *(Panurus biarmicus)*, these changes in morphology have profound implications for diet selection. Birds with a "spring and summer" gut adapted to an insectivorous diet cannot survive when switched to seeds, and vice versa (Spitzer 1972; T. Bairlein and P. H. Becker, unpub.).

Among mammals, studies of structural flexibility in the digestive tract

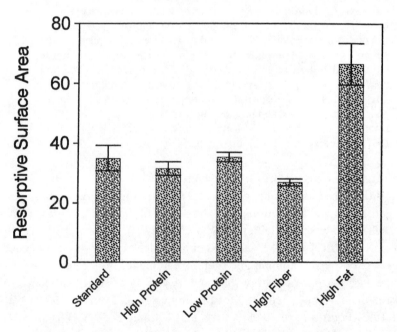

FIG. 6.2. Total resorptive surface area of mucosal epithelium (in 1,000 mm^2) in Japanese quail fed different diets, as calculated from the length of gut segments. (Adapted from Starck and Kloss 1995.)

have been restricted largely to rodents and ruminants. All have revealed substantial changes in morphology or physiology in response to changes in food quality and quantity (reviewed by Starck 1999a,b). Whereas primate data are currently unavailable, similar changes in gut morphology and physiology can be assumed to occur in primates. Such changes among seasons and habitats, within and between species, are likely to have profound implications for food selection and processing that have not yet been considered.

Physiological Changes in Digestive Efficiency

Associated with morphological reconstructions are physiological changes in the digestive tract. P. Lahann, U. Walbaum, and J. U. Ganzhorn (unpub.) investigated digestive efficiency in two lemur species *(Propithecus verreauxi* and *Lepilemur ruficaudatus)* during the dry and wet seasons in Madagascar. Concentrations of various chemical components in food and fecal samples were measured throughout the year. Digestive efficiency was defined as the ratio of concentrations in the food to those in the feces. In most of these comparisons, digestive efficiency differed significantly between the dry and wet seasons (fig. 6.3; higher concentrations of certain chemicals, such as acid detergent fiber, in feces than in food are due to the digestion, and thus removal, of other substances). The causes and consequences of these changes in digestive efficiency remain unclear, and strictly controlled studies will be required to understand them. But the work indicates that even species, such as *Lepilemur,* that live on a leaf-flower-fruit diet year-round modulate their

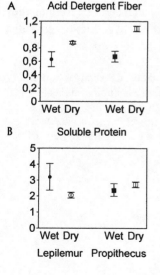

FIG. 6.3. Changes in the efficiency of digestion of *(A)* acid detergent fiber and *(B)* soluble protein in two lemur species, *Propithecus verreauxi verreauxi* (squares) and *Lepilemur ruficaudatus* (circles) between dry seasons (open symbols) and wet seasons (solid symbols) in Madagascar. Digestive efficiency is indexed by the ratio of concentrations in the food to concentrations in the feces (values are means ± 95% confidence intervals). (Data from P. Lahann, U. Walbaum, and J. U. Ganzhorn, unpub.)

digestive physiology across the annual cycle. The effect ought to be more pronounced in species engaging in major seasonal dietary switches (e.g., from animal to plant material, and vice versa), much as with the bird examples provided above. In addition to seasonal variation within species, interactions among behavior, anatomy, and physiology result in interspecific variation unaccounted for in most approaches (Lambert 1998).

Reduction of Energy Expenditure during the Lean Season

If, as argued above, primate populations are constrained by limitations during the rich season, not by food shortage during the lean season, we have to ask how primates survive the lean season. To face this unfavorable environmental condition efficiently, many animals gain weight before the lean season. There are alternative strategies, however. Heldmaier (1989) showed, for example, that many nonhibernating mammals from severe climates reduce body mass in anticipation of the season of food shortage, instead of increasing mass and accumulating large stores of body fat (fig. 6.4).

This finding seems counterintuitive and does not conform to Bergmann's rule, which states that endothermic animals with a large geographic distribution are often larger in colder areas because large animals have relatively low metabolic rates (Begon, Harper, and Townsend 1996). However, Heldmaier (1989) demonstrated that small mammals (< 1 kg), especially nonhibernators, benefit from weight reduction because they are not large enough to accumulate large fat deposits. Another nonintuitive aspect of this pattern is the increased mass-specific metabolic rate that results from small body mass (Schmidt-Nielsen 1997) and the consequent need to forage constantly during the lean season. Despite this, as mass decreases, so does the amount of tissue that needs to be heated. Consequently, reduction of body

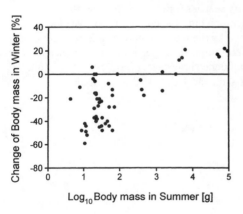

FIG. 6.4. Seasonal changes in the body mass of nonhibernating mammals prior to the lean season (called winter in the original work). (Adapted from Heldmaier 1989.)

mass by 30–50% in shrews and voles significantly reduces their total energy requirements during winter (Heldmaier 1971).

The situation for primates may be very similar. Malagasy lemurs have evolved various mechanisms for coping with seasonal changes in climatic conditions and in the resources on which they feed. Madagascar has a pronounced island-wide environmental seasonality that differs distinctly from that of other tropical landmasses (Dewar and Wallis 1999). As a result, lemurs have evolved adaptations to save energy under unfavorable conditions (the "energy conservation hypothesis"; summarized by Wright 1999), which have been studied more intensively and in more detail than those of other primates.

Pereira and colleagues (1999) investigated the metabolic and ecological adjustments of ringtailed lemurs *(Lemur catta)* and red-fronted brown lemurs *(Eulemur fulvus rufus)* in relation to normative seasonal changes in ambient temperature and food supply. They showed that under semi-free-ranging conditions, as well as when the lemurs were kept under constant temperature and food supply conditions but under changing light regimes, the lemurs decreased their food intake, used energy from adipose stores, and reduced their body mass during the photoperiod corresponding to the latter, more severe half of Madagascar's dry season. Hair growth also ceased during this period, possibly reflecting reduced rates of metabolism.

Because food was not limiting in these studies (Pereira et al. 1999), these results support the idea that, today, reduction of body mass might be an adaptation for reducing energy requirements for maintenance rather than a consequence of food shortage. Certainly, in evolutionary terms, it is most likely that food shortage was the ultimate factor that caused the evolution of this adaptation, while today it is triggered by changes in photoperiod. These findings demonstrated that nonhibernating lemurs are able to reduce their activity levels, and probably their metabolism, during the dry season.

There are other energy-saving mechanisms that need to be considered and tested under natural conditions. Pereira and colleagues (1999) have shown that in captive lemurs, concentrations of two hormones known to affect metabolism in humans and other mammals—insulin-like growth factor 1 (IGF-1) and thyroxine (T_4)—vary seasonally (Nagy, Gower, and Stetson 1995; Tomasi and Stribling 1996). Other studies have linked low levels of these hormones with high adiposity. In *Lemur catta* and *Eulemur fulvus,* circulating levels of IGF-1 and T_4 were low during the photoperiod corresponding to Madagascar's wet season, which presumably facilitates fat deposition in nature at the appropriate time of year. As mentioned above, fat deposits were then reduced during the photoperiod corresponding to the dry season. Thus, more than one mechanism (accumulation of energy reserves for the beginning of the lean season and reduction of body mass to reduce energy requirements late in the lean season) might be involved in helping lemurs to survive the lean season. Again, to be understood in evolutionary terms, the precise interplay of these possible adaptations needs to be studied under natural conditions.

During the lean dry season in Madagascar, lemurs show a variety of other features that significantly reduce energy requirements by decreasing metabolic rate and body temperature. The most extreme adaptation is realized in the small cheirogaleids, *Microcebus* and *Cheirogaleus,* known for their ability to enter daily torpor or hibernation under the low ambient temperatures and food scarcity conditions typical of Madagascar's cool dry season (Martin 1972; Pagès and Petter-Rousseaux 1980; Ortmann et al. 1997; Schmid 2000; Schmid, Ruf, and Heldmaier 2000; Dausmann, Ganzhorn, and Heldmaier 2000; Perret and Aujard 2001). Daily torpor is characterized by a dormancy bout duration of less than twenty-four hours, whereas hibernation is characterized by a sequence of prolonged torpor bouts with an average bout duration of two weeks (Twente, Twente, and Moy 1977; French 1982). Whereas torpor can significantly reduce energy expenditure by lowering body temperature and metabolic rate, these adjustments may not actually affect a given daily energy budget significantly. Schmid and Speakman (2000) measured the daily energy expenditure (DEE) of free-living gray mouse lemurs *(Microcebus murinus)* during the dry season. They found that both female and male *Microcebus* enter torpor spontaneously over a wide range of ambient temperatures during the dry season. Female mouse lemurs saved about 25% of their energy via daily torpor, whereas the DEE of torpid males was 5% higher than that of normothermic males (fig. 6.5A). Mean water turnover was significantly lower in torpid males and females, however, compared with that of normothermic animals (fig. 6.5B). Thus, a benefit of torpor for both males and females was water conservation, possibly reducing water requirements.

Daily torpor and hibernation do not seem to occur in primates other than lemurs (Hudson 1973; Müller 1983; Genoud, Martin, and Glaser 1997). Nevertheless, all endothermic animals need to produce heat to maintain body temperature, and most of their energy is spent on thermoregulation (Schmidt-Nielsen 1997). One heat-producing mechanism is nonshivering thermogenesis (NST), which occurs in brown adipose tissue (BAT). BAT is essential for the maintenance of euthermia in small and newborn mammals and for rewarming after hibernation. BAT thermogenesis results from high proton conductance by the inner mitochondrial membrane. This unique

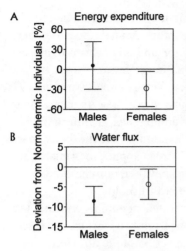

FIG. 6.5. Mean values of *(A)* daily energy expenditure (kJ/day), and *(B)* water flux (ml/day) in torpid male (solid circles) and female (open circles) gray mouse lemurs *(Microcebus murinus)*. The percentage of deviation from normothermic individuals is shown; values represent mean ± standard deviation. (Adapted from Schmid and Speakman 2000.)

proton translocation is due to the uncoupling protein (UCP1), which is exclusively expressed in BAT (Klaus et al. 1991; Pond 1998). By acting as a proton channel, UCP1 uncouples the respiratory chain from ATP synthesis. The oxidative energy gained from respiration and translocation of protons is no longer converted to ATP, but is rather released as heat. BAT is specific to mammals, and UCP1 expression has been detected in several orders (Rodentia, Lagomorpha, Carnivora, Primates, Chiroptera, and Artiodactyla; reviewed in Klaus et al. 1991; Brandner, Keith, and Trayhurn 1993).

The capacity to produce heat by NST is a function of body mass, and this mechanism is most effective in animals with a body mass lower than 3–5 kg (Heldmaier 1971). Norepinephrine (NE) stimulates nonshivering thermogenesis and increases oxygen consumption in relation to body mass. After injection of NE, heat production increases up to 10 times over basal metabolic rate in very small mammals (body mass ≤ 10g), but it fails to increase heat production in animals with a body mass of more than 3–5 kg (Heldmaier 1971; Wunder and Gettinger 1996). This finding leads to the assumption that animals above 3–5 kg may have different mechanisms for maintaining body temperature. At this time, we do not know to what extent NST is utilized by primates. Previous studies did not find BAT in adult *Lemur catta* and *Eulemur* spp. (Pereira and Pond 1995; M. E. Pereira, pers. comm.), though BAT is present in other primates (Trayhurn 1993). Thus, our understanding of adaptations to seasonality may benefit from systematic investigations of BAT occurrence and utilization among free-living primates, such as by measuring heat production following NE injection.

Animals that use NST as a heat-producing mechanism are adapted differently to lean-season conditions than are other animals because thermoregulation via NST is unrelated to locomotion and other forms of muscle movement. In contrast, animals without NST must move, shiver, or use other forms of heat production or conservation. Since animals without NST must spend additional energy for muscle movement, they ought to incur higher costs for temperature regulation. Assuming such a difference between animals with and without NST, and assuming that the body mass threshold of 3–5 kg found by Heldmaier (1971) also applies to primates, we could expect different allometric relationships between population density and body mass for species above and below that threshold.

The threshold of about 3–5 kg found for other mammals is within the range of body masses that separates folivorous from frugivorous and insectivorous primate species (Terborgh 1992). Inspection of the allometric relationships between body mass and mean population density for primates from different continents indicates steeper regression lines between primate body masses and population densities at higher body masses (Ganzhorn 1999). This finding suggests that larger primates are linked more tightly to primary production (and thus are regulated more by bottom-up processes) than smaller species that are distanced from primary production by foraging at higher trophic levels (e.g., arthropods). With this possibility in mind, it might be worth investigating whether the differences in the allometric relationship of body mass and density between large and small primate species are related to dietary differences or different mechanisms of thermoregulation.

There has been considerable debate about whether a primate's diet can be used to predict its basal metabolic rate relative to body mass (Müller 1983, 1985; Kurland and Pearson 1986; McNab and Wright 1987). According to the available data, there is no strong evidence that diet is directly linked to BMR, although low relative BMRs are found in species with folivorous diets (Ross 1992a). It may be worth reconsidering these analyses, repeating them with data collected under natural conditions.

Bottom-up Effects: Responses to Pulsed Food Resources

Variations in food availability, such as irregular or regular superabundant food supplies, offer "natural experiments" with which to study the effects of biotic resources on consumers (bottom-up effects), such as those postulated above for large primate species. Most frugivorous or omnivorous animals respond to mast fruiting with increased reproduction and population growth (Ostfeld and Keesing 2000). In contrast, the fat dormouse *(Glis glis),* a hibernating rodent from central Europe, failed repeatedly to reproduce in years following mast seeding (Bieber 1998; Schlund, Scharfe, and Ganzhorn 2002). This dramatic reproductive failure coincided with a lack of food resources in the following autumn; namely, fruiting and masting oak trees.

Generally, oaks show mast fruiting only in alternate years. In mast fruiting years, immature and adult *Glis* feast, fatten, and successfully overwinter at high rates. In following years, however, many trees produce few seeds. Given the high probability of scant food supplies in autumn (prior to hibernation), the dormice do not reproduce in spring. From an adaptive perspective this strategy seems reasonable, because prospective young would have difficulty finding adequate food to prepare for upcoming hibernation, and females themselves might often recover insufficiently well from raising those young. The question of underlying mechanisms remains for all such cases in which animals do not reproduce despite being in an immediate condition that would support reproduction. Reproduction occurs instead when times are tough, but also when the probability of increased food availability in the near future is high.

Documentation of a similar phenomenon may be under way for ringtailed lemurs *(Lemur catta)* in Berenty, Madagascar. There, the ringtails bear and raise relatively few offspring in years of good rain following drought (Jolly et al. 2002). These are years when tamarind trees produce little fruit. In contrast, tamarind trees fruit abundantly during years of drought, when there is a high chance of fruit failure in the subsequent year. Thus, under conditions that superficially appear to be unfavorable, the animals produce greater numbers of viable offspring.

Summary and Conclusions

In this chapter we provide evidence that, within limits, primate populations in seasonal habitats are not constrained by food scarcity as much as previously thought. They tend to have higher rates of natural increase in more seasonal habitats (Ross 1992b) and reach higher biomasses in seasonal dry than in wet forests. The reason for these phenomena may be that the quality of leaves decreases with increasing life span. According to this analysis, the availability of high-quality food during parts of the year, followed by low food abundance during the subsequent months, in deciduous forests allows higher primate population densities than in evergreen forests, where lower-quality food is available year-round. Further support for the idea that food limitations during dry seasons are not as important as previously assumed comes from new studies using doubly labeled water to measure energy expenditure in free-ranging mouse lemurs. Here, contrary to expectations, torpor appeared to be more important in saving water than in reducing energy expenditure during the dry season. Energy savings may not be the function of torpor in mouse lemurs.

Since there is less food available in most regions during the lean season than during the rich season, primates need adaptations for surviving the lean season without losing all the advantages gained during the rich season. Studies addressing the mechanisms facilitating survival in lean seasons are scant in primatology, especially beyond the behavioral level. There is a large literature describing physiological and morphological changes between seasons in nonprimate taxa, however. Nonprimate patterns offer predictions that can be applied to primate work to better reveal the limiting constraints that have shaped extant life histories. Whereas limiting constraints during the wet season may take the form of limited resources, physiological constraints may also intervene. Adaptations for coping with the season of limited resources, for example, might absolutely preclude more efficient exploitation of resources during the rich season.

Our conclusions coincide with the plea of Pereira and Leigh (chap. 7, this volume): We appeal for more interdisciplinary approaches, especially in field primatology, for greater awareness of findings from outside of primatology, and for a broader temporal perspective on the ontogeny and evolution of traits used in comparative life history analyses. In addition to integrating new technical tools and methodologies, we should remain mindful that evolutionarily significant events might have happened months or years before the appearance of the states we consider relevant traits shaped by evolution. Just as this caution applies generally to development (Pereira and Fairbanks 1993; Pereira and Leigh, chap. 7, this volume), it applies to our understanding of ecological constraints and adaptations as responses to environmental variation.

Acknowledgments

We thank Peter M. Kappeler for his invitation to participate in the "Freilandtage." He, M. E. Pereira, C. Ross, and an anonymous reviewer provided valuable comments on a previous draft of the manuscript.

PART TWO

Development

Organismal life histories are fundamentally about development. From birth, individuals grow, develop patterns of behavior, engage threats to their survival, strive to reproduce one or more times, senesce, and die. Since the inception of life history theory (Medawar 1952; Cole 1954; Williams 1957), comparative analyses have illuminated how and why life histories unfold rapidly in some species and more slowly in others. Analyses of relationships among life history traits have grown increasingly sophisticated, especially insofar as contemporary work routinely controls for effects of phylogeny, ruling out statistical violations pertaining to pseudoreplication and thereby elucidating evolutionary patterns more definitively. Detailed treatments have uncovered regularities among diverse taxa with regard to life span, age at maturation, body and brain sizes, sizes of neonates and litters, interbirth intervals, and even dietary and ranging patterns.

Analyses of mammalian life histories have not yet fully come to grips with development itself, however, as information about the actual ontogenetic trajectories traveled by individuals has been largely left aside (cf. Case 1978; Starck and Ricklefs 1998; Ricklefs and Starck 1998). At the root of this problem lie the complexities of growth phenomena and the difficulties of amassing detailed longitudinal data for members of many mammalian species.

The chapters of part 2 provide glimpses of various ways in which close attention to development will enrich our understanding of life histories and their evolution, especially in relation to socioecology. Pereira and Leigh (chap. 7) illustrate how inattention to juvenile primates results in the neglect of important life history variation. Particular adult states can be attained via divergent developmental paths, and the functional significance of different ontogenetic trajectories has hardly begun to be explored. In addition, early conditions can induce the development of alternative adaptive suites of morphological, physiological, and behavioral traits. Demographic or social conditions, for example, can modulate the timing of maturation, levels of reproductive effort, offspring sex ratios, or even an adult individual's sex. Some such effects are known for primates, but Pereira and Leigh emphasize that we yet recognize only the tip of this iceberg. They reason that plasticity itself should have been a target of selection among primates, given the length of developmental phases in these taxa relative to the chances for change in living conditions. Within and among taxa, divergent modes of development result from dissociation of phenotypic elements, and Pereira and Leigh feature many examples in which timings, rates, or extents of emergence have been modulated for traits including brain size, body size, dentition, play behavior, and natal dispersal.

Godfrey and colleagues (chap. 8) add forcefully to this theme with an extensive and detailed analysis of dental development. Whereas dentition constitutes a direct link to foraging competence, little previous work has examined relations among foraging ecology, dental development, and ontogeny of behavior among primates. Godfrey et al. begin by surveying immature and adult skulls for eleven strepsirrhines and twenty-nine haplorhines, including eighteen folivorous species among seven families of primates. Exploiting a variety of statistical methods, they corroborate that dental development proceeds most rapidly among primates with small cranial capacities, but especially those whose diets have heavy foliage components. In addition, across clades, species that wean infants at relatively late ages are found to show advanced dental development at weaning. Godfrey et al. note that acceleration of dental development and delay of weaning are alternative routes to high dental endowment at weaning, while only the former mode, decoupling dental and skeletal growth, endows youngsters with near-adult food processing capacities without significantly expanding costs, and therefore risks, to mothers. Contrasting adult survivorships in sympatric primates during a recent period of severe food shortage are offered as testimony to links among strategies of maternal investment, reproduction, and early development. In sum, the dissociation of dental and skeletal development provides a compelling example of how selection differentially modulates aspects of development to promote the success of individuals at particular life stages.

While Godfrey et al. show how interrelations among diet, development, and maternal strategies elucidate primate life histories, Hawkes, O'Connell, and Blurton Jones (chap. 9) suggest that the evolution of human life histories was vitally affected by these same interrelations across not just one generation, but two. Human life history features are well predicted by the primate regression of age at maturity against adult life span, whereas solely in humans does life span include a substantial postreproductive phase (menopause). Hawkes and colleagues note that the human data should fit the general primate pattern only if the entire adult life span devotes its production capacity to descendants. Williams (1957) suggested that selection favored premature reproductive senescence in human females. Hawkes et al. disagree, pointing out that age at last birth is similar in chimpanzees and humans, and suggest that the basic life history feature unique to humans is extreme longevity. A variety of data are marshaled to support the premise that protracted adult life spans have characterized human life histories for more than a million years.

How, then, to explain the long postreproductive phase of life in human females? Hawkes et al. build on diverse evidence that *Homo* emerged just as ecological changes would have required significant shifts in foraging strategy for this lineage of large-bodied primates. Greater seasonality and more open plant communities may have favored exploitation of previously little-used resources, such as tubers, that provided substantial returns to adult foragers while remaining inaccessible to young juveniles. Such conditions are suggested to have opened a new role for aged females: grandmothering. Unburdened by their own infants, aging females would have augmented their daughters' fertility, enabling earlier weaning and shortening interbirth intervals. Thereupon, selection would have favored greater vigor (delayed senescence) following last births (menopause). The extended human life span would, in turn, have favored further delays of maturation, augmenting juvenile dependence and increasing adult size. In sum, grandmothering is suggested to have been an adaptive shift that, from the outset of the Pleistocene, has linked an array of life history features distinguishing humans within the general primate pattern. Appreciably predating very large brain size, it may have been the pivotal factor enabling the geographic spread and success of our genus.

Every chapter in part 2 makes clear the value of attending to the details of development while wedding the perspectives of life history theory and socioecology. In each case, results on heretofore neglected but readily measured aspects of morphology, physiology, and behavior offer to deepen our understanding of well-known adult characters in extant or even extinct species. These works suggest that it is time for primatology to take full advantage of the available synergy among life history theory, developmental biology, and socioecology.

7 Modes of Primate Development

MICHAEL E. PEREIRA AND STEVEN R. LEIGH

Exceptional life history patterns keep development a major issue for primate research. Yet studies of juvenile primates had been rare up until ten years ago (Pereira and Fairbanks 1993a), and since then the trend has only grown worse. The life history literature states only that the juvenile period constitutes species-typical "slow" or "fast" development, with little consideration of actual trajectories traveled. The lack of work on behavior is similarly striking: only four studies of social play—the hallmark of juveniles—have appeared in the world's four leading primatological journals *(American Journal of Primatology, International Journal of Primatology, Folia Primatologica, and Primates)* over the past decade. Some morphologists have included juveniles among their subjects, but overall, juvenile primates remain severely understudied. This chapter seeks to make clear why diverse data on primates' prereproductive life histories are needed to understand these relatives of ours.

Initial investigation of any taxon is appropriately focused on its adult male and female representatives as they strive to reproduce before succumbing to predation, disease, starvation, or old age. Reproductive success is not determined solely during adulthood, however; there is much more to life histories. Following parental provisioning or care, individuals in many taxa must negotiate juvenility, the succeeding life stage that lasts until sexual maturation and the onset of reproduction (Pereira 1993b). Failures in this first great task of independent life do not even appear among adult study subjects (e.g., Altmann 1998), making research on adults the study of solely the relatively successful. Indeed, given primates' slow development and reproduction (Case 1978; Charnov and Berrigan 1993; Ross and Jones 1999b), immatures' struggles may constitute the single most important factor ultimately structuring primate societies (cf. Kappeler 2000; van Schaik and Janson 2000).

An understanding of life histories and behavior requires information on development and its adult endpoints. As smaller, less experienced, and unestablished individuals, juveniles face challenges that are different from those faced by adult males and females, and immature and adult primates each operate in ways designed to safeguard or detract from immatures' welfare (Pereira 1988b; Janson 1990; van Noordwijk et al. 1993; van Schaik and Kappeler 1997; Palombit 1999). As juvenile primates develop, they target and become targeted by particular adults for the development of animosities and alliances (Silk, Samuels, and Rodman 1981; Pereira 1988a, 1995; Fairbanks 1993), and shifting collaborations, conflicts, and compromises between juveniles and adults constitute a major dimension of primate social structure (Pereira and Fairbanks 1993a).

Also important is the observation that individual experience shapes phenotypic expression and that different conditions commonly invoke the development of different adaptive suites of morphological, physiological, and behavioral traits among individuals (Roff 1992; Stearns 1992; Schlichting and Pigliucci 1998). Members of many animal species respond to the presence of predators, for example, by developing particular shape-color-behavior modules that are adaptive under those conditions (table 7.1). Others undertake particular phenotypic conversions in response to certain demographic or social circumstances (e.g., Reavis and Grober 1999); such cases include known examples from primates (e.g., Stammbach 1978; van Schaik and Hrdy 1991; Maggioncalda, Czekala, and Sapolsky 2000).

Primate researchers should seek to describe not only "average" juvenile and adult males and females, but also individuals' responses to varied circumstances throughout development. Such descriptions are important because similar adult forms or careers can be achieved via divergent developmental trajectories, and because plasticity itself is especially likely to have been the target of selection wherever developmental phases are long relative to chances for change in living conditions. In this chapter we attempt to illustrate the importance of such ontogenetic research by describing modes of development in primates. Modes are alternative forms and ways of operating. In organisms, they are ontogenetic trajectories, combining particular timings, rates, and extents of codevelopment of integrated traits such as brains, somatic size, body parts (e.g., dentition), and types of behavior, including play, dispersal, agonistic intervention, and targeted aggression and amicability.

First, we review how natural and sexual selection have contributed to

Table 7.1 Examples of adaptive developmental conversions in animals

Phylum	Class	Common name	Species	Known socioecological cues or correlates	Developmental conversion	References
Mollusca	Gastropoda	Whelks	*Nucella lapillus*	High-impact shoreline	Mature early at small size; quadruple litter size	Etter 1989
Mollusca	Gastropoda	Freshwater snails	*Physella virgata*	Predator presence	Develop 100% larger adult size	Crowl and Covich 1990
Arthropoda	Crustacea	Acorn barnacles	*Chthamalus anisopoma*	Predator presence	Develop attack-resistant shape	Lively 1986a,b
Arthropoda	Arachnida	Filmy dome spiders	*Neriene radiata*	Autumnal photoperiod	Delay maturation, grow large, and seek winter hibernacula	Wise 1976
Chordata	Amphibia	Gray tree frogs	*Hyla chrysoscelis*	Predator presence	Develop attack-resistant shape, size, and behavior	van Buskirk, McCollum, and Werner 1997
Chordata	Osteichthyes	Blue-banded gobies	*Lythrypnus dalli*	Loss of reproductive male; dominance over peers	Change sex (female)	Grober 1998; Godwin et al. 2000
Chordata	Mammalia	Meadow voles	*Peromyscus maniculatus*	Autumnal photoperiod	Withhold infanticidal aggression; direct paternal behavior to pups	Parker 2000
Chordata	Mammalia	Ringtailed lemurs	*Lemur catta*	Autumnal photoperiod; attainment of high mass	Use adipose energy stores; suppress growth rate until spring equinox	Pereira 1993c; Pereira et al. 1999
Chordata	Mammalia	Papionins	*Papio, Macaca* spp.	High- or low-dominance status; food supply	Overproduce daughters or sons	van Schaik and Hrdy 1991
Chordata	Mammalia	Orangutans	*Pongo pygmaeus*	Alpha male presence	Males develop primary while suppressing secondary sexual traits	Maggioncalda, Czekala, and Sapolsky 2000

underappreciated diversity in morphological development across primates. Then, we present cases in which developmental conversions may have contributed to variation in behavioral development and early reproduction within species. Throughout the chapter, matters are contextualized in terms of developmental reaction norms, which comprise ranges of possible phenotypic expression for particular genotypes, depending on diverse effects of individuals' internal and external environments throughout life (Roff 1992; Stearns 1992; Schlichting and Pigliucci 1998). The reaction norm perspective maintains that ontogenies themselves are principal targets of selection, which results in functional changes in traits' developmental sequences, patterns of allometry, and degrees of plasticity. Insufficient genetic information still limits research into primate reaction norms, but progress in understanding relationships between primate life histories and socioecology will require incorporation of the reaction norm *perspective* into primatology—that is, its consistent consideration during observational, experimental, and theoretical work.

To shed light on the puzzle of primate development, we must examine the ways in which the pieces can relate. Development entails modularity (Wagner 1996)—that is, dissociability among functionally integrated character complexes and their recombination to express alternative modes (Needham 1933; Gould 1977). Such new character combinations characterize many evolutionary divergences (e.g., Schaefer and Lauder 1996). The protracted development of primates augments prospects for modular plasticity within populations and for taxonomic diversification (Leigh 1995). Distinguishing dissociable from more firmly coupled elements of ontogeny, within and among taxa, will extend our understanding of primate life histories appreciably beyond the slow-fast dichotomy.

Morphological Variation
Body Mass Ontogeny

Body mass is a basic life history variable with tangible links to every morphological, physiological, and behavioral trait (Schmidt-Nielsen 1984). Wide variation in adult body mass occurs among primates, and underlying ontogenetic patterns contribute to divergent modes of immaturity. Unfortunately, we know little as yet about primate growth outside of captive circumstances. But, assuming that captive animals reveal general patterns, nonparametric regressions of mass against age and related growth velocity measures (Leigh 1996) identify at least two major ways in which growth varies within and among primate species.

First, postnatal body mass growth can exhibit a persistent decay in rate, gradually decelerating until adult size is attained. This pattern occurs in

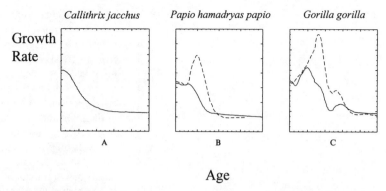

Fig. 7.1. Growth rate curves (kg/year) for selected primate species: *(A)* Common marmoset *(Callithrix jacchus)*. *(B)* Guinea baboon *(Papio hamadryas papio)*. *(C)* Western lowland gorilla *(Gorilla gorilla gorilla)*. In parts B and C, dashed lines represent male trajectories; solid lines represent female trajectories. In part A, rate curves for the two sexes are combined. Axes are unscaled to emphasize shape differences in trajectories (see Leigh 1996 for details).

many small primates, such as common marmosets *(Callithrix jacchus)* (fig. 7.1A), and in females of some larger species, including baboons (fig. 7.1B) and common chimpanzees. In contrast, this pattern is rare among males of larger, more sexually dimorphic species. A second pattern involves slow growth early in development followed by sharp accelerations, or growth spurts, just prior to attainment of adult size. Male baboons (fig. 7.1B) and both male and female gorillas (fig. 7.1C) illustrate this second pattern.

Combinations of these two basic growth patterns have generated diversity within and among species. Some measures of growth spurts, such as peak velocity, often scale with positive allometry (Leigh 1996). Thus, larger taxa tend to show disproportionately large growth spurts. Nonetheless, relations between growth rate and adult size remain complex. Female chimpanzees *(Pan troglodytes)*, for example, lack an obvious, single-peaked growth spurt, whereas female bonobos *(Pan paniscus)* show a distinct growth spurt, despite their smaller size. Such contrasts between closely related species suggest that mass growth patterns are not heavily constrained phylogenetically and are responsive to selective forces (Leigh 1996; Leigh and Park 1998).

Growth spurts underscore one problem in conceptualizing primate life histories simply as relatively slow or fast: changes in growth rates generate life histories that are both slow and fast. Human mass growth is extremely slow initially, for example, but later growth is consistent with expectations

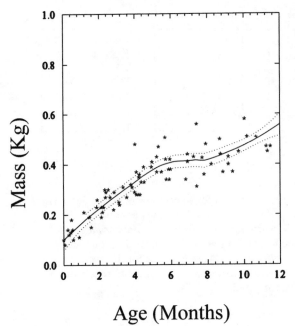

FIG. 7.2. Body mass growth in squirrel monkeys *(Saimiri sciureus)* from birth to twelve months. Regression lines are lowess fits based on cross-sectional treatment of data. Dotted lines represent 95% confidence intervals on position of slope. No detectable mass growth is evident from six to eight months of age.

Age (Months)

from comparative analyses (Leigh 1996; Leigh and Park 1998). Squirrel monkeys, even when provisioned in captivity, characteristically exhibit weeks or months of depressed growth rates following rapid mass growth throughout infancy (fig. 7.2). Such species-typical modulations of growth rate are known in all major primate radiations (e.g., Crockett and Pope 1993; Pereira 1993c; Hamada et al. 1999).

Because mass growth rates do not correlate strongly with other ontogenetic variables, rates and durations of growth contribute independently to adult size. After controlling for size and phylogeny, nearly all correlations among early growth rates, late growth rates, and duration of growth (see Leigh 1996), for both males and females, are below 0.80 (table 7.2). Particular examples also emphasize that adult size is not linked very tightly to age at maturation. Female Hanuman langurs grow faster than female siamangs to achieve a similar adult size, maturing at least one year earlier (fig. 7.3A). Male Diana monkeys attain comparatively small sizes after growing longer than male black-and-white colobus monkeys (fig. 7.3B). Growth rates of Diana males also appear more variable than those of male colobus. Conversely, some similar sizes are reached at similar ages by primates expressing divergent trajectories. Male blue monkeys radically accelerate growth at about five years of age, for example, more than doubling their mass in the last 25% of the growth period, whereas male bonnet macaques show a

Table 7.2 Reduced-major axis residuals of ontogenetic variables regressed on average species body mass

Raw Pearson correlations

Males ($n = 21$ species)

	TOV	PV	ARTOV
Take-off velocity	1.000		
Peak velocity	0.949	1.000	
Age at return to take-off velocity	0.457	0.618	1.000

Females ($n = 13$ species)

	TOV	PV	ARTOV
Take-off velocity	1.000		
Peak velocity	0.972	1.000	
Age at return to take-off velocity	0.505	0.643	1.000

Size-adjusted Pearson correlations, based on the ratio of growth variables to adult body mass

Males

	TOV	PV	ARTOV
Take-off velocity	1.000		
Peak velocity	0.638	1.000	
Age at return to take-off velocity	0.224	0.054	1.000

Females

	TOV	PV	ARTOV
Take-off velocity	1.000		
Peak velocity	0.964	1.000	
Age at return to take-off velocity	0.326	0.400	1.000

Phylogeny-adjusted correlations based on size-adjusted data [a]

Males

	TOV	PV	ARTOV
Take-off velocity	1.000		
Peak velocity	0.766	1.000	
Age at return to take-off velocity	0.787	0.732	1.000

Females

	TOV	PV	ARTOV
Take-off velocity	1.000		
Peak velocity	0.841	1.000	
Age at return to take-off velocity	0.200	−0.005	1.000

Note: Data for correlations are derived from growth rate curves presented by Leigh (1996), and include measures of early growth rates (take-off velocity, TOV), peak velocity during subadult growth spurt (PV), and age at return to take-off velocity (ARTOV, a measure of growth duration). Leigh (1996) provides definitions of these variables.

[a] Phylogeny adjustment utilizes independent contrasts methods described by Garland and Adolf (1994).

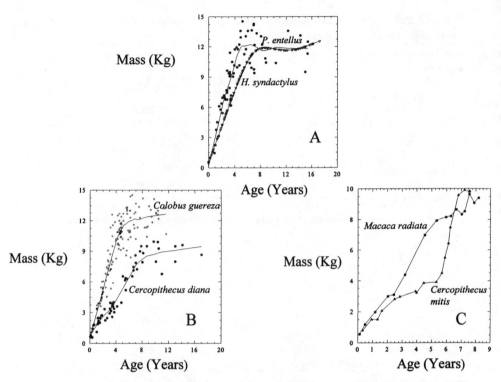

Fɪɢ. 7.3. Comparisons of body mass growth curves for selected species. *(A)* Cross-sectional data for female Hanuman langurs *(Presbytis entellus)* and lowess-predicted values for female siamangs *(Hylobates syndactylus)*. Although both taxa are largely folivorous and reach similar adult sizes, siamangs are expected to reach adult mass up to three years later than langurs. *(B)* Lowess-estimated mass growth curves for male black-and-white colobus monkeys *(Colobus guereza)* and male Diana monkeys *(Cercopithecus diana)*. Despite their smaller adult size, Diana monkeys grow for a longer period of time, and possibly at faster rates during limited phases of development. *(C)* Longitudinal growth curves for a male bonnet macaque *(Macaca radiata)* and a male blue monkey *(Cercopithecus mitis)*. These two individuals are comparably sized as adults, but reach adult mass via different trajectories. Quantitative comparisons of these two species corroborate this illustration based on individuals' growth histories.

consistent mass increase throughout the growth period (fig. 7.3C). New World monkeys and the large-bodied Malagasy lemurs corroborate the significance of growth rate variation across primate radiations, showing prominent adult size diversity without major shifts in age at maturation (fig. 7.4).

All such ontogenetic divergences presumably reflect life history adaptations to contrasting socioecologies (Leigh and Shea 1995; Pereira 1995),

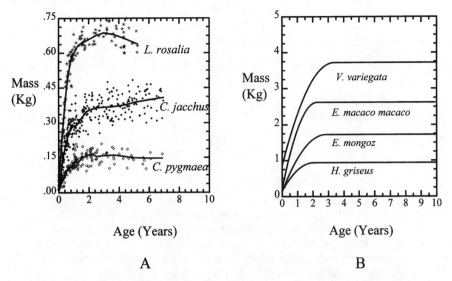

Age (Years)   Age (Years)

A   B

FIG. 7.4. Interspecific differences in rates of growth across similar growth durations. *(A)* Lowess fits for cross-sectional data from golden lion tamarins *(Leontopithecus rosalia rosalia)*, common marmosets *(Callithrix jacchus)*, and pygmy marmosets *(Cebuella pygmaea)*. *(B)* Piecewise regression fits for growth curves in selected lemurid species: black-and-white ruffed lemur *(Varecia variegata)*, black lemur *(Eulemur macaco macaco)*, mongoose lemur *(Eulemur mongoz)*, and gentle gray lemur *(Hapalemur griseus)*. Sexes are combined, given the lack of sexual dimorphism in mass at any point in ontogeny (see Garber and Leigh 1997; Leigh and Terranova 1998).

but they are invisible to life history analyses that summarize postnatal development with age at maturation alone (e.g., Charnov and Berrigan 1993; Pagel and Harvey 1993). Especially because separate components of growth trajectories—early rates, late rates, and duration—seem independently responsive to selection, information on actual growth rates should be incorporated into comparative analyses (e.g., Starck and Ricklefs 1998). Juvenile development can be both slow and fast, male and female patterns often diverge, and individual variation can be expected to differ among species and environments. Research on sex differences and individual variation in development under diverse ecological conditions is an important next step for primatology.

Modular Development

Mass growth schedules provide a platform from which to explore the ontogenetic coupling and dissociability of other traits (cf. Godfrey et al., chap. 8,

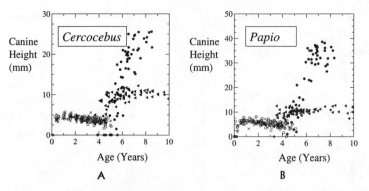

FIG. 7.5. Canine tooth height growth *(A)* in mangabeys *(Cercocebus atys)* and *(B)* in baboons *(Papio hamadryas anubis)*. Canine height is measured from the tip of the left canine to the gum line along the buccal aspect of the tooth. Deciduous teeth are denoted by small crosses for females, open circles for males. Adult teeth are denoted by filled triangles for females, filled circles for males. Female mangabeys show a rapid transition to adult condition, whereas female baboons show a variety of "character states" (absence of tooth, deciduous tooth present, or adult tooth present) prior to completion of adult tooth eruption.

this volume). Here, we present examples of variable dental eruption and brain development patterns among related species sharing similar adult sizes and schedules for somatic growth.

Dental eruption. Modes of development among papionin monkeys entail a dissociation of maxillary canine emergence relative to overall growth and maturation, with female mangabeys *(Cercocebus atys)* suddenly developing relatively large canine teeth despite rates of body mass growth below those of baboons (Leigh 2000). In all papionins, deciduous maxillary canine teeth are nondimorphic and decline steadily in size through wear following rapid eruption (fig. 7.5). Female mangabeys show a rapid change of state at about four years of age (fig. 7.5A). There is little overlap in age ranges between individuals with deciduous and with adult teeth, and few erupting female canines have been measured at small sizes. By contrast, female baboons *(Papio hamadryas anubis)* shift from juvenile to adult canine condition gradually, and on more individually variable schedules, increasing canine size for adult life relatively little despite higher rates of somatic growth than mangabeys (fig. 7.5B). Adult male canines generally appear later than female canines, emerging at similar ages and rates in the two species, though male mangabeys are more prone to a phase lacking canines (zero height in fig. 7.5A).

Canine development may have dissociated from somatic growth schedules in response to divergent selection for agonistic capacity early in development (cf. Plavcan and van Schaik 1997). The rapid emergence and large size of adult canines in female mangabeys would support independent acquisition of dominance by females between three and five years of age (see Gust and Gordon 1991, 1994), in contrast to the social "inheritance" of maternal rank characteristic of the closely related baboons and macaques (Chapais, Girard, and Primi 1991; Pereira 1992, 1995).

Modular recombinations within and among anatomical systems are also evident in hominid evolution. *Australopithecus africanus,* as represented by the Taungs child, shows dental development that deviates from that of both chimpanzees and modern humans (Smith 1992). Relations between dental development and epiphyseal fusion have shifted further over more recent human evolution, given contrasts between *Homo ergaster* (evidenced by the KNM-WT 15000 Nariokotome skeleton) and modern human standards (B. H. Smith 1993; but see Clegg and Aiello 1999).

Brain growth. Brains are expected to modulate primate life histories because they are the command center for behavior and because brain tissue is metabolically expensive (Leonard and Robertson 1994; Aiello and Wheeler 1995; Deaner, Barton, and van Schaik, chap. 10, this volume; Dunbar, chap. 12, this volume). A long-standing hypothesis, derived from comparative analyses, is that brain growth constrains somatic growth and the timing of maturation (Sacher 1975; cf. Prothero and Jürgens 1987; Martin 1990; Janson and van Schaik 1993). Recent analyses have begun to focus on the development of brain components (e.g., Barton and Harvey 2000; Dunbar, chap. 12, this volume). Whereas brain ontogeny at the cellular level remains poorly understood, phylogenetic and sex differences in life history and behavior presumably relate to differential patterns of proliferation and pruning of neurons and synapses (see below; see also Leigh 1996; Leigh and Park 1998).

Here we feature patterns of postnatal brain growth whose dissociation from somatic growth generates divergent modes of immaturity among closely related primates. Contrasting brain ontogenies within clades provide compelling examples of important modularity in the evolution of primate development.

Squirrel monkeys *(Saimiri sciureus)* are born with relative brain volumes comparable to those of other primates, about 55% of total adult

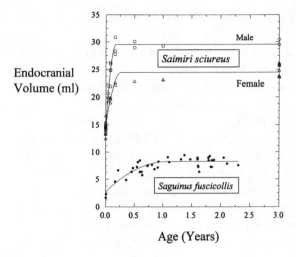

FIG. 7.6. Endocranial volume growth in squirrel monkeys *(Saimiri sciureus)* and saddleback tamarins *(Saguinus fuscicollis)*. Squirrel monkey data are from Manocha 1979. Endocranial volumes (ml) are calculated from Manocha's brain mass data by multiplying mass (g) ×1.036 (Gelvin, Albrecht, and Miller 2000). Piecewise regression fits to these data indicate that squirrel monkeys reach adult endocranial volumes within about three months after birth, while saddleback tamarins reach adult values much later.

volume. By two months of age, however (ca. 4% of the postnatal somatic growth period), squirrel monkeys attain more than 90% of adult brain volume (~25 ml, or 3.7% of adult female body mass: Manocha 1979). By contrast, despite a smaller body size and a shorter somatic growth period, saddleback tamarins *(Saguinus fuscicollis)* grow their brains for as much as one year (ca. 55% of the somatic growth period) to attain much smaller absolute and relative brain sizes (8 ml, or 1.9% of adult female body mass: Garber and Leigh 1997) (fig. 7.6).

These differences in brain ontogeny correspond to species differences in infant care. *Saimiri* mothers bear the entire metabolic burden of infant brain growth, nursing from four (Costa Rican taxa) to twelve months (Peruvian taxa) (Boinski and Fragaszy 1989). Callitrichid mothers absorb only a portion of brain growth costs, by contrast, weaning their offspring at about three months of age, whereupon other caregivers help to supply food until the immatures reach about nine months of age. Developing immatures themselves support only the latest, slowest period of brain growth.

Such species contrasts are neglected by comparative analyses of solely adult data. Age at maturation, adult size, and absolute and relative adult brain sizes in squirrel monkeys all exceed counterpart values for saddleback tamarins, generating the simple impression of strong positive correlations

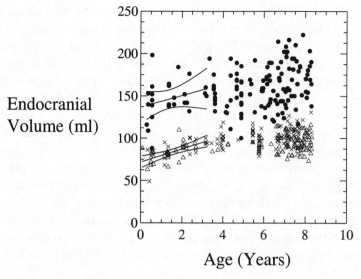

FIG. 7.7. Comparisons of postnatal endocranial volume (ml) growth trajectories among papionin primates (filled circles = baboons: all *Papio* subspecies; small crosses = mangabeys: *C. torquatus torquatus, C. t. atys,* and *C. galeritus;* open triangles = black mangabeys: *Lophocebus aterrimus* and *L. albigena*). Baboons reach maximal adult endocranial volumes very early in the postnatal period. whereas other papionins experience increases in volume throughout the first two postnatal years. Confidence intervals (95%) on least-squares regressions for ages 0–3 years suggest no significant growth in baboons, whereas the growth trend for mangabeys is statistically significant.

among these variables. In fact, brain growth is substantially dissociated from somatic growth and cannot be conventionally linked to longer immaturity in this comparison (cf. Deaner, Barton, and van Schaik, chap. 10, this volume).

Dissociation of brain growth from somatic and dental development also appears to have contributed to a distinct mode of immaturity in baboons. Cranial capacity data from wild-collected papionin skulls suggest that baboons reach adult brain volumes soon after birth, whereas other members of the papionin clade *(Macaca, Cercocebus, Lophocebus)* do so only about a year later (fig. 7.7). Least-squares regressions on data from all baboon subspecies detect no changes in cranial capacity across the first few years of life (estimated via dental eruption), and no variation deriving from subspecific differences is in evidence. Fossil evidence suggests that the unique baboon pattern was derived long after the divergence of *Papio* from the other lineages (Buchanan 2000).

Given the similarities of infant care across papionins, the baboon mode of development seems costly, as mothers produce neonates in the group's characteristic mass range, but with relatively large and rapidly growing brains (Ardito 1975; Smith and Leigh 1998). It is tempting to speculate on the possible adaptive advantages of this mode of development. Socioecological pressures unique to *Papio* have presumably selected for divergent neurological ontogeny, and we note Altmann's (1998) finding that foraging success around weaning predicted lifetime reproductive success among female savanna baboons in Amboseli, Kenya. Potential relations between brain development and immature foraging ecology cannot be evaluated definitively until weanlings are studied in detail at other field sites and then monitored throughout their lives. Already, however, baboons' unusual schedule for postnatal brain growth, alongside Altmann's (1998) findings, underscore the importance of looking beyond the traditional slow-fast maturational dichotomy to understand primate life history and behavior.

Coupling of traits: Functional integration. The coupling of traits allows integrated codevelopment of complex phenotypes. Aspects of neuromotor development, for example, appear to be firmly coupled with play behavior in mammals. Byers and Walker (1995) emphasized that cerebellar synaptogenesis and differentiation of fiber type in skeletal muscle can be influenced significantly by experience only soon after birth in mammals and that changes at this stage seem lifelong in effect. Demonstrating close correlations between ontogenetic schedules for play behavior and these aspects of physiology in mice, rats, cats, and giraffes, they argued that play behavior is designed specifically to promote the development of coordinated, responsive locomotion.

Fairbanks (2000) extended these findings to primates, showing that the times of solo, object, and social play in vervet monkeys *(Cercopithecus aethiops)* coincide with periods of maximal responsiveness to experience in development of the neocortex. Rates of play peak at the same time as synaptic densities in the motor and visual cortices. At five to six months, a sensitive period for development of vision (Harwerth et al. 1986), dendritic pruning begins to outpace synaptogenesis in rhesus monkeys, and visual input guides which synapses will be retained and lost. Peaking during the first year, social play continues stimulating the neuromotor connections activated by solo play and adds the complexity of coordination and strategic assessment between play-fighting partners while connections in the prefrontal cortex are being selected.

Fairbanks (2000) suggested that play promotes adult competence in locomotion, food handling, and fighting through early, permanent effects on the developing nervous system. By providing especially males with frequent and varied stimulation of the neuromotor pathways involved in fighting precisely when those pathways are being selected and myelinated, social play may promote the development of sociocognitive assessment abilities (see also Miller and Byers 1998; cf. Pellis and Pellis 1998). Moreover, play seems scaled to immaturity among the primate radiations—the longer the span of dental immaturity, the longer the period of play, without changes in the ontogenetic sequencing of play types. The extension of play behavior from prosimians through monkeys into apes probably reflects an extension of pruning and synaptogenesis. By evolving specifically in concert with neuronal development, play behavior may contribute importantly to phylogenetic differences in behavior (cf. Siviy 1998).

Linking Pattern with Process: Variation within Populations

Primate ontogeny is clearly a dynamic and evolvable process. Information on variable body mass growth, trait couplings, and dissociability of modular subsystems offers to enhance comparative analyses and our understanding of individual populations. Indeed, more information on covariation between development and socioecological circumstances must accumulate before today's "species-typical" patterns can be interpreted with confidence as reflecting taxonomic divergences. The rates, magnitudes, and timings of developments (e.g., pubertal growth spurts) often shift with environmental conditions, and individually variable plasticities of response to environmental input may *most often* be the raw material for evolutionary divergences (Schlichting and Pigliucci 1998, p. 315; emphasis theirs).

Primates' long lives and large social groups ensure that individuals will encounter a range of conditions during development, promoting the evolution of adaptive plasticity. We should explore individual variability across development as a putative manifestation of different portions of reaction norms (Roff 1992; Stearns 1992)—that is, as different modes expressed facultatively. Over time, long-term projects can correlate codeveloping morphology, physiology, and behavior with particular social and ecological circumstances. Modular phenotypic adjustments that occur rapidly are developmental conversions (e.g., maturation, sex change, dispersal, targeted social behavior) whose advantages over some graduated phenotypic modulation (Smith-Gill 1983) include efficient genetic regulation and facilitated preparation for upcoming conditions (Schlichting and Pigliucci 1998).

The following three examples from two long-term study sites portray primates responding to personal circumstances by adjusting major traits, including timing of dispersal and offspring sex. The examples' shortcomings enhance the value of the exercise by highlighting avenues for future research. Despite the duration and detail of the observations, uncertainties remain due to the general difficulty of primate research—outside of extensive collaborations, only modest numbers of primates can be studied at a time because their life spans far exceed those of most individual research projects. Together, however, the case studies show that longitudinal studies documenting primates' entire lives are needed to illuminate interactions between primate life histories and socioecology (cf. Goodall 1986; Altmann 1998). If longitudinal research can be conducted under a variety of captive conditions and at diverse protected field sites, progressively more detailed descriptions of developmental trajectories will accumulate for types within types of individuals, enabling primate studies to contribute, alongside integrative research on organisms more tractable in other ways, toward a deeper understanding of phenotypic evolution (Stearns, Pereira, and Kappeler, chap. 13, this volume).

Early Onset of an Alternative Reproductive Strategy

Our first example comes from research on the young and old juveniles in two baboon groups of moderate size (*Papio h. cynocephalus;* ca. twelve and eighteen adult females, respectively) at the long-term study site in Amboseli, Kenya (Alberts and Altmann, chap. 4, this volume). The three old juvenile males (3–5.5 years old) in the larger group had grown up playing with one another and, less often, with five older immature males, while the one old juvenile male in the other group, HOM1, grew up with no male age peer and only two older immature males, one partially crippled in his hind limbs.

Play data suggested that these old juvenile males valued their play relations with young juveniles (1–2.5 years old), especially the males among them. Each old juvenile male initiated play more gently with the males among these young groupmates than with the females (fig. 7.8A), despite the fact that the young juvenile males initiated play more often and more roughly with these older partners than their female counterparts did (Pereira 1984; Pereira and Altmann 1985). Lacking male age peers, HOM1 initiated play most gently of all with his younger partners, including the females. HOM1 also initiated 15% more of his play bouts, and played at more than twice the rate of his counterparts in the other study group, possibly compensating for his lack of variety in play partners.

If solely rates of play had been examined, HOM1 would have appeared to be a particularly "male" male (Symons 1978; Fagen 1981). His distinction by the details of his play interactions was furthered by a similar evaluation of his grooming interactions with unrelated adult females. Whereas all other juvenile males, young and old ($n = 6$), initiated the large majority of their grooming episodes with adult females by soliciting grooming (rather than beginning the work themselves), HOM1 began nearly 60% of his episodes by grooming his adult female partner first (fig. 7.8B). HOM1 was also the only male to conduct more than 50% of his bouts of grooming with adult females (Pereira 1984).

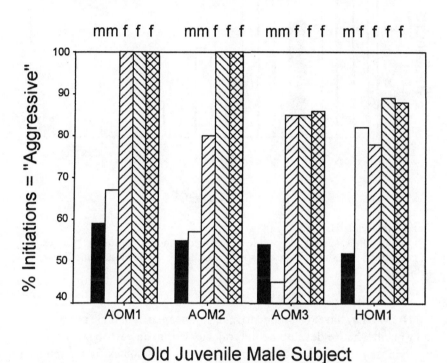

Old Juvenile Male Subject

A

FIG. 7.8. *(A)* Percentage of old juvenile males' (OM) initiations of rough-and-tumble play with young juveniles that were "aggressive" (e.g., slap head, grab coat, jump on) rather than gentle (e.g., touch-withdraw, duck head, jump over; see Pereira 1984). The three OM in Alto's group (acronyms with A) are listed left to right by descending dominance; HOM1 was the sole OM in Hook's group. Old juvenile males initiated play more gently with the most challenging of young juvenile partners, the males (young juvenile sexes are shown atop the bars). This was especially true of HOM1 and the bottom-ranking OM in Alto's group, and these two OM also played relatively gently with young juvenile females.

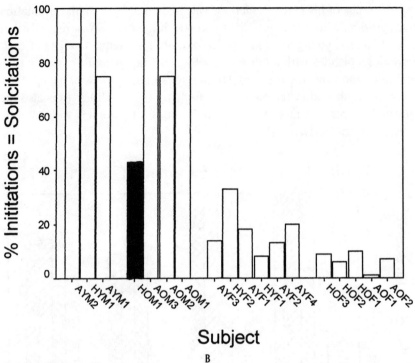

FIG. 7.8 (*continued*)
(B) Only HOM1, among all juvenile males, young and old, initiated episodes of grooming with adult females relatively unselfishly, as did all juvenile female subjects.

HOM1 may have responded to his extreme dearth of male peers by pursuing an alternative developmental trajectory from an early age, one featuring atypically balanced and friendly relations with females. While male baboons entering the late stages of adulthood commonly reduce their rates of group transfer and depend on friendly relations with females for mating privileges (Altmann 1980; Smuts 1985), HOM1's dearth of well-matched play partners throughout his early development may have evoked this developmental conversion unusually early in life. This interpretation was reinforced during later research, when these males were adolescents and young adults (Alberts 1992). HOM1 completed sexual maturation at a very late age and initiated his reproductive career without dispersing (Alberts and Altmann 1995). He continued to stand out as having several female "friends," whom he groomed and defended, and he engaged in many sexual consortships (S. C. Alberts, pers. comm.).

Developmental Trajectories in Male Ringtailed Lemurs

Patterns of juvenile social behavior, male group transfer, and mating behavior were also subjects of long-term study in two forest-living groups of ringtailed lemurs at the Duke University Primate Center (Pereira and Weiss 1991; Jolly et al. 2000). In 1994, focal animal observations were paired with fecal sampling for analysis of testosterone levels, deepening our exploration of male development and reproduction (Cavigelli and Pereira 2000).

Levels of testosterone, aggression, and mating behavior on females' individual days of estrus (Jolly 1966; Evans and Goy 1968) were strongly correlated and clearly distinguished "fighter" and "nonfighter" males (fig. 7.9;

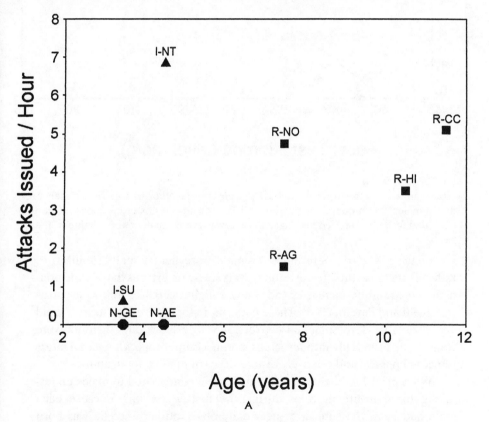

Age (years)

A

Fig. 7.9. *(A)* Mean rate of attacking other males exhibited on five days of estrus by each mature male in group Lc2 in 1994. Resident males AG, NO, HI, and CC had transferred from the neighboring social group (Lc1) three years earlier; males NT and SU had arrived in this group just months before mating season, and natal males GE and AE had not yet dispersed. Neither age nor residence status predicted rates of aggression among non-natal males.

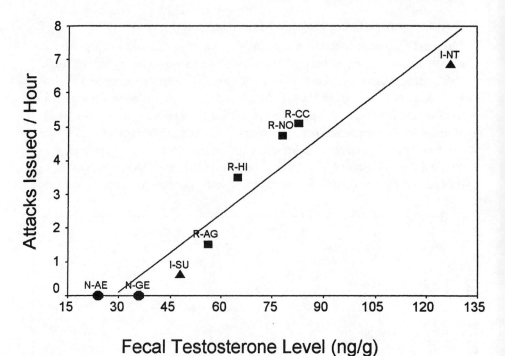

Fecal Testosterone Level (ng/g)

B

FIG. 7.9 (*continued*)
(B) Mean levels of testosterone in fecal samples produced on estrous days closely predicted males' mean rates of aggression. All males whose testosterone concentrations
exceeded 65 ng/g exhibited high aggression toward other males (were "fighters").

$r^2 = 0.93$; $p < .001$). Still more dramatic were fighter males' abilities to
radically increase their testosterone levels solely on estrous days, while nonfighters showed no increases, and showed slight decreases following estrus
(Cavigelli and Pereira 2000). These patterns, in themselves, illustrate a rapid
and reversible developmental conversion. On estrous days, fighters were
transformed suddenly into ferocious mating competitors, while nonfighters
remained passive and relatively disinterested in mating opportunities.

What prior life history factors might have contributed to males undertaking these divergent roles during the mating season? Because each
male had been the subject of near-daily observations since he was born
years earlier, we could construct a matrix of candidate factors to investigate
(table 7.3). This effort further illuminated some things that were already
known. Female ringtailed lemurs are generally disinterested in mating
with related males, for example (Pereira and Weiss 1991), and the two natal males studied in 1994 maintained the lowest testosterone levels by far

Table 7.3 Life histories of eight "nonfighter" and "fighter" males of group Lc2 during the 1994 mating season

Subject	Nonfighters				Fighters			
	AE	GE	SU	AG	HI	NO	CC	NT
Testosterone (ng/g)	24	36	48	56	65	78	93	127
Estrous-day attacks/h	0	0	0.62	1.51	3.51	4.75	5.09	6.85
Residence	Natal	Natal	Immigrant	Resident	Resident	Resident	Resident	Immigrant
Body mass (g)	2,346	2,091	2,027	2,502	2,608	2,693	2,410	2,466
Age (years)	4.5	3.5	3.5	7.5	10.5	7.5	11.5	4.5
Age at transfer	5	4	3.5	3	NA[a]	3	NA[a]	4.5
Age dominates first adult	4.5	3.5	3.5	3	3.5	3	3.5	2.75
Age acquires mid-high rank among males	>5	>4	3.5	3.5	3.5	3.5	3.5	2.5
Rank in birth cohort[b]	1 (2)	3 (3)	1 (4)	5 (5)	1 (2)	2 (5)	Unknown	2 (3)
Rank among juveniles[b,c]	None	3 (3) → None	1 (4) → 1 (2)	5 (5)	1 (2)	2 (5)	Unknown	2 (3) → 1 (2)
No. play partners at 2 years	1	0	2	9	8	4	9	3

[a] Timing of colony management decisions in male's life precluded observation of voluntary natal transfer between groups.

[b] Individual's dominance rank in birth cohort; size of birth cohort is shown in parentheses.

[c] Arrows indicate rank changes due to death of one or more birth peers. "None" indicates that individual was sole survivor of its birth cohort.

(fig. 7.9B). The degree to which males had achieved high body mass may also have played some role, but the two most aggressive fighters had body masses comparable to those of the heavier nonfighters.

Given the patterns observed in the baboon case study, we also examined number of play partners and dominance rank prior to maturation. Fighter males had developed in the company of about six play partners on average, whereas nonfighters had generally had far fewer (table 7.3). Also, all fighters for whom data were available had maintained high rank among their peers from infancy until after puberty. This was true of only one of the four non-fighters, SU, who actually supported the hypothesis that high immature rank enhances male development by rising in rank over the course of the mating season, even as a young immigrant among well-established adults.

Two measures of developmental rate—age-specific play rate and age at natal dispersal—correlated with dominance rank in four pairs of male birth peers. As in other primates (Fagen 1993), rates of play decline with age among immature ringtails (M. E. Pereira, unpub). Supporting our prediction that top-ranking males experience accelerated development, all four top-rankers initiated play less often than did their lower-ranking peers late in juvenility and, especially, early in adolescence (fig. 7.10A). In three of the four pairs, the high ranker also transferred out of his natal group at a younger age (fig. 7.10B). Adding four males that grew up without age peers after assigning them the lowest existing social rank in this analysis (4) left the correlation between immature rank and age of dispersal unchanged.

Prior analyses had suggested that development in ringtailed lemurs is influenced by early social experience. Ringtailed infants compete for dominance prior to weaning to maximize access to food, growth rate, and fat storage before characteristic autumnal reductions of growth rate and metabolism (Pereira 1993c, 1995). Late-born infants' conversion to extremely rapid growth corroborated the importance of maximizing growth before the austral winter (Pereira 1995). If high rank among peers accelerates development over the longer term as well, as suggested by our play and dispersal data, then age at first reproduction as well as first-winter survival is at stake during the dominance competitions of infant ringtailed lemurs.

Ultimately, this is one small data set in which several important factors are confounded, but significant patterns may be peeking through this blurry analysis. Note that these patterns emerged even with no differences in nutrition among weanlings (due to provisioning). In other words, the experience of top rank among infant and juvenile peers may "fast-track" development in young males of expanding ringtailed lemur populations, especially given many play partners (see also Fairbanks 2000). Alternatively, unidentified variables may affect both rank and developmental rate. These possibilities could be readily investigated with a modest number of captive groups, as birth orders are easily manipulated and size determines dominance rank among infants in this species (Pereira 1993a).

Age-Dependent Effects of Social Status on Offspring Sex Ratio

Unfortunately, we were unable to explore similar aspects of development for female ringtailed lemurs because infanticide and predation reduced the number of female subjects to one during the period of long-term data collection. Earlier data, however, including some from prior projects (Taylor and Sussman 1985; Taylor 1986), showed that female life histories are influenced by individual agonistic capacity. Females did not "inherit" maternal rank; young females in large groups were targeted by older females for evic-

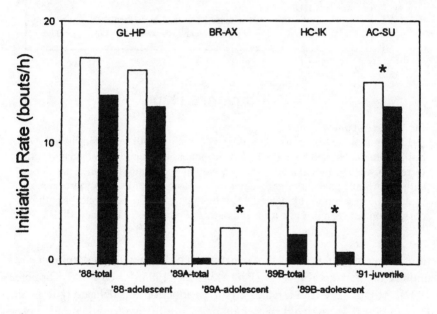

Group and Age Class

A

FIG. 7.10. *(A)* Among immature ringtailed lemurs, dominant males (solid bars) played less often after weaning, especially after puberty, than their subordinate male peers (open bars). Median monthly rates of play initiation are shown for three pairs of male birth peers (1988, 1989A, and 1989B) through the combined juvenile and adolescent periods ("total") and through only the latter portion of that phase ("adolescent"); rates are also shown through the juvenile (preadolescent) phase for another pair (1991; no adolescent data were available for these males). Asterisks indicate significant Wilcoxon signed-ranks tests contrasting monthly values for a pair of males in a given period.

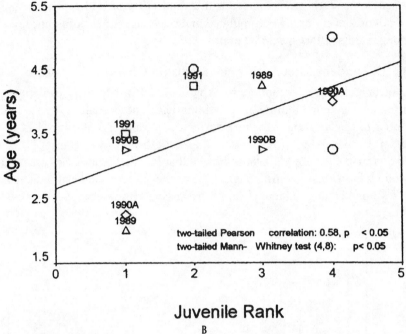

Juvenile Rank

B

FIG. 7.10. (*continued*)
(B) In three of the four pairs (identified by common birth years and shared symbol), the male that had been the higher ranking as a juvenile dispersed and transferred between groups at a younger age. Adding three males that had developed without juvenile peers after assigning them bottom rank (4) and a fourth male that had been second-ranking before his birth peer died (open circles) left the correlation between juvenile rank and age at group transfer unchanged.

tion, and primarily females that overturned older females in dominance remained in their natal groups (Pereira 1993c, 1995; Pereira and Kappeler 1997). At Duke, evicted females' first reproduction was delayed (Nunn and Pereira 2000). In the wild, the challenges of effective foraging and reproduction are exacerbated for evicted females by the difficulties of establishing a new territory between those of other groups (Koyama 1991, 1992; Hood and Jolly 1995; Jolly 1998; Jolly and Pride 1999).

Social play patterns different from those of species in which female social status is "inherited" matrilineally support the view that fighting skill is important in female ringtailed lemurs. Whereas juvenile female cercopithecines play less often, and with fewer regular partners, than their male counterparts (Symons 1978; Pereira 1984; Fagen 1993), juvenile ringtailed lemurs exhibit no such sex differences (M. E. Pereira, unpub.).

Females' lifelong records were investigated for evidence of reproductive adjustments in response to age and social circumstances (Nunn and Pereira 2000). Specifically, we tested predictions on the effects of targeted aggression and female territoriality on offspring sex ratios. Our prediction that members of large groups would produce a disproportionate number of sons was not supported. But individual females accompanied by several mature daughters did overproduce sons, while founders of new groups produced a disproportionate number of daughters, as predicted (Nunn and Pereira 2000).

Of particular interest is evidence of a reaction norm affecting offspring sex ratio that is apparently sensitive to social status only during the earliest phase of a female's reproductive career. More than 80% of the offspring produced by young adult females targeted for aggression during the weeks prior to the mating season were daughters, whereas neither the offspring sex ratios of nontargeted age peers nor those of older females targeted for aggression significantly diverged from 50:50 (fig. 7.11). These results parallel

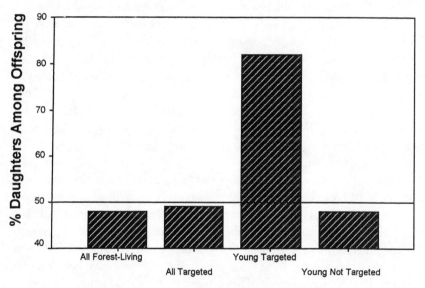

Female Class

FIG. 7.11. Effects of age and social status on offspring sex ratios. Bars indicate the percentages of offspring that were female born to all mothers studied in forest enclosures at the Duke University Primate Center ($n = 133$); all mothers targeted by adversaries for intense aggression in the weeks before the mating season ($n = 45$), young mothers (< 6 years old) targeted for such aggression ($n = 16$), and young mothers that were not targeted for such aggression ($n = 33$).

those from other bird and mammal systems in which mothers benefit early in adulthood from the support of maturing daughters (Nunn and Pereira 2000; see esp. Holekamp and Smale 1995).

Summary and Conclusions

Primates travel diverse ontogenetic trajectories, manifesting great variation in duration of juvenility, overall growth rate, growth rate changes prior to maturation, and modular phenotypic flexibilities within and among populations. Comparisons within and between the sexes, as well as across taxa, indicate that selection often affects early and late developmental phases differently. The significance of long-term changes in rates of growth within species has as yet gone largely unexplored, however, as has modular adjustment of ontogeny and allometry among morphological and behavioral subsystems.

Modes of juvenility should shift rates and timings of trait emergences to minimize risk early in primate development. Extending Janson and van Schaik's (1993) risk aversion hypothesis, we see juveniles responding to risk by changing not only the gross rate but also various "shapes" of development. Janson and van Schaik proposed that group living prolongs juvenility in haplorhine primates both by allowing slow growth, through protection from predators, and by requiring it, through strong feeding competition. In this same context, developmental trajectories whose more detailed modular features best mitigate juveniles' costs (e.g., growth of large brains, adipose stores, social status) by shifting them to caregivers, by shifting them apart from one another in timing, and by augmenting immatures' competitive capacities, should be favored.

Testing of Janson and van Schaik's (1993) basic model and the extension we propose will require detailed ontogenetic data from diverse populations alongside estimates of age, size, and group size effects on predation risk, foraging efficiency, and competition for food, dominance, and social partners. These are exciting goals, particularly because existing data provide some support for the model (cf. Godfrey et al., chap. 8, this volume). For example, assuming greater abundance and lower variability in food supply, immature folivores would be expected to face relatively low feeding competition. As expected, cercopithecoid folivores grow faster and mature earlier than counterpart frugivores (Leigh 1994a). The Callitrichidae may constitute exceptions that prove the rule (cf. Garber and Leigh 1997), insofar as alloparenting and food sharing deflect to other group members some of immatures' earliest risks pertaining to food supply, rapid growth, and predator pressure. Lemurs' sharp but predictable seasonal changes in food supply (Richard

and Dewar 1991) provide a supportive natural experiment. Juvenile lemurids radically suppress their growth rate throughout their first long dry
season, even when low winter nutrition and temperature are removed in
captivity (Pereira 1993b; Pereira et al. 1999; see also Godfrey et al., chap. 8,
this volume). Precocious dental development, dissociated from somatic
growth, additionally safeguards immature lemurs by mitigating foraging
challenges (Samonds et al. 1999; Godfrey et al., chap. 8, this volume). Finally, captive *Saimiri* essentially suspend growth for more than two months
around weaning (Garber and Leigh 1997). The generalized risk aversion
model predicts that weanling squirrel monkeys face acute challenges in assuming juvenile independence, and this seems likely for monkeys known to
depend on high quantities of high-quality foods that are difficult to process
and capture.

Competition for social status, as well as for food, can influence the rate,
duration, and shape of development under the right conditions. In groups
of ringtailed lemurs, for example, all infants are born within one or two
weeks of one another (Pereira 1991) and grow rapidly until autumn (Pereira
1993b), at least partly because social dominance among weanlings is determined by their relative sizes (Pereira 1993c, 1995) and high immature
rank conveys one or more important life history advantages. The baboon-
mangabey differences in adult female canine development also merit further exploration, as mechanisms of rank acquisition remain undescribed for
mangabeys (cf. Plavcan, van Schaik, and Kappeler 1995). Intrasexual selection also can accelerate or extend male growth (bimaturism: Jarman 1983),
and particularly great sexual dimorphism occurs where both effects are
seen. Whether acceleration or bimaturism is predominant presumably depends on the particular selection pressures affecting males. Slow growth, for
example, may relieve developing males of competition and aggression from
older males, functioning as a form of "crypticism" (Jarman 1983; Leigh
1995) while also promoting the quality of other behavioral or morphological developments (e.g., Deaner, Barton, and van Schaik, chap. 10, this volume). Virtual postponement of growth until growth spurts at late ages
should reflect one or more challenges that are potentially extreme early in
development.

In sum, primatology needs much more research on immatures. Juvenile
patterns of development will help to shed light on aspects of sexual selection, natural selection, and social structure in primates and other mammals.
Given primates' long lives, there is an obvious need for longitudinal research—both in captivity, where variables can be manipulated, and at long-
term field sites—as circumstances early in life can be expected to influence

aspects of development and behavior measurably years later (e.g., Altmann 1998).

Genuinely longitudinal research collaborations on known individuals under naturalistic conditions is one of the most important next steps for primate studies. Behavioral sampling methods can be readily implemented to evaluate individuals at fine scales, and because allometries functionally covary suites of traits, they can be organized to identify morphological and physiological candidates for investigation of putative developmental modes across settings and experiments. Alternative strategies should be characterized more definitively by documenting developmental conversions entailing particular couplings and dissociations of traits. Associations between developmental modes and socioecological conditions will illuminate adaptive dynamics within systems while also enriching comparative analyses.

Such detail on immatures will simultaneously sharpen our portrayals of "average" males and females, immigrants and residents, high- and low-rankers, and so forth. Ideally, collaborative projects will increasingly pair careful behavioral work with hormonal and genetic analyses. The necessity of studying animals of particular genotypes under diverse conditions to properly illuminate norms of reaction suggests judicious translocations of primate groups and individuals among populations as a way to enhance conservation efforts and life history research simultaneously.

8 Dental Development and Primate Life Histories

LAURIE R. GODFREY, KAREN E. SAMONDS,
WILLIAM L. JUNGERS, AND
MICHAEL R. SUTHERLAND

In 1989, B. H. Smith observed that the pace of dental development is strongly correlated with adult cranial capacity. Smith argued that the timing of dental eruption might therefore provide a good metric for the overall pace of primate life histories. Until recently, dental development played a rather negligible role in cross-species analyses of primate or mammalian life histories (e.g., Clutton-Brock and Harvey 1979; Harvey, Martin, and Clutton-Brock 1987; Promislow and Harvey 1990; Martin 1990; Ross 1991, 1992b). However, because of its promise as a vehicle for reconstructing aspects of the life histories of extinct species, dental development became a focus of life history research in the late 1980s and the 1990s (Bromage and Dean 1985; Dean 1987; Beynon and Wood 1987; B. H. Smith 1989, 1991b, 1993; Dirks 1998; Reed et al. 1998; Kelley 1999; Schwartz 2000). Tools were developed to assess the biological age of individuals from dental microstructure, with the expressed purpose of reconstructing the life histories of extinct primates (for reviews, see Macho and Wood 1995; Morbeck 1997).

Despite the strong correlation across primates between the absolute pace of dental development and adult cranial capacity, there are significant interspecific differences in the timing of dental eruption that are surprisingly uninformed by variation in either adult body size or cranial capacity. This chapter addresses the linkages among diet, phylogeny, life history parameters (e.g., age at weaning), and the absolute and relative pace of dental development across the order Primates.

With a few exceptions (e.g., Leigh 1994a), there has been little comparative research on the ontogenetic correlates of diet in primates. Dental development, including its outliers, is generally considered within the context of variation in the overall pace of growth and development (Smith, Crummett, and Brandt 1994). Whereas it is recognized that the relative timing of life history landmarks is highly variable (Lee 1996; Lee and Kappeler, chap. 3, this volume; Pereira and Leigh, chap. 7, this volume), how that variability relates to dental development is poorly understood.

Several researchers have noted that folivorous primates tend to grow and develop more rapidly than like-sized frugivores. Malinow et al. (1968) noted the rapid eruption of the permanent teeth in *Alouatta caraya,* one of the largest and most folivorous platyrrhines. Glander (1980) and Froehlich, Thorington, and Otis (1981) reported rapid growth and development in *Alouatta palliata.* Eaglen (1985) related variation in the pace of dental development in lemurs to diet: dental development in the more folivorous *Propithecus* proceeds more rapidly than in the more frugivorous *Lemur, Eulemur,* and *Varecia.* Smith, Crummett, and Brandt (1994) examined some of Eaglen's data within the context of data on other primates; *Propithecus verreauxi* appeared as an outlier on log-log plots of age at the eruption of selected teeth versus adult brain or body size. *Propithecus* exhibits rapid dental development, while *Papio* and *Macaca* are slow to mature dentally. Leigh (1994a) showed that diet affects the overall pace of growth of anthropoid primates, and he postulated a relationship between the overall pace of growth and maturation of the digestive system. He interpreted the rapid growth and development of folivorous anthropoids as support for Janson and van Schaik's (1993) ecological risk aversion hypothesis—that variation in intraspecific competition (due to differences in resource distribution and availability) affects the pace of growth and development in primates. However, Leigh did not examine dental development.

In some taxa, the pace of dental development is far more rapid than might be expected on the basis of craniofacial or skeletal growth (see especially Samonds et al. 1999; Godfrey et al. 2001; King, Godfrey, and Simons 2001; Godfrey, Petto, and Sutherland 2002). Whereas the risk aversion hypothesis might explain a tendency for folivores to accelerate dental development as part and parcel of an overall acceleration of growth and development, it cannot explain the *extreme* acceleration of dental (but not skeletal) development exhibited by some taxa. Our approach here is to evaluate dental development as one of a series of inherently dissociable life history components, under the potential influence of different selective forces. We wish to better understand why dental eruption is relatively precocious in some clades and delayed in others, and how such variation relates to attainment of ecological competence by weanlings. How might the study of dental development enhance our understanding of primate life history strategies?

Materials and Methods

Ideally, a comparative analysis of the effects of diet on dental development requires a sample that is taxonomically and ecologically heterogeneous. Unfortunately, dental developmental schedules are well known for relatively few primate species (B. H. Smith's 1989 analysis was based on twenty-one species) and particularly few folivores. There is also a dearth of information on dental development in species belonging to certain families or subfamilies, including the Colobinae, Atelidae, and most Malagasy lemurs (Smith, Crummett, and Brandt 1994). To remedy this situation, we conducted a survey of the skulls of immature and adult individuals belonging to over forty species of primates, concentrating on folivores and other poorly known taxa. We used multiple sources of information to reconstruct previously undocumented dental developmental schedules. For some colobines and New World monkey species that undergo marked changes in coat coloration from infancy to juvenility, skins provide information from which biological age can be estimated. For strepsirrhine species with strong reproductive seasonality, collection dates provide a means to estimate biological ages of immature individuals. Body mass data are also useful when growth curves are known (e.g., Kirkwood and Stathatos 1992; Leigh 1994a; Leigh and Terranova 1998). In addition, some museum specimens (particularly captive individuals) have records of birth as well as death dates. From these data, along with previously published dental eruption schedules (especially Smith, Crummett, and Brandt 1994), we reconstructed at least partial dental eruption schedules for forty species.

To assess the absolute pace of dental development, we measured dental precocity at four months and one year of age. We define "dental precocity" as the ratio of the number of postcanine teeth (deciduous and permanent) that have generally erupted at any given age to the total number of postcanine teeth (again, deciduous and permanent). To evaluate ecological competence at weaning, we measured both dental precocity and dental endowment at weaning. We define "dental endowment at weaning" as the ratio of average weanling postcanine occlusal area to average adult postcanine occlusal area. Occlusal area for each tooth was estimated as the product of mesiodistal and buccolingual crown dimensions. Teeth that normally have been replaced before weaning are not included in this calculation. Dental precocity and dental endowment are the variables we seek to predict.

Life history data (cranial capacity, neonatal mass, adult female mass, age at weaning, and gestation length) were compiled from the primary literature. Several sources of information contributed greatly to this effort (e.g., Miller 1997; R. J. Smith and Jungers 1997; R. J. Smith and Leigh 1998), but numerous additional original sources were consulted, as described elsewhere (Godfrey et al. 2001). When values for cranial capacity were unavailable, they were measured directly from museum specimens by one of the authors (W. L. J.). Adult female cranial capacities were used for all sexually dimorphic species and preferred for all species, although pooled-sex cranial capacities were used for some sexually monomorphic species when necessary. All size variables (including adult postcanine occlusal area) were log-transformed. In all, we had complete data sets for forty species, including eighteen folivores in seven families (table 8.1). Eleven of the forty species were strepsirrhines and twenty-nine were haplorhines.

To examine effects of diet, a crude foliage consumption scale was constructed. Species consuming little or no foliage (basically, "insectivores") were assigned the value of 1. "Frugivores" consuming varying amounts of foliage were assigned the value of 2, and "folivores" were assigned the value of 3. Because the percentages of fruit and foliage in the diets of many primate species can vary markedly by season, locality, and year, we relied on a variety of cues to assign species to the "frugivore" versus "folivore" categories, including the anatomy of the gastrointestinal tract and physiological adaptations, dietary adaptive strategies under resource crunches, and normative elements in the diet. Colobines, with their enormous sacculated stomachs for leaf fermentation, were considered "folivores" despite the fact that many also consume large amounts of ripe and unripe fruit and seeds (and might be classified on this basis as "frugivores"; see Kirkpatrick 1999). Other species considered "folivores" here (all indriids, *Lepilemur, Alouatta,* and *Hapalemur*) are cecal or ceco-colic fermenters with enlarged and sometimes coiled ceca and colons; they rely heavily on poor-quality foods under resource crises. All folivorous seed predators were classified as "folivores," as it is increasingly recognized that seeds are important components of the diets of many folivorous primates (Hemingway 1996, 1998; Overdorff and Strait 1998; Kirkpatrick 1999), and dentitions modified for masticating seeds are also excellent for processing foliage (Happel 1988; Lucas and Teaford 1994; Yamashita 1998a,b).

We tested the variables in our database for phylogenetically correlated evolution by examining the significance of differences between family means. We also calculated R. J. Smith's (1994) degrees of freedom adjustment for phylogenetic nonindependence. Highly significant ANOVAs, as well as marked differences between Smith's "effective" (or "discounted")

Table 8.1 Samples of measured crania

Family	Genus and Species	Immature individuals	Mature individuals
Indriidae	*Propithecus tattersalli*	1	1
	Propithecus verreauxi	21	25
	Propithecus diadema	13	30
	Avahi laniger	8	17
Lemuridae	*Varecia variegata*	11	6
	Hapalemur griseus	11	7
	Lemur catta	16	11
	Eulemur macaco	6	8
	Eulemur fulvus	20	8
Lepilemuridae	*Lepilemur ruficaudatus*	6	14
Galagidae	*Galago moholi*	15	14
Callitrichidae	*Callithrix jacchus*	7	6
	Saguinus nigricollis	8	9
	Saguinus fuscicollis	2	7
Cebidae	*Saimiri sciureus*	4	6
	Cebus albifrons	14	10
	Cebus apella	12	10
	Aotus trivirgatus	4	9
Atelidae	*Alouatta palliata*	9	10
	Alouatta caraya	5	7
	Ateles geoffroyi	16	17
Cercopithecidae	*Chlorocebus aethiops*	16	8
	Macaca fascicularis	28	15
	Macaca mulatta	9	6
	Macaca nemestrina	14	7
	Papio cynocephalus	8	6
	Papio anubis	12	9
	Kasi vetulus	9	14
	Semnopithecus entellus	30	8
	Trachypithecus françoisi	2	9
	Trachypithecus cristata	17	14
	Trachypithecus obscura	18	36
	Nasalis larvatus	11	8
	Piliocolobus badius	11	6
	Colobus guereza	23	17
Hylobatidae	*Hylobates lar*	16	9
	Hylobates syndactylus	5	7
Pongidae	*Pongo pygmaeus*	8	9
Hominidae	*Pan troglodytes*	13	8
	Gorilla gorilla	6	7

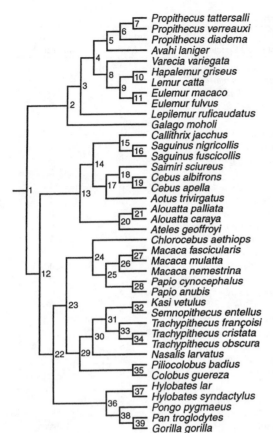

FIG. 8.1. Cladogram used in this analysis. (Data from Razafindraibe, Montagnon, and Rumpler 1997 and Fleagle 1999.)

sample sizes and the traditional sample sizes, indicate strong phylogenetic inertia.

To control further for effects of phylogenetic correlation, we also ran our statistical analyses on phylogenetically independent contrasts (Felsenstein 1985). Contrasts are *differences* between observed or reconstructed variable values at nodes in a specified cladogram (Felsenstein 1985; Harvey and Purvis 1991; Garland, Harvey, and Ives 1992; Pagel 1992, 1993). To create our independent contrast data, we constructed a cladogram for the forty species in our database (fig. 8.1), assumed equal branch lengths, and specified an evolutionary model of Brownian motion. Under such a model, with a fully resolved phylogeny, species differences are independent and can be used in regression and correlation analysis with virtually no loss of statistical power. Beginning with N species values, $N - 1$ independent contrasts can be generated. The signs of contrasts are indeterminate until the researcher selects some criterion on which to order species differences to ensure consistency and interpretability of principal components. We ordered species differences by making contrasts in adult female body mass always positive. This procedure automatically determines the signs of contrasts in other variables (e.g., contrasts in age at weaning will also usually be positive as long as evolutionary increases in size are associated with evolutionary increases in age at weaning).

We analyzed both species mean and independent contrast data because we believe that concordance between the two provides the best evidence in favor of a given relationship among variables. Neither method is error-free, and each is prone to different types of error. Analyses based on species mean data are prone to type I error (falsely rejecting the null hypothesis of no relationship) due to pseudoreplication of values sampled in closely related taxa (see Felsenstein 1985; Harvey and Purvis 1991; Garland, Harvey, and Ives 1992; Garland, Midford, and Ives 1999). R. J. Smith's (1994) degrees of freedom correction adjusts for this possibility and has some other nice properties (Nunn 1995), but it suffers from the same "negative variance estimate" problem that sometimes plagues variance components analysis. To avoid the negative variance problem, we used the maximum likelihood solution to variance components available in SAS ("proc varcomp"). Analyses based on independent contrast data are strongly affected by assumptions hidden in the methodology itself (e.g., that the cladogram is correct, that evolution proceeded in a certain manner; see Ricklefs and Starck 1996; Pagel 1999). Price (1997) suggests that the two types of analytic techniques address different types of questions: independent contrast data reflect evolutionary history (the pattern of diversification), whereas species mean data describe current relationships. He also argues that independent contrasts do not necessarily eliminate the problem of phylogenetic inertia (see also Martins and Hansen 1996).

Multiple regression and correlation techniques were employed to model dental precocity at four months, at one year, and at weaning, using both species mean values and independent contrasts. As is generally recommended for analyses of contrasts (see Garland, Harvey, and Ives 1992), we forced each contrast regression through the origin (y-intercept of zero). All analyses based on species mean values were checked for statistical significance of the regression F statistic under Smith's (1994) reduced or effective sample size. We followed Smith's recommendation of selecting the smaller of two choices: (1) the effective N calculated for the dependent variable or (2) the mean of the effective Ns calculated for all variables. To determine the best explanatory model for each dependent variable, we used a hierarchical step-down regression procedure, first including all explanatory variables and then successively eliminating the poorest until the adjusted R^2 peaked and all variables accounted for significant portions of the variance. The predictive power of contending explanatory variables could then be evaluated by dropping them (individually or in various combinations) from the larger models initially suggested by the full correlation structure. This procedure guards against model misspecification, which occurs when the constructed model contains either too few or extraneous elements. We note that the value of the adjusted R^2 can increase as variables are dropped from the explanatory pool, as it takes into consideration both the degrees of freedom in the model and the change in the unadjusted R^2.

Two-sample t tests were used to test the significance of dental developmental differences between frugivorous and folivorous primates, and one-tailed probabilities were calculated for both traditional and discounted degrees of freedom. Primate species consuming large amounts of insects or animal matter were not included in this suite of comparisons. The same suite of tests was performed for two subsets of our sample: strepsirrhines and cercopithecoids. Discounting was not done here, as the sample sizes were prohibitively small and we were more interested in the direction of differences than their statistical significance.

Finally, principal component analyses were run on the correlation matrices generated for both species mean values and independent contrast data. The former allows the investigator to identify clusters of taxa that "behave" in a similar manner (have similar projections on factor axes), whether because of common phylogenetic heritage or convergence. The latter allows the investigator to explore the historical pattern of evolutionary change and phylogenetic stasis. Strongly positive or negative projections on contrast PCA plots indicate marked differences between immediate descendants at the designated nodes, whereas examples of stasis (little difference between immediate descendants at any given node) will cluster around zero. Projections of particular contrast nodes on factor axes can be interpreted if one evaluates the signs of the raw contrast values (see below).

Results
Is Phylogenetic Correlation Important?
One-way ANOVAs of dental development by family demonstrate significant differences at four months, at one year, and at weaning (table 8.2), suggesting that dental development is strongly affected by phylogenetic inertia.

Table 8.2 Analysis of variance of dental precocity and endowment by family

Family	N of species	Dental precocity at four months: Mean (SD)	Dental precocity at one year: Mean (SD)	Dental precocity at weaning: Mean (SD)	Dental endowment at weaning: Mean (SD)
Indriidae	4	0.60 (0.18)	1.00 (0.00)	0.73 (0.09)	0.71 (0.08)
Lemuridae	5	0.34 (0.02)	0.56 (0.04)	0.37 (0.05)	0.41 (0.07)
Lepilemuridae	1	0.67 (—)	1.00 (—)	0.67 (—)	0.70 (—)
Galagidae	1	0.88 (—)	1.00 (—)	0.72 (—)	0.91 (—)
Callitrichidae	3	0.46 (0.07)	1.00 (0.00)	0.38 (0.00)	0.42 (0.04)
Cebidae	4	0.28 (0.06)	0.54 (0.25)	0.40 (0.05)	0.56 (0.09)
Atelidae	3	0.30 (0.06)	0.48 (0.06)	0.46 (0.08)	0.51 (0.18)
Cercopithecidae	14	0.20 (0.09)	0.38 (0.10)	0.37 (0.13)	0.37 (0.16)
Hylobatidae	2	0.14 (0.00)	0.29 (0.00)	0.36 (0.10)	0.35 (0.13)
Pongidae	1	0.00 (—)	0.29 (—)	0.57 (—)	0.72 (—)
Hominidae	2	0.00 (0.00)	0.29 (0.00)	0.43 (0.00)	0.44 (0.01)
All	40	0.30 (0.20)	0.56 (0.27)	0.44 (0.15)	0.47 (0.18)
F, (d.f.)		17.21 (10,29)	23.29 (10,29)	6.43 (10,29)	4.53 (10,29)
P		< .0001	< .0001	< .0001	< .001

Table 8.3 Discounted or effective sample size

Variable	Discounted N
Dental endowment at weaning	22
Foliage consumption	21
Dental precocity at weaning	18
Log adult postcanine occlusal area	14
Log adult female mass	12
Gestation length	12
Log adult female cranial capacity	10
Dental precocity at one year	11
Age at weaning	11
Age at female first breeding	11
Dental precocity at four months	9
Log neonatal mass	8

Note: Unadjusted $N = 40$.

R. J. Smith's (1994) effective sample sizes (table 8.3) confirm important phylogenetic correlation for all variables in our database. Smith's adjustment reduces the sample size from a traditional N of 40 to effective Ns between 8 and 22, depending on the variable tested.

Predicting the Pace of Dental Development

Tables 8.4 and 8.5 show the prediction models for dental development generated by step-down multiple regression. All regressions, whether based on species mean values (with traditional or discounted Ns), or independent contrast values, have strongly significant F values. The latter regressions yield consistently lower adjusted R^2 values than their species value counterparts, but the results of the two methods are highly concordant: the same explanatory variables tend to be influential, sometimes in the same order of importance.

Our analyses support B. H. Smith's (1989) observation that the absolute pace of dental development is best predicted by cranial capacity (table 8.4): the smaller the brain, the more rapid the pace of dental development on absolute scales. That correlation is particularly strong during early postnatal development (at four months). At one year, cranial capacity explains less of the variance in dental development, and more variables affect dental development significantly. For both species value and contrast regressions, the signs of the contributions of adult female mass and age at first breeding are positive once effects of cranial capacity are controlled. Species whose cranial capacities are small given their adult female mass and/or age at first breeding are likely to be dentally advanced at one year. Age at weaning is

Table 8.4 Multiple regression analysis: Dental precocity at four months and one year

Dental precocity at four months, species mean data

$N = 40$	$F = 85.61$	$P > F = .0000$	d.f. $= (2,37)$	Adj. $R^2 = .81$
N eff. = 9		**$P > F = .0000$**	**d.f. $= (2,6)$**	

Parameter	Coefficient	t	$P > \lvert t \rvert$
Log cranial capacity	−0.72	−7.24	.000
Log female adult mass	0.30	3.68	.001
Constant	1.32	10.94	.000

Dental precocity at four months, independent contrasts

$N = 39$	$F = 49.94$	$P > F = .0000$	d.f. $= (2,37)$	Adj. $R^2 = .72$

Parameter	Coefficient	t	$P > \lvert t \rvert$
Log cranial capacity	−0.43	−9.17	.000
Gestation length	0.47	2.07	.045

Dental precocity at one year, species mean data

$N = 40$	$F = 47.41$	$P > F = .0000$	d.f. $= (4,35)$	Adj. $R^2 = .83$
N eff. = 11		**$P > F = .0001$**	**d.f. $= (4,6)$**	

Parameter	Coefficient	t	$P > \lvert t \rvert$
Log cranial capacity	−1.19	−8.46	.000
Log adult female mass	0.40	3.76	.001
Age at female first breeding	0.04	3.04	.005
Gestation length	0.59	2.04	.049
Constant	1.86	9.84	.000

Dental precocity at one year, independent contrasts

$N = 39$	$F = 23.88$	$P > F = .0000$	d.f. $= (3,36)$	Adj. $R^2 = .64$

Parameter	Coefficient	t	$P > \lvert t \rvert$
Log cranial capacity	−0.96	−5.26	.000
Log adult female mass	0.40	3.02	.005
Age at female first breeding	0.02	2.03	.05

Note: Probabilities and degrees of freedom for discounted Ns are given in boldfaced type.

not an important predictor of the *absolute* pace of dental development, once effects of body mass and/or cranial capacity are taken into account. Contrast data suggest an important role for length of gestation: increases in gestation length are associated with more advanced dentitions in early development.

Our explanatory variables cannot account for as much of the variance in dental development at weaning as at four months or one year (table 8.5). Cranial capacity is still an important predictor of dental development at weaning, and, just as for four months and one year, whenever adult female mass enters the regression, it is secondary to cranial capacity and has the opposite sign. Species with small brains for their body size are likely to show dental advancement at weaning. Neonatal mass is strongly correlated with

Table 8.5 Multiple regression analysis: Dental development at weaning

Dental precocity at weaning, species mean data

$N = 40$	$F = 12.71$	$P > F = .0000$	d.f. = (3,36)	Adj. R^2 = .47
N eff. = 15		**$P > F = .0006$**	**d.f. = (3,11)**	

Parameter	Coefficient	t	$P > |t|$
Log cranial capacity	−0.38	−5.78	.000
Age at weaning	0.12	4.54	.000
Foliage consumption	0.12	4.40	.000
Constant	0.64	7.55	.000

Dental precocity at weaning, independent contrasts

$N = 39$	$F = 7.17$	$P > F = .0007$	d.f. = (3,36)	Adj. R^2 = .32

Parameter	Coefficient	t	$P > |t|$
Age at weaning	0.11	3.92	.000
Log cranial capacity	−0.60	−3.41	.002
Log adult female mass	0.30	2.23	.032

Dental endowment at weaning, species mean data

$N = 40$	$F = 9.93$	$P > F = .0001$	d.f. = (3,36)	Adj. R^2 = .41
N eff. = 15.5		**$P > F = .0018$**	**d.f. = (3,11.5)**	

Parameter	Coefficient	t	$P > |t|$
Log neonatal mass	−0.35	−5.44	.000
Age at weaning	0.13	4.25	.000
Foliage consumption	0.09	2.58	.014
Constant	−0.11	−0.89	NS

Dental endowment at weaning, independent contrasts

$N = 39$	$F = 10.69$	$P > F = .0002$	d.f. = (2,37)	Adj. R^2 = .33

Parameter	Coefficient	t	$P > |t|$
Age at weaning	0.15	4.30	.000
Log cranial capacity	−0.33	−4.05	.000

Note: Probabilities and degrees of freedom for discounted Ns are given in boldfaced type.

cranial capacity and can substitute for it as the best explainer of dental development at weaning.

Two other explanatory variables appear in our regression analyses of dental precocity and endowment at weaning: age at weaning and foliage consumption. Age at weaning correlates with dental development at weaning independently of, and in a direction opposite to that of, cranial capacity; this relationship holds both for contrasts and for species mean values. Small-brained late weaners are likely to be the most dentally advanced at weaning. Without any change in the absolute pace of dental development, delaying the age at weaning can provide weanlings with a better masticatory toolkit.

Foliage consumption is the third most important explanatory variable

Table 8.6 One-tailed t tests on dental development by diet for distinct clades as well as for all thirty-five primate frugivores and folivores

Sample	Variable	Frugivores Mean (N, SE)	Folivores Mean (N, SE)	t	d.f.	P
Primates	Dental precocity at four months	0.20 (17, 0.03)	0.34 (18, 0.05)	−2.52	33	.008
					6	**.023**
Primates	Dental precocity at one year	0.41 (17, 0.04)	0.59 (18, 0.06)	−2.38	33	.012
					6	**.027**
Primates	Dental precocity at weaning	0.37 (17, 0.02)	0.50 (18, 0.04)	−2.84	33	.004
					16	**.006**
Primates	Dental endow- ment at weaning	0.41 (17, 0.04)	0.51 (18, 0.04)	−1.86	33	.036
					19	**.039**
Strepsirrhini	Dental precocity at four months	0.35 (4, 0.01)	0.57 (6, 0.07)	−2.34	8	.024
Strepsirrhini	Dental precocity at one year	0.55 (4, 0.02)	0.94 (6, 0.06)	−4.70	8	.001
Strepsirrhini	Dental precocity at weaning	0.38 (4, 0.02)	0.66 (6, 0.07)	−2.98	8	.009
Strepsirrhini	Dental endow- ment at weaning	0.42 (4, 0.04)	0.65 (6, 0.06)	−2.81	8	.011
Cercopithecoidea	Dental precocity at four months	0.14 (6, 0.03)	0.25 (8, 0.02)	−2.78	12	.008
Cercopithecoidea	Dental precocity at one year	0.31 (6, 0.02)	0.44 (8, 0.03)	−3.29	12	.003
Cercopithecoidea	Dental precocity at weaning	0.29 (6, 0.00)	0.43 (8, 0.05)	−2.42	12	.016
Cercopithecoidea	Dental endow- ment at weaning	0.25 (6, 0.01)	0.45 (8, 0.06)	−2.88	12	.067

Note: Discounted degrees of freedom and associated one-tailed probabilities are given in boldfaced type.

in our species value regressions for both dental precocity and dental endowment at weaning. Contrasts in foliage consumption do not correlate with contrasts in dental advancement at weaning strongly enough to attain statistical significance. With insectivorous species removed from the analysis, two-sample t tests demonstrate a clear relationship between foliage consumption and dental development. Folivorous primates exhibit faster dental development on both absolute and relative scales than do frugivorous primates (table 8.6). Directional (one-tailed) tests of the significance of this relationship hold up under discounting. This difference is not driven by differences in body size, as the folivorous and frugivorous primates in our sample do not differ significantly in adult female mass. This pattern characterizes members of very different clades (including strepsirrhines and cercopithecoids). Small samples mitigate against meaningful tests for ceboids and hominoids, but we found no evidence of faster dental development in frugivores than in like-sized confamilial folivores.

Principal Component Analyses of Species Mean and Independent Contrast Data

Principal component analyses of species mean values and independent contrast data are remarkably concordant. For both, the first axis describes the *absolute* pace (or contrasts in the absolute pace) of dental development (dental precocity at four months and at one year) and correlated variation in the pace of the life history. The second axis describes the *relative* pace (or contrasts in the relative pace) of dental development (i.e., dental development and precocity at weaning) and correlated variation. The third axis describes variation in diet and its correlates.

Factor score coefficients (or beta weights) for the original variables on the first two principal component axes of the species value analysis are given in table 8.7. These axes explain 61.5% and 19.6% of the variance respectively (totaling 81.1%). A third axis explains an additional 9.8% of the variance. The variables influencing axis 1 (with beta coefficients stronger than ±0.25), in order of decreasing importance, are cranial capacity, neonatal mass, adult female mass, total postcanine occlusal area, dental precocity at four months, dental precocity at one year, age at first breeding, age at weaning, and gestation length. Influencing axis 2, in addition to dental precocity and endowment at weaning, are foliage consumption ($\beta = 0.25$) and (with

Table 8.7 Factor score coefficients for species value PCA

Variable	Axis 1	Axis 2
Log cranial capacity	0.36*	0.03
Log neonatal mass	0.36*	0.01
Log female adult weight	0.35*	0.16
Log adult postcanine occlusal area	0.33*	0.18
Age at female first breeding	0.30*	0.24
Age at weaning	0.29*	0.24
Gestation length	0.27*	0.22
Dental precocity at one year	−0.31*	0.24
Dental precocity at four months	−0.33*	0.22
Dental precocity at weaning	−0.14	0.59*
Dental endowment at weaning	−0.15	0.55*
Foliage consumption	−0.11	0.25*

Note: Axis 1 = life history pace (61.5% of the variance); axis 2 = relative pace of dental development (19.6% of the variance). Asterisked values are strongly positive (equal to or higher than +0.25) or strongly negative (equal to or lower than −0.25).

beta coefficients slightly under 0.25) reproductive parameters such as age at weaning and age at first breeding. Folivores and species with delayed weaning are likely to be dentally advanced at weaning. The third axis isolates variation in foliage consumption ($\beta = 0.77$) that is inversely related to age at weaning ($\beta = -0.37$) and age at first breeding ($\beta = -0.32$) and positively related to postcanine occlusal area ($\beta = 0.26$). Species consuming a lot of foliage are likely to exhibit early weaning and reproductive maturation and to have relatively large postcanine teeth.

The first, second, and third principal component axes generated by the contrast PCA explain 50.8%, 20.9%, and 11% of the variance, respectively. The beta coefficients for the first axis of the contrast PCA are surprisingly similar to those describing the corresponding axis of the species PCA. Except for slight changes in the order of importance of the cluster of variables measuring the absolute pace of dental development and the size of the postcanine teeth, and a reversal of the order of importance of the pair of variables measuring the relative pace of dental development, the two sets of beta coefficients bear identical signs and relative strengths. Axis 2 of the contrast PCA (like its counterpart in the species value analysis) describes variation in dental development at weaning that is positively correlated with age at weaning and age at first breeding. Clearly, one way to increase dental development at weaning is to delay weaning and first breeding. Contrasts in foliage consumption are virtually entirely allocated to the third contrast axis. The beta coefficient for foliage consumption (0.81) is considerably higher than that of any other variable on this axis, and no other variable has a beta coefficient stronger than ± 0.25. Nevertheless, this axis isolates interesting relationships among contrasts. Evolutionary increases in foliage consumed are associated with (in order of decreasing importance) earlier ages at weaning ($\beta = -0.237$), earlier ages at first breeding ($\beta = -0.235$), larger postcanine occlusal areas ($\beta = 0.225$), greater dental precocity at weaning ($\beta = 0.218$), and a longer gestation period ($\beta = 0.206$).

Both species value and contrast PCAs highlight the complex nature of the relationships among life history and dental developmental parameters in primates. The two analyses carry parallel messages. With regard to the relationship between diet and dental precocity at weaning, for example, the species PCA suggests that primates consuming large amounts of foliage are likely to exhibit advanced dental development at weaning. The contrast PCA suggests that evolutionary increases in foliage consumption are likely to be associated with evolutionary increases in dental development at weaning.

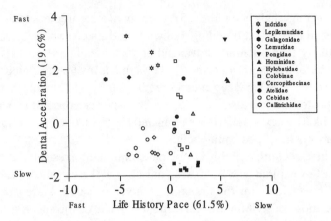

FIG. 8.2. Species PCA, showing projections of species mean values on axes 1 and 2.

Figures 8.2 and 8.3 show the projections of species values and contrast nodes (respectively) on axes 1 and 2 of the species and contrast PCAs. Projections on axis 1 (life history pace) of the species PCA offer no surprises (fig. 8.2). All strepsirrhines, all callitrichids, and the smallest of the cebids *(Aotus, Saimiri)* fall on the negative side; all other anthropoids fall on the positive side. *Galago moholi* (solid hexagon) has the strongest negative projection on this axis, and *Gorilla gorilla* (one solid triangle) has the strongest positive projection.

More interesting are the projections of taxa on axis 2 (dental development at weaning). The indriids (stars) have strongly positive projections on axis 2, signaling high dental precocity and endowment at weaning. *Galago moholi* (solid hexagon) and *Lepilemur ruficaudatus* (solid diamond) are fairly precocious at weaning, but not as markedly as the larger-bodied indriids. Lemurids (open diamonds) have strongly *negative* projections on this axis, signaling low dental precocity and endowment at weaning. Without exception, the colobines (open squares) lie above the cercopithecines (solid squares) on this axis, and the colobines with the strongest positive projections (and greatest dental precocity and endowment at weaning) are the African species and *Trachypithecus françoisi.* Atelids (solid circles) generally lie above cebids (open circles) on this axis, and siamangs above gibbons (both open triangles). The strongly positive projections of some taxa—*Pongo pygmaeus,* the African apes, and *Ateles geoffroyi* (with the most strongly positive projection of all atelids in the sample)—reflect a tendency toward very delayed weaning. The positive projections of the indriids, colo-

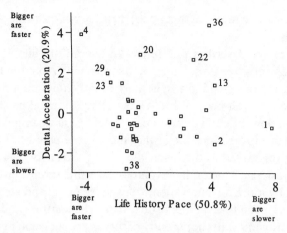

FIG. 8.3. Contrast PCA, showing projections of contrast nodes on axes 1 and 2. Nodes are labeled as in figure 8.1. To interpret the projections, the investigator must decipher which of the two taxa diverging at any node has been reconstructed as having increased in body size at that node. On the first axis (life history pace), nodes at which an evolutionary increase in size is associated with an increase in the life history pace have negative scores, and nodes at which an evolutionary increase in size is associated with a decrease in the pace of life history have positive scores. On the second axis (dental acceleration), nodes at which an evolutionary increase in size is associated with an increase in relative dental development have positive scores, and nodes at which an evolutionary increase in size is associated with a decrease in relative dental development have negative scores.

bines, and howler monkeys, on the other hand, are outcomes of a relative acceleration of the timing of dental eruption.

Of special interest is *Hapalemur griseus,* with its diet rich in bamboo and other foliage. This species does not distinguish itself from the frugivorous lemurids; rather, it falls squarely among them. We will return to its dental developmental adaptations below.

Figure 8.3 shows which contrast nodes experienced the most dramatic evolutionary shifts. Because we made all differences in body mass positive, we need only know the sign of the beta coefficient of each variable on each axis, and which of the two taxa diverging from any given node is the larger (or, more accurately, which has been *reconstructed* as having undergone an evolutionary increase in size at that node), to interpret the projections of the contrast nodes.

Contrast axis 1 describes differences in overall life history pace (see above). Strongly positive projections on this axis mean that the taxa reconstructed as having increased in size at these nodes exhibited a concomitant slowing of their overall pace of life history (larger cranial capacity, larger neonatal mass, slower dental development, later age at weaning, and so on). Strongly negative projections mean that the larger-bodied of two taxa diverging at a given node has the faster life history. Strongly positive projections on contrast axis 2 mean that evolutionary shifts toward larger size are accompanied by more advanced dentitions at weaning, and strongly negative projections mean that the opposite is true. The contrast nodes at which extreme evolutionary shifts apparently occurred are shown in figure 8.3; the nodes are labeled as in figure 8.1.

Contrast node 1 projects strongly positively on axis 1 but weakly negatively on axis 2. Haplorhines are generally larger than strepsirrhines, and an increase in body size in the basal haplorhine at contrast node 1 is inferred. The projections indicate a concomitant slowing of the absolute pace of life history (and dental development), but little change in dental development at weaning. Dental development in haplorhines is generally slower on an absolute scale than in strepsirrhines, but not on a relative scale.

Several contrast nodes project negatively on axis 1 but strongly positively on axis 2. The most extreme is contrast node 4, representing the divergence of indriids from lemurids. This projection shows that the indriids (inferred to have experienced a size increase at this node) are faster than the lemurids in their absolute pace of dental development and have more teeth at weaning. The projections of contrast node 29 indicate that the (larger-bodied) African colobines have faster dental development than Asian colobines on absolute and relative scales. Similarly, though to a slightly lesser extent, the ancestral colobine is reconstructed as having undergone an increase in body size and acceleration of absolute and relative dental development at contrast node 23.

Contrast node 20 has a near zero projection on axis 1, but a strongly positive projection on axis 2. The larger-bodied *Ateles* has more teeth at weaning than does *Alouatta* because its offspring are weaned at an advanced age.

Several contrast nodes project positively on axes 1 and 2. In each of these cases, a slowing in the pace of life history in the larger-bodied branch is accompanied by an increase in dental development at weaning due to a delay in age at weaning. This pattern characterizes contrast node 36 (great

Table 8.8 Dp4/M1 occlusal area ratios in strepsirrhines

Taxon	Mandibular dp_4/M_1 area index (as %)	Maxillary dp^4/M^1 area index (as %)
Avahi laniger	16.6	24.7
Propithecus diadema	18.3	25.5
Propithecus verreauxi	18.7	25.5
Lepilemur ruficaudatus	34.9	47.6
Galago moholi	59.4	58.3
Varecia variegata	75.7	64.8
Lemur catta	76.5	72.8
Eulemur fulvus	79.1	78.7
Eulemur macaco	86.5	89.5
Hapalemur griseus	101.0	98.7

apes develop more slowly than hylobatids but have more teeth at weaning), contrast node 22 (hominoids develop more slowly than cercopithecoids but have more teeth at weaning), and contrast node 13 (atelids develop more slowly than do cebids/callitrichids but have more teeth at weaning).

Contrast node 38 is distinguished by its very negative projection on axis 2 and its slightly negative projection on axis 1. African apes (inferred to have undergone a size increase at this node) tend to be slightly faster in their absolute pace of development, but less advanced dentally at weaning, than *Pongo.* This is apparently because age at weaning is much earlier in *Gorilla* (of the African ape clade) than in *Pongo,* and *Gorilla* has smaller neonates, more rapid dental eruption, a smaller brain size, and earlier breeding than might be expected on the basis of body size.

One contrast node is noteworthy for its lack of any significant difference in absolute or relative pace of dental development among its descendants. Contrast node 10 is located within the cluster of nodes with near-zero projections on both axes. *Hapalemur griseus* does not deviate from *Lemur catta* (or other lemurids) in the manner that might be expected of an extreme folivore. *Hapalemur* consumes far more foliage than is the norm for indriids such as *Propithecus,* yet its teeth develop at a far slower pace, and it is far less dentally precocious at weaning. *Hapalemur* lacks the developmental specializations that allow indriids, regardless of adult body size, to experience extraordinarily rapid dental development. The key to this puzzle is the size of the last deciduous premolar relative to the first permanent molar (table 8.8). The dp4/M1 occlusal area ratio has direct bearing on the relative space allocated to the crowns of the deciduous and permanent teeth during their development. It also (at least partly) reveals the extent to which dp4

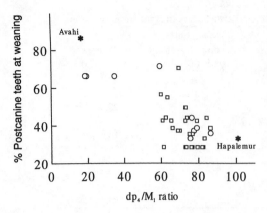

FIG. 8.4. Scatterplot of dental precocity at weaning (expressed as a percentage) versus the mandibular dp4/M1 occlusal area ratio in thirty-nine primate species. Open squares are anthropoids; open circles are strepsirrhines. The coefficient of correlation is −0.75 ($P <$ .0001 for 37 d.f.; with a discounted N of 15 and 13 d.f., $P <$ .0001).

can act as a mechanical substitute for M1 in weanlings. Indriids sacrifice the size of their deciduous teeth to allocate more space in their jaws to the developing crowns of permanent molars. Mandibular and maxillary dp4/M1 occlusal area ratios are excellent predictors of dental precocity at weaning in primates in general and in strepsirrhines in particular (fig. 8.4). To understand how *Hapalemur* copes with a diet of bamboo, one must examine this phenomenon in more detail.

In fact, rather than sacrificing the deciduous teeth to increase the speed of dental development, *Hapalemur* simply increases the size of its deciduous teeth (table 8.8; fig. 8.4). Its last deciduous premolar is not merely very like M1 in morphology, but is sometimes larger in absolute size than M1. *Hapalemur* also increases the *effective* occlusal area of its posteriormost deciduous premolar and permanent cheek teeth by increasing the lengths of their shearing blades (table 8.9; see Kay 1984; Covert 1986). It thus exhibits a considerably higher shearing quotient than does its more frugivorous relative, *Lemur catta*. A parallel contrast has been shown between cercopithecoid folivores and frugivores (Kay 1978).

Perhaps the most interesting species of the genus *Hapalemur* is *H. simus,* which, at 2.4 kg, is still smaller in adult size than *Lemur catta* (ca. 2.7 kg). Whereas *H. griseus* feeds on bamboo shoots, *H. simus* exploits the pith of giant bamboos (Tan 1999a). *Hapalemur simus* has an extraordinarily high molar shearing quotient (table 8.9) to cope with a diet much tougher than that of *H. griseus*. Furthermore, whereas *H. griseus* weans its offspring at four months, weaning does not occur in *H. simus* until eight months (Tan 1999b). If *H. simus* follows a *Lemur catta-* or *H. griseus*-like eruption schedule (and very limited data suggest that it does), then at eight months (or weaning) it would exhibit moderately advanced dental development (with

Table 8.9 Mandibular second molar shearing quotients in the Indriidae and Lemuridae

Genus and species	M_2 shearing quotient[a]
Hapalemur simus	28.79
Avahi laniger	14.77
Hapalemur griseus	12.51
Lemur catta	6.01
Eulemur macaco	3.95
Propithecus diadema	3.93
Eulemur fulvus	2.72
Propithecus verreauxi	1.31
Varecia variegata	−9.43

Source: Covert 1986.
[a]The shearing quotient is calculated as $100 \times$(observed shear−predicted shear)/(predicted shear). Baseline data are prosimians.

M1 certainly erupted, and probably M2, in addition to the full set of deciduous teeth). Thus, greater bamboo lemurs apparently achieve dental competence at weaning not by accelerating the pace of dental development, but by delaying weaning.

Discussion

Weaning and Dental Development

B. H. Smith (1991a) reported a tendency in primates for weaning to approximate the age of emergence of the first permanent molar. This finding suggests that dental development at weaning is more or less invariant across primates. In contrast, our data underscore just how variable dental development at weaning can be. One cannot assume that weanlings will exhibit comparable degrees of dental maturation, or that dental maturation will be linked in a consistent manner to skeletal growth or reproductive maturation.

This variability is an outcome of real differences in the *pace* of dental development and the *timing* of weaning vis-à-vis other life history parameters. In some anthropoid primate species (e.g., spider monkeys, orangutans, chimpanzees), the process of weaning is extremely prolonged, and weaning can occur at relatively advanced ages. In other primate species (particularly lemurs), weaning is more narrowly constrained to correspond in time to the occurrence of certain phenological events (Sauther 1991, 1998; Meyers and Wright 1993; Wright 1997, 1999). Other factors, including predation, climate, and cognition, also appear to affect the timing of weaning in primates (Wolfe 1986; Boinski 1987; Lee, Majluf, and Gordon 1991; Lee 1996). Furthermore, because weaning is a process, not an event, and because field researchers use different criteria to identify it, measurement error probably contributes significantly to interspecific variation in "age at weaning." Our inferences regarding the teeth that have erupted at weaning are sometimes based on scanty information, and measurement error is likely to be high here as well. Whatever the source of the error, marked intraspecific as well as interspecific variation makes any simple interpretation of dental development at weaning impossible. Therefore, it is not surprising that less of the variance in dental development at weaning than at selected biological ages can be explained, and that single parameters, taken alone, are poor predictors of weanling dental development. Nevertheless, large interspecific differences in relative timing of weaning and in normative patterns of dental eruption at weaning clearly exist.

Our data suggest that adult cranial capacity is the best predictor of dental development at weaning: small-brained (often small-bodied) species tend to be dentally precocious. Charnov's (1993) assumption that weaning occurs when the young reach a certain percentage of adult female mass has been shown to be incorrect for mammals; instead, the ratio of the mass of the weanling to that of the adult female is negatively correlated with adult female mass in mammals (Purvis and Harvey 1995). We have found that the same holds for primates (Godfrey et al. 2001). This pattern may at least partly account for the tendency for small primates to have advanced dentitions at weaning, but it does not explain why cranial capacity is a better predictor of dental development at weaning than is body size. B. H. Smith (1989) suggests that dental development may be pleiotropically tied to the development of the brain. Nevertheless, there are important ways in which adult cranial capacity fails to predict dental competence at weaning (e.g., strikingly different indriid and lemurid weanling dentitions are predicted by neither adult cranial capacity nor other size parameters). Other variables, such as diet or phylogenetic inertia, must be enlisted.

Diet, Phylogeny, and Dental Development

Janson and van Schaik (1993) linked variation in rates of growth and development explicitly to diet. They hypothesized that species that experience low intraspecific resource competition should exhibit relatively rapid growth and maturation. Leigh (1994a) saw the rapid growth of anthropoid folivores, relative to like-sized anthropoid frugivores, as support for Janson and van Schaik's hypothesis. However, neither Leigh (1994a) nor Janson and van Schaik (1993) examined folivorous strepsirrhines within this context.

Strepsirrhines differ from haplorhines in their patterns of prenatal and postnatal growth (Martin 1990; Leigh and Terranova 1998). Postnatal growth is more rapid in strepsirrhines than in haplorhines, and adult size of strepsirrhine postcranial skeletons may be achieved well before sexual maturation. In contrast, prenatal growth is slow in strepsirrhines. This pattern of growth affects the space available in the jaws of strepsirrhine neonates for development of deciduous teeth. This space is particularly limited when the development of the permanent dentition is so accelerated that well-calcified crowns of the first and second permanent molars occupy most of the jaw of neonates (as in Indriidae). The deciduous teeth of these species are effectively vestigial, and the dp4/M1 occlusal area ratios are very low.

The Janson-van Schaik hypothesis predicts wholesale developmental acceleration in folivorous species, and therefore also early weaning and sexual maturation. Our data support some of these predictions, but not others. The tendency for folivorous species to exhibit more rapid dental development (on an absolute scale) than like-sized frugivores is consistent with the expectations of this hypothesis. But variation in intraspecific resource competition cannot account for the tendency for *weanlings* of folivorous species to exhibit more advanced dentitions than like-sized frugivores, the striking differences in absolute pace of dental development between haplorhines and strepsirrhines, or the dissociation between dental and somatic development characterizing strepsirrhines.

Contrary to the expectations of the Janson-van Schaik hypothesis, folivorous strepsirrhines do not grow more quickly than frugivores, nor do they have more accelerated life histories (see Wright 1999; L. R. Godfrey et al., unpub.). Instead, dental development tends to accelerate in the more folivorous, seed-crunching species. Indriids are among the slowest growing of extant strepsirrhines; indriid weanlings are small in body mass, but dentally precocious. Lemurids grow faster and have relatively larger weanlings that are dentally retarded. Among lemurids, the most frugivorous genus, *Varecia,* has the most rapid rate of postnatal skeletal growth (Kappeler 1996) and the slowest pace of dental development.

Mature foliage and seeds are best processed by cheek teeth with large crushing basins and elongated shearing crests (Kay 1978). Given seasonal changes in the availability of fruits and other high-quality foods, the need for heavy-duty food processing equipment may increase precisely as food quality and availability drops to its annual low. Adults can be presumed to be mechanically and physiologically prepared to handle seasonal shifts in food properties, but recently weaned youngsters must process foods without the benefit of a mature digestive system or a full battery of adult teeth. The problem can be severe if prolonged facultative suckling could threaten the very survival of the mother. Only through developmental dissociation of sets of morphological traits can species with relatively slow growth and low maternal investment have weanlings that are prepared to process foods that might be expected to be accessible only to considerably larger (or older) individuals.

Life Histories and Dental Development

We have argued that the dental occlusal area present at weaning may be critical to the survival of mammals because it directly relates to food processing proficiency, and that high dental endowment at weaning can be attained through alternative routes. Acceleration of dental development is one; delaying weaning is another. How different developmental options affect population dynamics depends not on their adult products (or outcomes) per se, but on their consequences for the survival, reproductive value, and interactions of individuals at various life cycle stages (see Alberts and Altmann, chap. 4, this volume; Hawkes, O'Connell, and Blurton Jones, chap. 9, this volume; see also Godfrey, Petto, and Sutherland 2002). Increases in maternal investment, for example, can positively affect the survival and reproductive value of infants while decreasing the probability that the adult females themselves will survive. The contrast between indriids and lemurids serves as an intriguing example.

The growth rates of both indriids and lemurids show marked seasonal fluctuations (Pereira 1993c; Pereira and Leigh, chap. 7, this volume). Dental development is also episodic (Eaglen 1985; our own data). There are "waves" of rapid eruption followed by periods of stasis during which no teeth erupt. Indriids and lemurids differ in the timing of these waves and in which teeth erupt during each. In all indriids for which we have data, the entire deciduous dentition is fully erupted or erupting at birth. Indriids begin their second wave of dental eruption (during which permanent incisors and anterior molars emerge) in their second to fourth month (depending on the species). An eight-month-old sifaka possesses all of its permanent teeth except the upper canines, and a one-year-old sifaka is dentally mature, or nearly so. In lemurids, the deciduous teeth erupt over a several-month period. The second wave of dental eruption does not begin until between five and eight months (again, depending on the species); the third wave of eruption (during which permanent premolars, posterior molars, and upper canines emerge) generally begins at about fourteen months. Weaning occurs between four and eight months in both lemurids and indriids. There are clear differences in dental development at weaning and during the first post-weaning dry season for species in these two groups (Eaglen 1985; L. R. Godfrey et al., unpub.).

Recent research has confirmed that maternal investment tends to be low in large-bodied strepsirrhines (Tilden and Oftedal 1995, 1997; Wright 1999). It is particularly low in indriids, which produce weanlings with exceptionally low body mass. Age at first birth is also delayed in indriids; female sifakas, for example, may not give birth to their first offspring until they are five or six years old or older (Richard, Rakotomanga, and Schwartz 1993; Wright 1995; A. F. Richard, pers. comm.). Seasonal weight gain and loss can be pronounced, particularly in females, and reproductive success in female sifakas is inversely correlated with extent of weight loss (Richard et al. 2000). Low maternal investment (coupled with physiological tolerance of water shortage and ability to process highly fibrous foods) may help resource-stressed adult females to survive prolonged periods of food shortage. Indeed, A. F. Richard (pers. comm.) documented surprisingly low adult mortality among *Propithecus verreauxi* during the devastating 1991–1992 drought in southwestern Madagascar. Infants and juveniles, by contrast, experienced very high mortality. Decoupling of dental and skeletal growth allows immatures in this species to acquire food-processing competence quickly while minimizing maternal risk. Apparently, in sifakas and (probably) other indriids, a selective premium is placed on adult survival (Godfrey, Petto, and Sutherland 2002).

Ringtail and other lemurid infants grow more rapidly than do indriids, and they are weaned at much higher body masses. Accordingly, maternal investment is higher than in indriids. Females also begin reproducing earlier than do like-sized indriids. Under the dietary stress induced by severe drought, mortality is high among ringtail adults as well as youngsters (Gould, Sussman, and Sauther 1999). Populations are maintained less through the ability of adults to survive droughts and other natural catastrophes than through the rapid reproductive rebound of those that do survive. Here, the selective premium is on early sexual maturation, dietary diversity, and reproductive resilience (Gould, Sussman, and Sauther 1999).

Propithecus verreauxi and *Lemur catta* have different life history "strategies" (i.e., different means to population recovery following episodes of high mortality), yet they thrive in the same habitats. We suggest that differences in the ability of adults to survive severe food shortages may be linked to differences in maternal investment and reproductive strategies, and that these differences have developmental correlates.

Our data suggest that selection tailors dental developmental and life history schedules to meet the masticatory needs of immature individuals, and that alternative strategies for preparing weanlings for ecological independence exist. These developmental strategies have consequences for population dynamics. The dissociability of dental development and skeletal growth allows selection to regulate dental and skeletal developmental schedules independently so as to promote the survival of individuals at particular life cycle stages. More research on interrelationships between developmental profiles and life history strategies is warranted.

Summary and Conclusions

For forty species of primates at selected absolute ages (four months, one year) and at weaning, we measured dental endowment (postcanine occlusal area) and precocity (number of postcanine teeth erupted) as percentages of species' mean adult values. These data allowed us to test the influence of dietary and phylogenetic factors, as well as adult body and brain size, on dental developmental schedules.

1. Of the several size variables we considered (neonatal mass, adult female mass, postcanine occlusal area, and adult female cranial capacity), adult female cranial capacity is indeed the best predictor of the absolute pace of dental development. Adult cranial capacity is a better predictor of early postnatal than of late postnatal dental development.

2. Cranial capacity has a weaker correlation with dental precocity and endowment at weaning than with dental precocity at particular absolute ages. The absolute pace of dental development is only weakly correlated with dental development at weaning across the order Primates.

3. Diet influences both the absolute and the relative pace of dental development. When compared with their more frugivorous like-sized relatives, folivorous strepsirrhines resemble folivorous cercopithecoids in exhibiting accelerated dental development both on an absolute scale and relative to particular life history landmarks, such as weaning. Folivore/seed predators tend to exhibit faster dental development, as well as greater dental precocity and endowment at weaning, than like-sized frugivores. However, accelerating the relative pace of dental development is only one way to increase the masticatory competence of weanlings.

4. Delaying the age at weaning is an alternative strategy for increasing the food processing competence of weanlings. When weaning is delayed, the masticatory toolkits of weanlings are improved, as are their foraging skills. Other strategies for increasing the ecological sophistication of weanlings include enlargement of the dp4 occlusal

area and modification of the morphology of the teeth available to the weanling (particularly the anterior molars and/or dp4). All of these strategies (or combinations of strategies) are manifested within the order Primates (indeed, even more narrowly, within the Strepsirrhini). The potential dissociation of dental development and skeletal growth allows selection to regulate dental and skeletal developmental schedules independently so as to promote the survival of individuals in particular life cycle stages.

5. Phylogeny has an important influence on the particular strategy adopted by species. These strategies may have consequences for population dynamics.

9 Human Life Histories: Primate Trade-offs, Grandmothering Socioecology, and the Fossil Record

KRISTEN HAWKES, J. F. O'CONNELL,
AND NICHOLAS G. BLURTON JONES

Human life histories differ from those of other animals in several striking ways. Recently Smith and Tompkins (1995, p. 258) highlighted the combination of "slow" and "fast" features of human lives. Our period of juvenile dependency is unusually long, our age at first reproduction is late, and we have the maximum life span of the terrestrial animals. Yet we wean babies relatively early, and we space births closely. We also have (midlife) menopause. Smith and Tompkins predicted that the evolution of our life cycles would be explained by a combination of developments in life history theory with increasingly sophisticated techniques for extracting information from the fossil record. Their prudent guess was that "no new sunburst theory—in which all human characteristics are drawn from one adaptive shift—is likely" to emerge (1995, p. 274).

Here we use work since that review to confirm Smith and Tompkins's optimism about the explanatory gains of combining life history theory with paleoanthropology. We focus on the framework provided by Charnov's symmetry approach, combining Charnov's mammal model with a grandmother hypothesis about the socioecology of *Homo erectus (ergaster)*. This combination links several otherwise contradictory features of human life histories, showing that our remarkable longevity, late maturity, and relatively high fertility could all be a systematic consequence of a single adaptive shift from an australopithecine life history like that of modern apes. If ancestral life spans were similar to those of modern chimpanzees, with few females surviving past the age of menopause, an ecological change that increased maternal provisioning of juveniles would alter the optimal allocation to somatic survival (Williams 1966; Schaffer 1974; Kirkwood 1977). If mothers shared more food with juvenile offspring, then help from aging females who were not nursing infants themselves would have a large, novel effect on their daughters' fertility. The increased fitness payoffs for adult survival would favor longer average adult life spans and, in turn, would alter the optimal age of first reproduction. This hypothesis is a partial "sunburst." It links some of our life history features to one another, but does not tie those features to the greatly expanded *Homo sapiens* brain or to increased meat eating and the contribution of males to subsistence—features long assumed to be foundations for our late maturation and longevity (e.g., Kaplan et al. 2000).

We conclude by consulting the paleoanthropological record for the ecological context and dates of life history changes, specifically to see how those changes are associated with the evolution of modern encephalization and with changes in the meat fraction of hominid diets. The evidence reviewed is consistent with the proposition that distinctive features of our life histories may have come before large brains and increased carnivory, and may have been responsible for the character and spread of the first widely successful and longest enduring members of our genus.

Life History Theory

A framework that links variation in different life history features is especially valuable for investigating human life history evolution for (at least) two reasons. First, all the other species in the hominid radiation(s) are known only from the fossil record. While paleoanthropologists can estimate some life history features from skeletal specimens, the difficulties of doing so are not trivial, and other features may never be indexed directly in the fossils. A framework that points from more easily measured features to features that are less easy to measure directly can help to extract more information from fossil data. It can leverage more from the paleoanthropological data—the only line of evidence we have for the timing, context, and order of appearance of the modern human patterns.

Second, a framework linking variation among life history features is valuable for investigating our own evolution because of the wide developmental and behavioral gulf between contemporary humans and our closest living primate relatives. There are so many differences between modern people and the (other) great apes that each feature of human life histories can be attributed to a wide array of possible causes. Yet some fundamental trade-offs that apply to other primates apply to us as well. A framework that can generate expected life histories for a model primate with our age-specific mortalities or fertilities will show how many of the human patterns could result from the same trade-offs that we share with other members of the order.

Is There a Framework?

The search for a general model of life history strategies has been fueled over the past few decades by some tantalizing regularities. MacArthur and Wilson (1967; Pianka 1970) proposed r and K selection to account for some of them. They hypothesized that r-selected species evolve in the face of repeated ecological disruptions, reasoning that population crashes with periodic opportunities for rapid population growth should favor life history tactics that give high maximal rates of population increase (r): early maturity at small size, with all effort expended in producing many small offspring, and thus early death. Species of the K-selected type were hypothesized to evolve in "saturated" environments. With population densities near carrying capacity (K), they hypothesized that selection would favor life histories that maximize competitive capacities: late maturity at large size and the production of a few large, well-developed offspring over long adult lives. This theory was especially influential because the logic was simple and compelling, and it gained wide use as a description of relative differences. But subsequent empirical and theoretical work showed that density-dependent mortality has different effects on optimal strategies depending on which age classes suffer the mortality (Gadgil and Bossert 1970; Charnov and Shaffer 1973; Stearns 1977), and that both r and K clusters of characters can evolve and persist in stable populations (see Stearns 1992 for review). Life history features do correlate generally with body size, but many features are even more strongly correlated with one another when the effects of body size are removed (Harvey and Read 1988; Harvey, Read, and Promislow 1989).

The value of a framework that explains links among life history features depends on whether the model's assumptions about fundamental trade-offs capture enough of the actual trade-offs in the empirical world to explain what really happens. The failure of r and K selection on that score has stimulated pessimism about the possibility of finding broad regularities. Stearns (1989b) enumerated a long list of trade-offs that can all be important, and, in the face of so many variables, addressed life history evolution one trait at a time in his (1992) text. In a recent discussion of life span variation, Finch (1997, p. 247) concluded that the range is so enormous (a millionfold in eukaryotes) that it "implies the absence of evolutionary limits in life-history schedules."

Yet, many relationships among life history variables are quite robust within particular taxonomic groups. Stearns labeled these relationships "lineage specific effects" (1992, chap. 5), and Finch (1997, p. 247) noted that lim-

its are set by "the physiological architecture of the species." Within taxonomic groups, larger-bodied species do live "slower" lives, but some taxa are
relatively slower, and others faster, at similar body sizes. During the last decade, Charnov, building on his own earlier work and that of many others (see
Purvis et al., chap. 2, this volume, for additional review) has focused on only
a very few trade-offs, elaborating a symmetry approach to explain both aspects of this interspecific variation (Charnov 1991; Charnov and Berrigan
1991). In his "dimensionless" model of mammalian life histories, he adopts
the widely used assumption that adult mortality risk selects age at maturity,
because growing longer increases maternal size (and consequent production capacity) but trades off against the risk of dying before reproducing.
Age at maturity, in concert with taxon-specific growth (production) capacity,
sets adult size. Trade-offs in offspring quantity versus quality determine optimal weanling size and combine with the growth assumptions to set fecundity. Elaborations and criticisms of Charnov's mammal model (Charnov
1993; Kozlowski and Weiner 1997; Harvey and Purvis 1999; Purvis and Harvey 1995, 1996; Purvis et al., chap. 2, this volume) do not undercut the main
elements of this framework (see below). To the extent that something like
Charnov's model is generally correct, it provides a powerful guide to developing and evaluating hypotheses about the evolution of human life histories.

Charnov's Model

Patterns of growth vary widely between and within species (e.g., Leigh 1996;
Pereira and Leigh, chap. 7, this volume). Charnov's mammal model uses a
simplified growth model based on the assumption that the energy a mammal
can put into production (of itself or of offspring) is captured by a characteristic average rate for size. He divides growth into two parts. The first is conception to weaning, which is assumed to be fueled by production from the
mother. The second is weaning to maturity, assumed to be under juvenile
control. At maturity, production previously allocated to growth is redirected
to offspring. Growth rates are an allometric function of body mass (W) and
a characteristic production coefficient (A). Individual production rates take
the form $dW/dt \sim AW^c$, where the exponent c is ~0.75 (Kleiber 1932; West,
Brown, and Enquist 1997). Adult size at maturity (W_x) and production available for offspring both vary directly with A, which is characteristically low
in primates compared with other mammals (Charnov and Berrigan 1993).
In a recent elaboration of the model, Charnov (2001) makes within-species
growth sigmoid rather than a power function and includes a cellular maintenance rate that is adjusted by selection.

Charnov captures key features of mortality schedules with another simplification. He assumes an early burst of mortality that incorporates any density-dependent effects. Mortality then drops to a constant level before the age of first reproduction. Given that constant mortality risk, selection sets α (the period from weaning to maturity) according to the trade-off between the benefits of continued growth (and so larger adult size) versus reproducing sooner. Since adult production is a function of maternal size, it increases with age at maturity. The probability of dying before reproducing depends on the instantaneous mortality rate (M), which is unaffected by size. As that rate falls (average adult life span increases), selection favors delaying maturity to reap the benefits of larger size. Thus α and M vary widely, but inversely. Their product (αM) is approximately invariant.

Another set of trade-offs affects fecundity in this model. Weaning releases the mother to allocate production to the next offspring, so weanling size determines the rate of offspring production. But what determines weanling size? Charnov (1993, pp. 107–108) notes that a version of the offspring size-number trade-off (Smith and Fretwell 1974) easily leads to an optimum ESS. For a sample of mammals (cf. Purvis and Harvey 1995, 1996), *and for primates separately*, the ratio of size at weaning to adult size $(W_o/W_\alpha = \delta)$ is approximately constant (Lee, Majluf, and Gordon 1991; Ross and Jones 1999). That means that the (ESS) weanling size in these taxa scales isometrically with adult size. Since production scales allometrically with size (the exponent in the initial growth model is about 0.75, modified to include the cost of cell maintenance in the sigmoid model), the ESS size of weanlings goes up faster with maternal size than does the production the mother can put into them. Consequently, the number of daughters produced per year (b) goes down as age at maturity (α) goes up. Larger mothers produce larger but fewer babies, making αb another approximate invariant.

Kozlowski and Weiner (1997) and Harvey and Purvis (1999) suggest that Charnov's model may be crucially flawed because later maturity means both a greater risk of dying before reproducing and lower fecundity, so that there can be no optimal α. But the model treats these as independent adjustments, one to an ESS α in which weanling size is assumed fixed (the same assumption used by Kozlowski and Weiner 1997), and one to an ESS weanling size, which takes W_α as fixed (since the ESS W_α is independent of W_o: Harvey and Nee 1991).

These assembly rules for mammalian life histories identify links that could account for an array of observed covariates. Other variables, other trade-offs, clearly play important roles as well (Stearns 1989b, 1992; Pereira and Leigh, chap. 7, this volume), but the general fit of empirical patterns to the model predictions (confirmed since Charnov 1993 on other, larger data sets: Purvis and Harvey 1995, 1996) suggests that the trade-offs in the model are close to the real trade-offs shaping the broad variation in mammalian life histories.

Charnov's Model and Primate Life Histories

Charnov's model captures similarities between primate and nonprimate mammals in the relationships among life history features as well as some distinctive features of the primate order (Charnov 1993; Charnov and Berrigan 1993). The "mouse lemur to gorilla" curve looks like the "mouse to elephant" curve. In both cases, larger bodies are associated with longer adult lifetimes, later ages at maturity, and lower fecundity. Not only are these features correlated within each group, but the relationships among adult life spans, age at maturity, and fecundity remain the same across transformations of body size. Primates and other mammals share the same αM invariant. But primates, growing more slowly, are smaller at a given α. The "production factor" (A) accounts for both lower growth rates (and thus smaller sizes for a given age at maturity) and lower fecundities for size in primates. This variable (A), approximately 1 in mammals, is generally less than half that in primates (Ross and Jones 1999). (It is even lower in modern humans [Hill 1993] than in the average primate—about the same for us as it is for chimpanzees.) Within each taxonomic group, smaller maternal size is associated with relatively higher fecundity (b). The lower primate b for a given size offsets the higher α for a given size, so that primates and other mammals share the same αb invariant.

Charnov (1993, p. 104) plotted the relationship between age at maturity (α) and average adult life span $(M^{-1}$, i.e., the inverse of the instantaneous rate of adult mortality) for fifteen primate subfamilies using data from Harvey and Clutton-Brock (1985). In this scatter, one point holds a much higher value on both variables than any other, but is not an outlier. It falls almost exactly on the best-fit regression line (correlation coefficient = 0.95, intercept not different from zero). That point represents the Hominidae, and there is, of course, only one species in the sample: modern *Homo sapiens*.

The fact that modern humans show the primate relationship between adult life span and age at maturity highlighted by Charnov's framework is initially astonishing. Unlike that of other primates (and most other mammals), the average adult life span of women includes a substantial postmenopausal component, a part of the life span often characterized as "postreproductive." If the symmetry captured in the αM invariant really depends on selection setting α according to the trade-off between the risks and the benefits of delaying maturity, and if the total expected benefits depend on the duration of time over which the gains for waiting will accumulate (Charnov 1997), then the human case should fit only if the whole life span is spent putting "production" into making descendants.

Postmenopausal Longevity as a Species Characteristic

This deduction points directly to questions about postmenopausal survival. Is it really typical of humans, and if so, why did it evolve? These are classic evolutionary puzzles. Williams (1957), noting that selection cannot maintain "post-reproductive" function, suggested that human midlife menopause might have evolved with increasing offspring dependence on maternal care. As the probability of seeing the next baby through to independence declined, and pregnancies became increasingly dangerous to aging mothers, those who "stopped early" and allocated late reproductive effort to the welfare of children already born would leave more surviving descendants. This hypothesis takes long human life spans as given and assumes that selection favored "premature reproductive senescence." It continues to stimulate useful work (Hill and Hurtado 1991, 1996, 1999; Rogers 1993; Peccei 1995; Shanley and Kirkwood 2001). But age at last birth is quite similar in humans and chimpanzees, suggesting that we may share our age at menopause with all descendants of our common ancestor. The derived feature that distinguishes us from our nearest living relatives is not our age at menopause, but our extreme longevity (Hawkes, O'Connell, and Blurton Jones 1997, 2000; Hawkes et al. 1998; Kaplan 1997; Kaplan et al. 2000).

The claim that extended longevity—relatively slowed senescence—is a general human characteristic is not without challenge (e.g., Austad 1997a; Olshansky, Carnes, and Grahn 1998). The extremely high life expectancies of many (but not all) contemporary human populations are clearly a very recent novelty. In most of the United States, Europe, and Japan, newborns can now expect to live about eighty years, while historical demography and population profiles in high-mortality settings show life expectancies of less than four decades. But life expectancies at birth are averages and are strongly affected by rates of infant and juvenile mortality (so that a life expectancy of thirty years does not mean that few adults live to forty) (e.g., Bailey 1987; D. W. E. Smith 1993; Lee 1997). French historical demography provides an instructive set of comparisons (table 9.1). Life expectancy at birth was only 39 in 1850, compared with double that in 1985. The largest source of difference between these two time periods is the rate of death in the juvenile age classes. Even in 1850, anyone who lived to adulthood had the prospect of a long life ahead. As table 9.1 shows, most women lived past the age at last

Table 9.1 French historical demography

	At birth: Average life expectancy (e_x0)	At maturity: Probability of living past age of last birth (45)	At age of last birth: Average additional life expectancy (e_x45)
France 1985	79	.96	36
France 1950	69	.92	30
France 1926	57	.84	27
France 1900	48	.75	25
France 1850	39	.72	24

Source: Keyfitz and Flieger 1968, 1990.

birth, and the average number of years remaining for anyone who reached that age was more than two additional decades.

Nineteenth-century France had an agrarian economy, something relatively recent in human experience. Where people depend only on wild foods, however, survival curves are very similar. In all four of the best-studied cases of modern hunter-gatherers, the !Kung (Howell 1979), the Ache (Hill and Hurtado 1996), the Hadza (Blurton Jones et al. 1992), and the Agta (Early and Headland 1998), representing populations in different environments with distinct recent genetic histories, age-specific survival is very like that recorded in nineteenth-century France (table 9.2). Although life expectancies at birth are less than four decades, this does not mean that people live only into their thirties. Most women live past the age of last birth, and those who do have an *average* of more than twenty years of life still ahead.

The inference from history and ethnography that substantial post-menopausal longevity is a usual characteristic of human populations has

Table 9.2 Contemporary hunter-gatherers

	At birth: Average life expectancy (e_x0)	At maturity[a]: Probability of living past age of last birth	At age of last birth: Further life expectancy (e_x45)	Women past age of last birth (%)	Source
!Kung	31	.66	20	31	Howell 1979
Ache	38	.79	22	36	Hill and Hurtado 1996
Hadza	33	.71	21	29	Estimated from Blurton Jones et al. 1992
Agta	24	.59		36	Early and Headland 1998

[a]If maturity is at 20, and last birth at 45 (40 for the Agta).

been challenged by age distributions observed in archaeological skeletal assemblages (Weiss 1973). Remains from individuals estimated to be over sixty at death are rare, supporting skepticism about the generality of the demographic patterns found among living populations (Austad 1997a; Trinkhaus 1995). But new sources of error are introduced in constructing population profiles from archaeological assemblages. Where historical records have provided independent evidence of the ages of individuals interred, two especially important sources of bias have come to light. The bones of the old and the young are disproportionately unlikely to sustain long preservation, and the ages of adults are systematically underestimated (Walker, Johnson, and Lambert 1988; Paine 1997). Standard aging techniques applied to samples of known ages illustrate the pervasiveness of this problem (Bocquet-Appel and Masset 1982; Key, Aiello, and Molleson 1994). Attempts to model sustainable populations using parameters estimated from cemetery profiles have repeatedly shown that the usual estimates are quite unrealistic (e.g., Howell 1982; Bermudez de Castro and Nicolas 1997).

Biases in the other direction, novel features of the modern world that might extend longevities in ethnographically known foraging populations, can also be explored. Living people are now everywhere affected to some degree by global networks of interaction (e.g., Wolf 1982; Schrire 1994; Blurton Jones, Hawkes, and O'Connell 1996 discusses interactions with the Hadza and the !Kung). We have tried to assess the effects that some access to Western medical care and interaction with neighboring farmers and herders might have on Hadza demography (Blurton Jones, Hawkes, and O'Connell 2002). Even the most generous estimates of regional medical services and the investigators' own possible effects make only negligible differences in the population parameters initially reported (Blurton Jones et al. 1992).

Comparisons with other mammals clearly show the unusual longevity of humans (Pavelka and Fedigan 1991). Table 9.3 lists the percentage of females who, having reached maturity, live past the age of last birth in three other primate species for whom vital rates have been monitored outside captivity. In macaque and baboon populations (Pavelka and Fedigan 1999; Packer, Tatar, and Collins 1998), reaching the age of last birth is the luck of only a very few individuals (whereas it is the norm among humans even under high-mortality conditions; see table 9.2). The living primates most closely related to us, the chimpanzees, have slower life histories than smaller-bodied macaques and baboons, and a few more females may live past childbearing age (table 9.3; Caro et al. 1995). But in all three of these nonhuman species, only a very small fraction of the adult females in a living

Table 9.3 Nonhuman primates

	At maturity: Probability of living past age of last birth	Adult females past age of last birth (%)
Macaques[a]	<.05	<2
Baboons[b]	.04	<1
Chimpanzees[c]	.17	<6

[a]*M. fuscata*, Texas (Pavelka and Fedigan 1999).
[b]Gombe (Packer, Tatar and Collins 1998).
[c]Composite of five study sites (Hill et al. 2001).

population are past the age of last birth. Even in high-mortality human populations (e.g., the hunter-gatherer examples in table 9.2), about a third of the adult females are beyond childbearing age.

None of these findings should obscure the novelty of the increasingly larger proportions of adults in senior age ranks in many contemporary human populations. This large fraction of oldsters presents economic, medical, and social challenges that can hardly be overestimated. But neither should the novelty and importance of those challenges obscure the strength of the evidence that long adult lives are normal for humans. Based on the arguments summarized below, they may be a feature of our lineage much older than *Homo sapiens*.

Lessons from the Hadza

In high-mortality circumstances, there is a characteristic human pattern of age-specific survival and fertility. Most women live long past the age of last birth. The additional years of life are several multiples of average birth intervals, which mark the time mothers devote to one offspring before turning to the next. Yet humans mature unusually late, as expected for our unusually long adult life span; thus, human life histories preserve the αM invariance. This fit to the broad primate (and more generally mammalian) pattern would be expected if women actually continued to produce descendants during those postmenopausal years. There are good reasons to think that they do.

The Hadza, hunter-gatherers in the arid Tropics of northern Tanzania (Woodburn 1968; Blurton Jones et al. 1992), provide an instructive lesson. Here, postmenopausal females have clear effects on the production of descendants, and the basis for those effects points to aspects of socioecology that could be key to an adaptive shift in ancestral life histories.

Young Hadza children are energetic foragers. Those between the ages of five and ten years supply half their own nutrient requirements in some seasons (Blurton Jones, Hawkes, and O'Connell 1989). Mothers take advantage of their children's foraging capacities, choosing to focus on foods the children can handle efficiently when those resources are in season (Hawkes, O'Connell, and Blurton Jones 1995). But the year-round staple in this habitat is deeply buried tubers, which young children are not strong enough or skilled enough to handle effectively. Senior Hadza women, long-experienced gatherers, spend even more time acquiring food than do women of childbearing age (Hawkes, O'Connell, and Blurton Jones 1989). The extra time is largely devoted to digging those deeply buried tubers, which they acquire at rates equivalent to those of younger adults (Hawkes, O'Connell, and Blurton Jones 1989).

The seasonal and age-related variation in Hadza foraging highlights some likely causes and consequences of human economic interdependence that could have large evolutionary implications. Other primate juveniles must be successful enough at feeding themselves to support their own survival (Altmann 1998). In humans, unlike other primates, weaned offspring depend on other individuals to supply a substantial component of their nutrition (Bogin and Smith 1996; Bogin 1999; Kaplan 1997; Kaplan et al. 2000). In one sense, this dependence seems to make human children a greater burden on their mothers. Yet, if human children depended on their own foraging, their mothers, like mothers in other primate species whose weanlings remain in their close company for other reasons, would face increased offspring mortality from extended foraging in habitats where the youngsters could not feed themselves. Food sharing allows adults accompanied by young offspring to invade habitats and exploit foods they otherwise could not.

Some mother-offspring food sharing occurs in many primate species (Feistner and McGrew 1989), but shared food accounts for at most a small component of juvenile diets in all living primates but one. When mothers supply a substantial fraction of their weanlings' nutrition, this sharing opens a novel opportunity for senior females to have large effects on their own fitness. An aging female, unencumbered by nurslings of her own, can provision a just-weaned grandchild so that its mother can allocate less effort to that child and can produce the next child more quickly. In this way, more vigorous perimenopausal females can have larger effects on the fertility of younger kin. These effects would strengthen selection against senescence in elder females, lowering adult mortality and increasing average adult life spans.

The interrelationships described in this scenario are evident among the Hadza (Hawkes, O'Connell, and Blurton Jones 1997). The foraging effort of a mother who is not encumbered with a nursing infant has a measurable effect on the nutritional welfare of her children. At the birth of a newborn, its mother's foraging effort declines. The link between a mother's effort and the nutritional welfare of her weaned children disappears. Now the weight gains of those children depend on the foraging effort of their grandmother (Hawkes, O'Connell, and Blurton Jones 1997).

An Evolutionary Scenario

Imagine ancestral australopithecine populations with a chimpanzee-like life history. Ecological changes constrict the forests and the availability of fruits that young juveniles can handle. Increasing aridity and seasonality favor plants that cope well with dry seasons—for example, by holding starches in underground storage organs. Such resources can yield high return rates, but only to those with the strength and skill to extract and process them. Young juveniles cannot do it. To rely on these resources and succeed in these environments, mothers must provision offspring who are still too young to extract and process the tubers for themselves. If older females whose own fertility is declining feed their just-weaned grandchildren, the mothers of those weanlings can have shorter interbirth intervals without reductions in offspring survivorship. The more vigorous elders will thus raise their daughters' fertility. Under this scenario, normalizing selection would maintain menopause at about the age at which it usually occurred in the ancestral population. The fraction of females living beyond menopause would increase, but any who continued to have babies at later ages themselves (as a consequence of increased initial oocyte stocks or slowed rates of follicle loss) would thereby reduce their contribution to the fertility of their daughters. Continued childbearing would interfere with grandmothering, erasing the selective advantage of vigorous adaptive performance late in life and collapsing life histories back toward the chimpanzee-like pattern.

Grandmothering and the Human αM

The novel effects that aging females could have on their own fitness in the socioecological circumstances sketched above would strengthen selection against age-related declines in somatic performance. Senescence is an important source of mortality in any large-bodied primate (Ricklefs 1998). Enhanced selection for continued adult vigor would reduce this mortality, increasing average adult life spans. (Effects on most physiological systems— e.g., cardiovascular, renal, gastrointestinal, pulmonary—would likely be correlated across sex: longer-lived mothers would have longer-lived sons as well as longer-lived daughters.) Lower adult mortality rates and greater longev-

Table 9.4 αM for living hominoids

	Average adult life span (years) M^{-1}	Weaning to maturity (years) α	αM
Orangutans	17.9	8.3	.46
Gorillas	13.9	6.3	.45
Chimpanzees	17.9	8.2	.46
Humans	32.9	14.5	.45

Note: See Hawkes et al. 1998 for sources; Alvarez 2000 for discussion of alternative calculations of α and statistical evaluation.

ity would, in turn, favor delaying maturity. The risk of dying before reproducing would decline, and the greater productive capacity of larger adults could be exercised over a longer period. So, according to Charnov's symmetry arguments (1993, 1997, 2001), the lower M (the inverse of the average adult life span) would favor higher α, a longer juvenile period, and continued growth to a larger size before maturing. In fact, the αM products of four living hominoids are strikingly similar (table 9.4). Alvarez (2000) shows that the human αM falls well within the confidence interval for sixteen primate species.

A grandmothering socioecology, combined with the interrelated life history trade-offs expected for any large-bodied female primate, could account for our unusual longevity—with midlife menopause—as well as our late maturity. This framework can explain why humans take about "twice as long to reach adulthood, and live about twice as long as great apes" (Smith and Tompkins 1995, p. 260).

Grandmothering and the Human αb

This scenario also has implications for the duration of lactation, and thus for interbirth intervals. As with age at maturity and average adult life span, discussed above, the "mouse lemur to gorilla" curve is like the "mouse to elephant" curve for variation in annual fecundity (b) (Charnov 1993; Charnov and Berrigan 1993). Whereas age at maturity and adult life span rise with increases in adult body size across the mammals, fecundity declines. In general, the (ESS) weanling size is larger for larger mothers ($W_o/W_\alpha = \delta$, approximately a constant: Lee, Majluf, and Gordon 1991). In Charnov's mammal models, the trade-off between offspring size and number changes as the ESS α rises. The optimum for larger mothers is more investment per offspring; in other words, larger babies at longer intervals.

Table 9.5 αb for living hominoids

	Weaning to maturity (years)	Offspring per year (daughters)	
	α	b	αb
Orangutans	8.3	.063	0.52
Gorillas	6.3	.126	0.79
Chimpanzees	8.2	.087	0.70
Humans	14.5	.142	2.05

Note: See Hawkes et al. 1998 for sources; Alvarez 2000 for discussion of calculations and statistical evaluation.

Grandmothering, however, is allocation of production by senior females to the offspring of younger kin. Females nearing the end of their own fertility gain fitness by changing the size-number trade-off faced by their daughters. The annual baby production (b) of childbearers in a grandmothering species incorporates production by both mother and grandmother, so it should be higher during the childbearing years than expected for a grandmotherless species with the same age at maturity (α). Table 9.5 shows the αb products for four living hominoid species. Human interbirth intervals are slightly shorter than those of chimpanzees and similar to those of gorillas. The instructive comparison is the expected b given our α. Human fertilities are more than twice that expected for a grandmotherless primate with our late maturity. Alvarez (2000) has extended the comparison to include sixteen primate species, showing human fertilities to be well outside the confidence interval of grandmotherless primates.

A general expectation from life history theory is that organisms must trade off current and future reproduction, so that lower fertility is the price of greater longevity (Williams 1966; Kirkwood 1977; Rose 1991). Primates generally spend less energy on lactation than do other mammals (Oftedal 1984). Some measures suggest that human mothers, who are likely to be grandmothers and have long lives ahead, put even less energy into current reproduction than do other primates (Prentice and Whitehead 1987). The grandmother hypothesis assumes that effort in producing children comes not only from mothers, but from postmenopausal helpers as well, so that even with less allocation to current reproduction, human fecundity is high relative to that expected for such a late-maturing primate.

This hypothesis accounts for the surprising combination of late maturity with a short nursing period and short interbirth intervals. Bogin and Smith (1996, p. 703) phrased the puzzling combination of human life history features this way: The problem is to explain "how humans successfully combined a vastly extended period of offspring dependency and delayed reproduction with helpless newborns, a short duration of breast-feeding, an adolescent growth spurt, and menopause." A grandmothering socioecology, combined with Charnov's symmetry framework, explains almost all the items in the list.

Grandmothering and *Homo erectus*

The grandmother hypothesis and Charnov's framework link an array of distinctive features of human life histories as systematic adjustments of a general primate pattern. If these features are the result of a single adaptive shift that actually took place in an ancestral population, then the paleoanthropological record should be consistent with that scenario. Evidence linking this shift to the appearance of the first widely successful member of our genus, *Homo erectus (sensu latu),* can be found in the paleoecological record of the early Pleistocene in Africa, the geographic distribution of this taxon, and morphological and developmental characteristics indicated in the skeletal evidence (O'Connell, Hawkes, and Blurton Jones 1999).

Data from a wide range of sources (e.g., deep marine sediments, soil chemistry, pollen, fossil faunas) consistently indicate that the 1.9–1.7 mya time period bracketing the first appearance of African *H. erectus (ergaster)* was marked by ecological changes likely to alter foraging opportunities for a large-bodied primate. Africa saw an unusually pronounced shift toward cooler, drier, more seasonal conditions and a related trend toward open, less wooded plant communities (e.g., Cerling 1992; deMenocal 1995; Reed 1997; Spencer 1997). Extinctions among frugivorous primates in various East African localities after 1.8 mya may all be related to the restriction of closed forest habitats (e.g., Reed 1997).

These circumstances could have favored the exploitation of previously unused or little-used resources that provide predictable returns to an adult forager, although they are difficult for young juveniles to handle. Many resources meet these criteria, notably certain varieties of small game, shellfish, nuts, seeds, and the underground storage organs of plants. Though none of these foods are especially well represented in the archaeology of early *H. erectus,* this may reflect problems of preservation, the absence of attention to their recovery by archaeologists, or both.

The best (though not the only) prospective candidates for increased exploitation under these circumstances may be underground plant storage organs ("USOs" or, loosely, "tubers") (see also Hatley and Kappelman 1980;

Wrangham et al. 1999). They are encountered in many forms across a wide array of habitats, sometimes occurring at densities greater than one metric ton per hectare (Raunkiaer 1934; Vincent 1985; Thoms 1989). Data from African, Australian, and North American settings indicate returns to modern human foragers of 1,000–6000 calories per hour, which commonly translates to about 8,000–12,000 calories per forager-day, all with very low day-to-day variance—easily enough to support at least two consumers (e.g., Blurton Jones, Hawkes, and O'Connell 1999; Couture, Ricks, and Housley 1986; Hawkes, O'Connell, and Blurton Jones 1995; O'Connell, Latz, and Barnett 1983; Thoms 1989; see Schoeninger et al. 2001 for a contrary estimate of nutrient value and discussion of this estimate in Hawkes, O'Connell, and Blurton Jones 2001b).

These dependable return rates make USOs attractive food resources, but mechanical and chemical defenses limit consumer access (Coursey 1973; Thoms 1989; Wandsnider 1997). Children's primary acquisition of USOs among modern human foragers is limited to forms found close to the surface that require little or no processing (e.g., Blurton Jones, Hawkes, and O'Connell 1989; Hawkes, O'Connell, and Blurton Jones 1995). Tuber use among other primates is probably limited by the same factors: chimpanzees rarely take them; baboons do so only in highly arid environments, and even then target only forms that juveniles can handle on their own (McGrew 1992; Whiten et al. 1992; Altmann 1998).

Archaeological evidence of tuber exploitation is often limited and indirect because the activity itself leaves ephemeral traces. Still, several lines of evidence are consistent with increased use of USOs beginning with the appearance of *H. erectus*. Efficient exploitation of deeply buried USOs requires, at minimum, a digging tool. The earliest known examples of such tools date to about 1.7 mya (Brain 1988). Cooking is essential to the use of chemically defended tubers and important for the conversion of the complex carbohydrates they commonly contain to simpler, more readily digestible forms. Though controversial, the earliest dates for humanly controlled fires, suitable for tuber processing, fall in the range 1.4–1.6 mya (Bellomo 1994; Rowlett 1999).

The geographic range of *Homo erectus* indicates the capacity of this taxon to exploit a far broader range of habitats than any previous hominid. While earlier hominids are restricted to Africa, early *H. erectus* is found as far east as Java and as far north as 50° latitude (Dennell and Roebroeks 1996; Gabunia and Vekua 1995; Gabunia et al. 2000). This sharp change in distribution strongly implies access to new food sources. Tubers are a staple among ethnographically known hunters in continental habitats extending to approximately 50° N (Thoms 1989), and so may also have been important in the first expansion of the genus *Homo*.

Aspects of *H. erectus* morphology also indicate a shift in resource exploitation. Reductions in chewing architecture clearly point to increased use of resources that require less postconsumption processing (Aiello and Wheeler 1995; Klein 2000; Wood and Collard 1999a,b; Suwa, White, and Howell 1996), implying either a narrower range of foods exploited or increased investment in preconsumption processing. Tuber cooking is a good example of such processing.

Skeletal criteria also clearly identify early African *H. erectus* as the first hominid with a nonpongid life history (D. W. E. Smith 1993; Wood and Collard 1999a,b). Australopithecines have been characterized as "bipedal apes" (D. W. E. Smith 1993; Smith and Tompkins 1995; Klein 2000). With *H. erectus,* there is a substantial increase in body size (McHenry 1994; Kappelman 1996; Ruff, Trinkhaus, and Holliday 1997), an expected consequence of delayed maturity. In addition to adult size, other maturation measures (e.g., Smith 1991b; Tardieu 1998; Clegg and Aiello 1999) suggest that age at maturity may have been within the range of modern humans (but see Dean et al. 2001). Since age at maturity plays a central theoretical and empirical role in the life history ideas discussed here, this aspect of the fossil evidence is of primary importance.

Delayed maturity implies a change in mortality rates. But, given the character of the fossil record (including the age biases in archaeological assemblages discussed above), it is not possible to directly test the expectation that *H. erectus* life spans were substantially longer than those of australopithecines. Recent work on aging skeletal specimens has focused on characterizing the biases in standard methods (Paine 1997). Clearer appreciation of these biases is spurring the development of new techniques for aging adult skeletons. It may also eventually be possible to identify osteological traces of postmenopausal physiology (Ruff 1991; Bogin and Smith 1996). At present, all that can be said is that claims of *short* adult lifetimes in premodern (and early modern) human populations are not empirically warranted.

Another direct test of the proposed adaptive shift in *H. erectus* should be feasible. A grandmothering primate should have higher fertility during the childbearing years than expected for a grandmotherless primate with the same age at maturity because that higher fertility in young adults is the benefit that drives selection against senescence. The specific prediction is that if grandmothering *is* the adaptive shift of *Homo erectus,* weaning ages will be lower (implying higher annual fecundity during the childbearing years) than expected for a grandmotherless ape with the age at maturity estimated for this taxon. It may be possible to monitor age at weaning by reference to changes in trace element composition (especially ^{18}O and ^{13}C) in teeth formed across the weaning period (e.g., Wright and Schwarz 1998).

What About Brains?

This hypothesis about the evolution of human life histories gives no instrumental role to changes in brain size. But brains are central in many influential scenarios of human evolution. Finding maximum life span more strongly correlated with brain size than with body size in his analysis of variation across mammals, Sacher (1959) hypothesized that increases in brain size improved physiological regulation and consequently slowed rates of aging. Others (notably Allman 1999) have continued to pursue work along these lines. Many analysts have linked the provisioning of human juveniles to a delay in maturity, which they propose to be a consequence of selection for larger brains (e.g., Washburn and Lancaster 1968; Bogin 1999; Kaplan et al. 2000).

Above, we reviewed evidence suggesting that modern human life histories may begin with *Homo erectus*. If so, then brain size in this taxon is relevant to any hypothesized causal role for brains. Differential preservation of cranial remains should make brain size a much easier feature to estimate for fossil taxa than many life history variables (e.g., Leigh 1992a). But because larger mammals have larger brains, measures that remove effects directly attributable to body size alone are required (Jerison 1973; Eisenberg 1981; Armstrong 1985a; Martin 1983). A dimensionless index, EQ (encephalization quotient), which compares observed brain size to that expected for a reference animal of the same body size, solves this problem. But the brain size expected due to changes in body size alone depends on the scaling relationship assumed. Observed scaling relationships between body size and brain size vary among orders and with taxonomic level (Martin and Harvey 1985).

The lower the taxonomic level, the smaller the observed change in brain size for each change in body size—in other words, the flatter the slope of the regression (Martin and Harvey 1985). One hypothesis to explain this taxon-level effect proposes that evolutionary changes in body size can be more rapid than evolutionary changes in brain size. A direct test of this lag hypothesis using pairwise contrasts in living primate species found no support for it (Deaner and Nunn 1999). Body size data are inherently extremely noisy. Much of the taxon-level effect could be a systematic consequence of measurement error (Pagel and Harvey 1988b, 1989).

Difficulties in estimating body sizes for living species are compounded in fossil taxa. Calculating EQ requires not only a choice of allometry, but a body size to go with each brain size. Even with a full skeleton, body mass estimates are not straightforward (e.g., Smith 1996), and they are even less so when fossils represent immatures and when elements cannot be clearly assigned to single individuals. Moreover, the very small number of specimens associated with some fossil taxa and with some time periods makes estimates of within-species variability, and thus between-species differences, difficult to assess.

The work to date is especially impressive in light of these difficulties. Brain size in *Homo erectus* is about double that found in australopithecines and modern chimpanzees. McHenry (1994) estimated female body size to be about 60% larger in *H. erectus* than in australopithecines and so calculated encephalization as only slightly higher—much lower than his EQ for modern humans. Using a cranial measure to estimate body size, Kappelman (1996) found that EQ increased in *H. erectus* over australopithecines, then remained stable in the genus *Homo* throughout the lower and middle Pleistocene, increasing again only within the upper Pleistocene. Ruff, Trinkhaus, and Holliday (1997) applied other indices of body size and also found no increases in relative brain size over time in *H. erectus sensu latu,* but "archaic" *H. sapiens,* appearing in the late middle Pleistocene, was more encephalized.

Most recently, Wood and Collard have reconsidered the definition of genus *Homo.* Linking several morphological features to adaptive strategies (body size and form, locomotion, maturation rate, and chewing architecture), they sorted fossil hominids into two broad categories, australopithecines on one hand and members of genus *Homo* on the other. This sorting reassigns *H. habilis* and *H. rudolfensis* to *Australopithecus* spp. *H. erectus (ergaster)* aligns with modern humans on these criteria, although "relative brain size does not group the fossil hominins in the same way" (Wood and Collard 1999b, p. 203). As they say elsewhere, "Although there are twofold differences in the mean absolute brain size of early hominids, these differences are almost certainly not significant when body mass is taken into account. A notable effect of body-mass correction is that the absolutely larger brain of *H. ergaster* is 'cancelled out' by its substantial estimated body mass" (Collard and Wood 1999, p. 324).

New fossils, of course, as well as increasing attention to the bases for estimates of relative brain size (including within-taxon variation), will clarify the picture. While the timing and extent of changes in encephalization across hominid phylogeny continue to be disputed, there *is* clear consensus that the extremely high EQs of contemporary human populations are quite recent, no older than the late middle Pleistocene. *Homo erectus* emerged and spread 1.5 million years before that. If essentially modern human life histories did evolve with that taxon, then the grandmothering adaptive shift is distinct from, *and much older than,* the adaptive shift that gave us our very big brains.

The Evidence for Meat Eating

The appearance of *Homo erectus (ergaster)* at the beginning of the Pleistocene coincides with the dates of the best-known Plio-Pleistocene archaeological sites (Isaac 1997; Leakey 1971), all of which fall in the 1.5–1.9 mya range (Feibel, Brown, and McDougal 1989; White 1995). Sites of this age often contain the bones of large ungulates in close association with stone tools. Though opinions differ on whether these animals were hunted or scavenged, cut marks show that early humans certainly took meat from them (Bunn 1981; Potts and Shipman 1981). Following Isaac (1978), many argue that these sites were "central places," similar to the base camps occupied by modern hunters—spots to which meat and marrow were routinely transported, probably by hominid males, for distribution to mates and offspring (e.g., Rose and Marshall 1996; Oliver 1994). Paternal provisioning and increased meat eating are central elements of most versions of this argument (e.g., Kaplan et al. 2000). Even those who disagree with certain aspects of this hypothesis usually assume that meat was a significant part of early human diets and was crucial to the emergence and subsequent evolutionary success of *H. erectus* (e.g., Blumenschine 1991; Rogers, Feibel, and Harris 1994). Increased carnivory is linked to the invasion of more seasonal habitats (Shipman and Walker 1989). Furthermore, the nutritional quality of meat and, especially, provisioning by males (Lovejoy 1981) are assumed to allow mothers to produce more offspring that are more dependent on provisioning through a longer childhood (Kaplan et al. 2000).

This scenario has been supported by appeals to the importance of meat and paternal provisioning among modern hunter-gatherers. But quantitative data on modern human foragers show that big game hunting and scavenging are often highly unreliable food acquisition strategies, and are not aimed toward family provisioning, even in tropical Africa, where both practices are thought to have evolved. Among the Hadza, for example, whose foraging practices provide a stimulus to the grandmothering argument, men specialize in taking big ungulates. These excellent hunters operate in a game-rich savanna woodland habitat. Even with bows and arrows (modern projectile weapons that appear archaeologically only in the late upper Pleistocene), they succeed in acquiring large animal prey at an annual average of only one every thirty hunter-days (O'Connell, Hawkes, and Blurton Jones 1988; Hawkes, O'Connell, and Blurton Jones 1991). The animals are very large, and a success draws the interest of many. Ninety percent *or more* of the meat goes to claimants outside the hunter's own household (Hawkes, O'Connell, and Blurton Jones 2001a). Marked variation in hunters' success rates has no effect on the number or size of shares their household gets from the kills of other men. This wide sharing, with no evidence of "risk reduction reciprocity," makes big game hunting clearly inferior to many alternative strategies for provisioning families (Hawkes, O'Connell, and Blurton Jones 2001b). Annual average meat consumption is high, but weeks may pass with no meat in camp, even in camps with several active hunters. Everyday support for children must come from elsewhere.

Archaeological data are sometimes said to indicate higher rates of meat acquisition in the past. Bunn (1982; Bunn and Kroll 1986), for example, claims that the well-known assemblage from the Olduvai site named FLK Zinjanthropus, containing the remains of at least forty-eight large ungulates, accumulated in less than two years, implying a minimum carcass acquisition rate of more than twenty-five animals per year at this site alone. Assuming that early humans were primarily responsible for these remains, that most of the carcasses represented were taken in complete or nearly complete condition, and that this was only one of many large faunal assemblages created by the local hominid group involved, then a meat consumption rate much higher than reported for the modern Hadza is implied. As many have observed, however, the basis for Bunn's estimated carcass deposition rate is weak: other, more plausible interpretations imply a much longer period of accumulation, possibly up to several centuries (e.g., Kroll 1994; Lyman and Fox 1989; Potts 1988). If most of the carcasses represented at FLK Zinj and other sites from that time period had also been heavily ravaged by other predators by the time hominids took control (Blumenschine and Marean 1993; Marean et al. 1992), then a lower, even more sporadic large animal tissue intake than enjoyed by the modern Hadza is indicated— far too low to support effective offspring provisioning.

Another aspect of the emerging paleoanthropological record is inconsistent with the argument that dependence on meat is the key to the evolution of *H. erectus*. Whatever the pattern of carcass acquisition represented at Plio-Pleistocene Olduvai and East Turkana, it may have a much greater antiquity than previously suspected. Reports from Kanjera in southern

Kenya indicate that a hominid-created large faunal assemblage comparable to the one at FLK Zinj was deposited there as early as 2.2 mya (Plummer et al. 1999); similar assemblages in Ethiopia may have been deposited by 2.6 mya (Asfaw et al. 1999; Heinzelin et al. 1999; Semaw et al. 1997). If these early assemblages do indeed prove to have been deposited by the same hominid-related processes, at about the same rates, as those at the later sites, then either *H. erectus* appeared 400,000–800,000 years earlier than currently supposed, or australopithecines are the hominids implicated. If the archaeology *is* associated with australopithecines (Heinzelin et al. 1999), then it cannot represent a distinctive adaptation of genus *Homo*.

Summary and Conclusions

A combination of Charnov's model of the trade-offs that underlie the variation in mammalian life histories and a specific hypothesis about ancestral socioecology can explain why "humans live on a vastly extended time scale compared to most other mammals" (Smith and Tompkins 1995, p. 258). It can also explain why the extreme slowness of some aspects of our life history is combined with producing offspring faster than they can become independent foragers. Grandmothering is not just a hypothesis about the fitness payoffs of postmenopausal survival; rather, it is part of a framework that links several life history features to one another as aspects of a single adaptive shift.

This theory is only a partial "sunburst" (Smith and Tompkins 1995). Changes in brain size or paternal provisioning, which Smith and Tompkins also list among the features that distinguish us from other primates, are not included in it. The paleoecology, the geographic spread of *Homo erectus (sensu latu),* and an array of characteristics inferred directly from the fossils themselves are generally consistent with a scenario in which a systematic cascade of life history changes occurs in this taxon. But, as we have noted above, the fossil record does not show marked encephalization in younger taxa. Nor does the archaeology show any clear indication of increased meat eating with the appearance of this first long-enduring and widely successful member of our genus. Results of stable carbon isotope analysis on one set of fossils shows no more meat in *H. ergaster* diets than in the diets of contemporary australopithecines (Lee-Thorp, Thackery, and van der Merwe 2000). A combination of the life history framework and the paleoanthropology data points toward the possibility that for *most* of the history of genus *Homo* (the first million and a half years of the Pleistocene), the successful way to be human included life histories similar to ours. Our very large brains, however, are much more recent, and the increased consumption of meat that characterizes some modern human foragers may have no greater antiquity. Instead, grandmothers may have been the distinctive addition to the human story during the Pleistocene.

This hypothesis has implications for each of a series of distinctive features of human life history:

1. Our extended juvenile dependency. An ecological shift in tropical Africa at the beginning of the Pleistocene reduced the availability of foods young juveniles could handle for themselves and increased the prevalence of foods that gave high return rates to an essentially frugivorous, large-bodied primate old (large) enough to overcome the mechanical or chemical defenses of those foods. Young children had to depend on their mothers for these resources.

2. Our long life span. Maternal provisioning allowed older females whose own fertility was declining to have a large effect on the fertility of their younger kin. Without infants and weanlings of their own, older females could provision the weanlings of their daughters. These fitness benefits strengthened selection against senescence, altering the apelike equilibrium allocation to somatic maintenance and repair and so lowering rates of adult mortality and lengthening adult life spans.

3. Our unusual midlife menopause. Longer life spans allow the production of more descendants through higher fertility in young adults. With senescence delayed, the fraction of postmenopausal females increased. But any tendency to delay menopause would not be favored because older females who were occupied with their own weanlings could not help younger females, removing the advantage for increased somatic maintenance and repair in those lineages. Higher fertility at young ages in the grandmothering lineages would continue to select against delaying menopause beyond midlife.

4. Our late maturity. Longer average adult life spans would alter the optimal age at maturity. As the risk of dying before reproducing declined, selection would favor continuing to grow for a longer period, and so to a larger size, before maturing. Gains from the added productive capacity due to an extended period of growth would be enhanced as they accumulated over the longer duration of adulthood.

5. Our early weaning and short birth spacing. The assistance of grandmothers would lower the offspring survival costs of earlier weaning. This would alter the optimal size (age) at weaning otherwise expected for a primate of our size (age) at first reproduction.

Acknowledgments
We thank H. Alvarez, E. L. Charnov, M. Collard, A. Harcourt, P. Kappeler, M. Pereira, C. Stanford, and C. van Schaik for very helpful comments.

PART THREE

Evolution of Primate Brains

The chapters of this last major section examine the evolution of the primate brain and its various links with life history traits. Because the brain plays a regulatory role in so many vital processes—from basic maintenance functions to social relationships—and because its growth and maintenance require much energy, this organ is considered an important factor in life history evolution, and sometimes as a life history trait in its own right. Compared with other mammals, primates have relatively large brains, slow life histories, and complex social systems. To what extent these trait suites are interrelated and whether their major characteristics evolved independently constitute the overarching questions tackled by the contributors of these last three chapters, and their answers do not entirely agree.

Deaner, Barton, and van Schaik (chap. 10) review the literature on evolutionary relationships between brain size and life history traits, turning up seven major hypotheses while also revealing that the assumptions, implied mechanisms, and predictions of these hypotheses have yet to be rigorously evaluated. Controlling for body size and phylogeny in allometric analyses, Deaner and colleagues take the innovative step of calculating residuals for each variable of interest using an independent data set on estimated body masses to minimize effects of correlated measurement error. Their analyses demonstrate a robust positive correlation between brain size and life span, but no significant correlations either between brain size and other life history traits or between life history traits and socioecological variables. Inferring from these initial results that socioecological factors, including group size, have not likely influenced brain size and life history in parallel, Deaner et al. consequently focus on hypotheses that suggest direct rather than ecological links between brain size and life span in primates.

Deaner et al. find some support for four hypotheses that are not mutually exclusive. The maturational constraints hypothesis suggests that some types of complex behavior can be supported only by mature nervous systems, and that this constraint leads evolutionary increases in brain size to delay maturation. The cognitive buffer hypothesis holds that large brains, with their augmented behavioral flexibility and learning capacities, serve to reduce extrinsic mortality. All else being equal, if the larger-brained are better buffered against ecological dangers, they should evolve slower life histories. The authors themselves develop the brain malnutrition risk hypothesis, positing that the high energy demands and extreme developmental sensitivities of a large brain are most compatible with conservative brain and body growth trajectories. In this case, extensions of the duration of somatic growth that also entail decreased growth rates would permit evolutionary increases in brain size. Finally, the delayed benefits hypothesis offers that increases in longevity should increase the likelihood of increased brain size because long-lived animals have greater opportunities than short-lived ones to exploit learning to promote fitness. By identifying these hypotheses for which interspecific comparisons provide preliminary support, this chapter paves the way for better-focused evaluations of general explanations for the brain size-life history connection, including explorations of its theoretical implications.

Ross (chap. 11) examines the relationship between brain size and life history traits from a different angle, exploring relations among brain size, rates of development, and patterns of infant care, of which primates possess an underappreciated diversity. Unlike Deaner et al., in phylogenetically controlled analyses, Ross finds relative brain size to correlate significantly with age at first reproduction and group size and, subtracting ages at weaning from ages at first reproduction, she also finds brain size to correlate with juvenile growth rate and duration of juvenility. Next, Ross shows that species with substantial allocare of infants also exhibit relatively rapid infantile growth, but do not reach sexual maturity sooner relative to adult body size. Similarly, no relationship is detected between amount of allocare and relative brain size. Presumably, then, mothers gain the main advantage from allocare, accelerating reproductive rate by reducing interbirth intervals. This work uncovers another link between life histories and social systems.

Together, the chapters of Part Three make clear how much is at stake in careful selection and application of analytic methods and how much will be gained through widespread effort to quantify many socioecological dimensions for a broad diversity of primates. After Ross's apparent corroboration of the brain size-group size correlation cracks back open a door that Deaner et al. provisionally shut, Dunbar slaps it back against the wall, unwincingly asserting that haplorhines' large social groups selected for increased brain size and, requiring extension of brain growth, also extended life histories. Along the way, Dunbar also underscores that important effects of socioecology go beyond home range size, degree of frugivory, and group size to include factors such as degree of terrestriality, relative dependence on ripe fruit, and patterns of subgrouping and alliance formation.

Dunbar argues for a link between apes' very large brains and their social systems, extending the more general claim that the evolution of large brains in haplorhines and several orders of nonprimate mammals is driven by the cognitive demands of succeeding in complex societies. This explanation is contrasted with ecological and developmental explanations, for which Dunbar finds no supportive evidence, whereas the analyses by Deaner et al. suggest that large brain size can be favored by additional behavioral advantages as well as by deeply integrated allometric constraints. Dunbar then points out that hominoids differ markedly from other anthropoids with respect to relative neocortex volume, marshaling the argument that this difference, in particular, has likely functioned to support advanced sociocognitive abilities. Why apes differ from other primates in this respect — that is, why apes are so smart — remains to be determined conclusively, but Dunbar staunchly favors socioecological over other causation. Specifically, he proposes that dispersed social groupings, combined with increased terrestrialization, were responsible for the increase in neocortical volume. Other primates with dispersed social systems offer opportunities to test this hypothesis, which echoes a main theme of this volume — namely, that social, ecological, and life history traits are much more closely linked than has previously been recognized.

One fine young investigator of nonprimates contributing to the conference that launched work on this volume remarked at the time that the amount of discussion we primatologists were devoting to brains was shocking and clearly disproportionate. We did not agree then. We do not agree now. The chapters of this section help to make unmistakably clear that research on the manifold nature of brains' links to life history and socioecology will remain a central and most exciting topic for research on primates and other mammals for decades to come.

10 Primate Brains and Life Histories: Renewing the Connection

ROBERT O. DEANER, ROBERT A. BARTON,
AND CAREL P. VAN SCHAIK

Across primate species, brain size correlates with several life history variables, including maximum recorded life span, gestation length, and age at first reproduction (Harvey, Martin, and Clutton-Brock 1987; Austad and Fischer 1992; Allman, McLaughlin, and Hakeem 1993a; Allman 1995; Hakeem et al. 1996; Allman and Hasenstaub 1999; Barton 1999; Ross and Jones 1999; Judge and Carey 2000; Ross, chap. 11, this volume; see also Friedenthal 1910; Sacher 1959, 1975, 1978; Sacher and Staffeldt 1974; Mallouk 1975). These life history variables are themselves correlated, meaning that large-brained primates generally have slow, prolonged growth periods, late sexual maturation, and long lives (Harvey, Martin, and Clutton-Brock 1987; Charnov and Berrigan 1993; for nonprimates see Western and Ssemakula 1982; Stearns 1983; Read and Harvey 1989). To account for these patterns, researchers have offered a number of hypotheses about brain development, function, and degeneration (table 10.1). Thus far, however, these hypotheses have not been examined rigorously, especially with respect to their theoretical bases (but see Allman 1995; Hakeem et al. 1996). One reason for this neglect is uncertainty as to whether the brain size-life history correlations that suggested these hypotheses actually indicate direct evolutionary links between the variables. In particular, it has been proposed repeatedly that these correlations may be artifacts of statistical methodology or unconsidered evolutionary processes (Calder 1976; Economos 1980a; Prothero and Jürgens 1987; Harvey, Martin, and Clutton-Brock 1987; Harvey and Krebs 1990; Barton 1999).

Our goal in this chapter is to evaluate these hypotheses. Given the uncertainty regarding the correlations that gave rise to them, however, we begin by considering the factors that could have spuriously produced those correlations. In the first section, we show that there are tractable statistical

Table 10.1 Hypotheses generated by comparative studies of primate brains and life history

Hypothesis	Premise and claim	Hypothesis type
Physiological regulator	Larger brains allow greater precision of physiological regulation and thus longer life. Longer life can be achieved only with larger brains, and vice versa.	Strong constraint
Growth regulator	Brain growth governs body growth; larger brains grow for longer periods, and thus larger-brained animals exhibit prolonged somatic growth. Prolonged somatic growth can be achieved only with larger brains, and vice versa.	Strong constraint
Neuronal investment	The functioning population of neurons inevitably diminishes; the only way to maintain a functioning brain (and population of neurons) at advanced age is to grow a large brain initially. The evolution of increased longevity must *often* be accompanied by an increase in brain size.	Weak constraint
Maturational constraints	Brain and behavioral maturation is an inherently slow process that takes longer to complete for large-brained animals. If there is selection for increased brain size, the period of development must *often* be extended.	Weak constraint
Cognitive buffer	The behavioral flexibility permitted by larger brains reduces extrinsic mortality. An evolutionary expansion of the brain increases the likelihood of an evolutionary increase in life span.	Adaptive
Brain malnutrition risk	Developing brains require a constant energy supply to avoid irreversible damage; to keep risks of brain damage low in unpredictable environments, large-brained animals grow their brains slowly; slow brain growth requires slow somatic growth. An evolutionary extension of somatic growth period increases the likelihood of an evolutionary increase in brain size.	Adaptive
Delayed benefits	Brains, and the learning they allow, provide delayed benefits that continue to accrue throughout the animal's life; long-lived animals benefit more from large brains than short-lived animals. An evolutionary extension of life span increases the likelihood of an evolutionary increase in brain size.	Adaptive

methods for considering the most problematic potential confounding factors; namely, body size, phylogeny, and socioecology. We demonstrate that even when conservative methods are used to deal with these confounds, there remain indications of correlated evolution between brain size and life history. With this evidence in hand, we turn in the following section to the hypotheses proposed to explain the correlations. Although several of the hypotheses are mutually compatible, we evaluate the logic and premises of each separately. Although some of the hypotheses can be refuted a priori, most are at least plausible. For these, we derive additional predictions and test several of those predictions with data from primates and other mammals.

Are Life History and Brain Size Truly Linked?

The hypotheses in table 10.1 assume that brain size and life history variables have direct evolutionary links, such that when there is a change in either brain size or a life history variable, the other trait shows, or becomes more likely to show, some corresponding change, *regardless of other biological processes*. In other words, the challenge is to show that a given brain size-life history correlation is not an artifact of some other process that simultaneously affects both variables. Although there are numerous potential confounds, body size, phylogeny, and socioecology are thought to be most problematic.

Body Size

Body size is arguably the most informative single variable for understanding an organism's physiology, morphology, and life history (Schmidt-Nielsen 1984; Calder 1984). It is highly correlated with brain size and with virtually all aspects of life history; therefore, the influence of body size alone can be expected to produce brain size-life history correlations. Several methods have been employed to examine whether brain size and life history traits are linked independently of body size, but all of them have been plagued by the fact that measures of body size are influenced by numerous factors. Body mass, for instance—the most commonly used body size measure—is highly sensitive to nutritional and reproductive status (e.g., Pagel and Harvey 1988b; Dunbar 1992; Smith and Jungers 1997).

Sacher (1959) made the first attempt to show that life history is related to brain size independently of body size. He found that in a sample of sixty-three mammals, maximum recorded life span (hereafter life span) was more strongly correlated with brain mass than with body mass. Nevertheless, the fact that the brain mass-life span correlation statistically exceeds the body mass-life span correlation does not automatically demonstrate that a portion of the life span-brain mass correlation cannot be explained by body size. The reason is that body mass is more subject to sampling error than brain mass is (Pagel and Harvey 1988b); hence, body mass could be worse than brain size as an estimator of true body size (Lindstedt and Calder 1981; Harvey, Martin, and Clutton-Brock 1987; Harvey and Krebs 1990; Barton 1999). Economos (1980a; see also Prothero and Jürgens 1987) argued this point forcefully, showing that two other size-related organs, the liver and the adrenals, are more highly correlated with life span than is body mass.

Other investigators have asked whether brain size and life history variables are related once the influence of body mass is statistically removed from both variables. The procedures used have included the residuals method (or partial correlation), in which the variables of interest are first regressed on body mass to generate residual or "relative" values (e.g., Harvey, Martin, and Clutton-Brock 1987; Allman, McLaughlin, and Hakeem 1993a,b), and multiple regression (Sacher 1975, 1978; Barton 1999). Nevertheless, because these methods are closely related to each other, they are both sensitive to the same potential pitfall. The problem is that if species-specific body mass estimates are highly prone to error, then using them to remove the effects of size from any two variables of interest could lead to spurious correlations between those variables (Harvey and Krebs 1990; Barton 1999). We can show this most easily by considering the results obtained with the residuals method when body mass estimates overestimate or underestimate true body size (fig. 10.1). For any given species (or independent contrast: see below), the error in body mass will create a bias in the same direction in both residuals: if the body mass estimate is erroneously too large, both residuals will be smaller than they should be; if the estimate is too small, both residuals will be larger than they should be. If the error in body mass is substantial, the two residuals may end up showing a significant correlation, even if there is not a biological relationship. Multiple regression is prone to the same problem because multiple regression coefficients represent the relationship between two variables (e.g., brain size and a life history variable) once the effects of another predictor (e.g., body size) have been partialled out.

Without "correct" body size measures, there may be no completely satisfactory way to control for body size. Nevertheless, the problem of correlated errors can be ameliorated with the residuals method if independent estimates of size are used for calculating the residuals for each variable (Harvey and Krebs 1990). In this case, an over- or underestimated body size estimate will produce an over- or underestimated brain size residual; how-

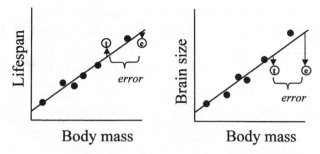

FIG. 10.1. The problem of correlated errors. Consider one species (open circles) in a hypothetical case in which true body size (t) and estimated body mass (e) are both known. If the residuals are generated from true body size, the life span residual will be positive and the brain size residual will be negative. In this instance, the hypothesis that the residuals are correlated will not be supported. On the other hand, if the residuals are generated from (over) estimated body size, both residuals will be negative, falsely supporting the hypothesis of correlation.

ever, because the life history residual is derived from a different body size estimate, it will not (generally) be over- or underestimated in the same manner.

To explore the use of independent estimates of size in primates, we first calculated brain mass residuals from the regression of $\log_{10}$ brain mass on $\log_{10}$ body mass, taking brain and body mass data from Stephan, Frahm, and Baron (1981; H. Stephan, H. D. Frahm, and G. Baron, unpub.; see Deaner and Nunn 1999).[1] Next, we calculated life span residuals from the regression of $\log_{10}$ life span on $\log_{10}$ body mass, taking life span as the maximum value in primary and secondary sources (Hakeem et al. 1996; Ross and Jones 1999; unpublished data from the Duke University Primate Center) and taking body mass from Smith and Jungers (1997). For the sake of comparison, we also repeated the analysis using the body mass data from Stephan, Frahm, and Baron to derive both brain mass and life span residuals.[2]

As shown in table 10.2, the correlation between brain mass and life span residuals was positive and significant, both when the residuals were derived from the same body mass values and when they were derived from different values. As expected, however, the relationship was stronger when the residuals were derived from the same values. When we performed these analyses with gestation length and age at first reproduction (age at first reproduction from Ross and Jones 1999; gestation length from Harvey, Martin, and Clutton-Brock 1987), we found highly significant relationships in all cases,

Table 10.2 Relationships between brain size residuals and life history residuals in primates when the residuals are derived from the same or different body mass estimates

	Same estimates			Different estimates		
	r	P	N	r	P	N
Life span	.46	.0002	56	.43	.0005	56
Age at first reproduction	.52	.0002	47	.47	.001	47
Gestation length	.38	.0087	46	.34	.0153	46

Note: All results are based on species values.

but again, the relationships were somewhat stronger when based on the same body size estimates. Thus, this brief investigation of correlated errors in body mass shows that using the same body mass estimates to calculate brain size and life history residuals may in fact lead to artificially strong associations between brain size and life history. Although much more work is needed on the issue, we provisionally recommend that, whenever possible, workers employ different body mass estimates for calculating each of the residuals that will later be compared.

Phylogeny

Early studies that documented statistical relationships between brain size and life history traits treated species values as independent data points (e.g., Sacher 1959; Sacher and Staffeldt 1974; Austad and Fischer 1991, 1992; Allman, McLaughlin, and Hakeem 1993a,b; Hakeem et al. 1996). Such analyses implicitly assume that the brain size and life history of each species represents an independent evolutionary event. This assumption is clearly incorrect, however: a species' brain size and life history, as well as most other aspects of its biology, are partially predictable from its phylogenetic history (for a discussion of the reasons, see Harvey and Pagel 1991). In primates, for example, relative brain size is largely a function of whether the species in question is a strepsirrhine or a haplorhine (e.g., Martin 1990; Barton 1999). The fact that trait expression is dependent on phylogeny means that statistical tests that treat species (or other taxonomic units) as independent data points greatly overestimate degrees of freedom, a situation that may lead to incorrect conclusions (Harvey and Pagel 1991; Martins and Hansen 1996; Purvis and Webster 1999).

There is a rapidly growing literature on the methods that are most appropriate for treating phylogenetically dependent data (reviewed by Harvey and Pagel 1991; Martins and Hansen 1996; Purvis and Webster 1999). At present, the method of independent contrasts (ICs) is considered best for calculating independent instances of evolutionary change in continuous variables (Purvis and Webster 1999). In brief, this method calculates evolutionary contrasts (differences between paired taxa) throughout a phylogenetic reconstruction of a character's evolution. The contrasts calculated at each node of the reconstruction should be fully independent of the other nodes and are therefore, in principle, suitable for standard statistical analysis (Felsenstein 1985; Harvey and Pagel 1991; Martins and Hansen 1996). Most importantly, because the ICs are representative of evolutionary change, a hypothesis of correlated evolution (e.g., that evolutionary changes in relative brain size are associated with evolutionary changes in life history variables) can be addressed explicitly.

We explored the associations between relative brain size and life span, age at first reproduction, and gestation length, employing Purvis and Rambaut's (1995) CAIC computer program and Purvis's (1995) phylogeny for calculating ICs. We followed standard practice in obtaining residuals after employing least-squares regressions, forced through the origin, on $\log_{10}$ values (Harvey and Pagel 1991; Garland, Harvey, and Ives 1992). There are theoretical reasons to expect that more recent contrasts could be misleading for variables that are sensitive to measurement or sampling error (Purvis and Rambaut 1995; Purvis and Webster 1999), and this problem could be acute for life history variables. Life span, in particular—a maximum measure—should be highly sensitive to the size and quality of the sample from which it is drawn (i.e., species that are poorly represented in captivity should often have appreciably underestimated life spans: Allman, McLaughlin, and Hakeem 1993a). Hence, we divided the contrasts into two groups of equal size, based on the corresponding nodes in the phylogeny (i.e., "young contrasts" and "old contrasts"). Then we repeated all of the tests with only the old contrasts (Purvis and Harvey 1995). The logic here is that because old contrasts usually incorporate data from several species, any error associated with a particular species estimate will have reduced influence.

Table 10.3 presents the results of our contrasts analyses and, for comparison, the results obtained when we considered species as independent data points (as presented in table 10.2). For all of these tests, we used different body mass estimates for calculating brain size and life history residuals (see above). Although these analyses indicate positive evolutionary relationships between the life history variables and brain size, the relationship reaches significance only for life span (fig. 10.2). For the analyses based only on old contrasts, the results are similar, although the life span-brain size relationship is no longer significant (but the correlation coefficient is slightly greater). These analyses confirm that life span is a genuine correlate of brain

Table 10.3 Relationships between brain size residuals and life history residuals in primates when residuals are derived from species, all independent contrasts, or old independent contrasts

	Species			All ICs			Old ICs		
	r	P	N	r	P	N	r	P	N
Life span	.43	.0005	56	.31	.03	52	.36	.07	26
Age at first reproduction	.47	.001	47	.20	.18	45	.27	.21	23
Gestation length	.34	.0153	46	.21	.16	44	.18	.43	22

size, but suggest that there might not be evolutionary relationships between age at first reproduction and brain size and gestation length and brain size, despite the apparent relationships when species were considered as independent data points. It is worthwhile to bear in mind that because life history variables may be highly error-prone, the absence of statistical correlations is not conclusive evidence that there are not biological relationships.

Socioecology

Several socioecological variables have been linked to differences in relative brain size (reviewed by Harvey and Krebs 1990; Barton and Dunbar 1997; Deaner, Nunn, and van Schaik 2000). There are also indications that some socioecological variables are associated with life history (reviewed by Ross and Jones 1999). It is possible, therefore, that brain size-life history correlations are by-products of socioecological factors (Harvey, Martin, and Clutton-Brock 1987; Ross and Jones 1999). For instance, a large group size (presumably meaning increased social demands: see Dunbar 1992; Dunbar, chap. 12, this volume) may select for both extended juvenility and larger brains, thereby producing a brain size-juvenility correlation (see Joffe 1997; Ross and Jones 1999).

The most direct way to evaluate whether socioecological factors are responsible for brain size-life history correlations is to test whether the same socioecological variables that correlate with brain size also correlate with

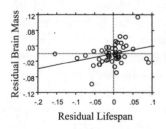

Fig. 10.2. Relationship between life span residuals and brain size residuals in primates, based on all independent contrasts. (See table 10.3 for statistics.)

Table 10.4 Relationships between life history residuals and socioecological variables in primates

	Group size			Home range residual			Frugivory		
	r	P	N	r	P	N	r	P	N
Life span	.19	.14	64	.24	.08	53	.10	.49	51
Age at first reproduction	.20	.15	54	−.34	.01	50	.13	.39	44
Gestation length	.01	.97	52	−.33	.03	45	.04	.78	40

Note: Life history residuals and home range residuals were calculated from independent body mass estimates. All results are based on all independent contrasts.

life history. Home range size, group size, and percentage of fruit in the diet have been previously shown to correlate with relative whole brain or neocortex size (Harvey and Krebs 1990; Barton and Dunbar 1997; Deaner, Nunn, and van Schaik 2000). Thus, we asked whether any of these factors correlate with life history variables. For these analyses, we took data on home range size and group size from Nunn and van Schaik (2002) and on frugivory from Barton (1999). We followed the procedures described above in calculating ICs, including generating residuals to remove the effects of body size from the life history variables. In addition, we used residual home range size, rather than absolute home range size, because we found this variable to be correlated with body mass. Again, for the reasons discussed above, we used different body mass estimates in generating the home range size residuals than we used when calculating the life history residuals.

We found that the relevant socioecological variables were not significantly correlated with life history variables (table 10.4; see also Harvey, Martin, and Clutton-Brock 1987; Ross and Jones 1999). The only significant findings were negative correlations between home range size and age at first reproduction and gestation length. Clearly, these *negative* correlations cannot explain why home range size should cause age at first reproduction and brain size to be *positively* correlated. Furthermore, Barton (1999; see also Allman 1999) has shown that in multiple regression, ecological and life history variables independently explain variation in brain size. Thus, it is highly unlikely that brain size-life history correlations are products of the common influence of socioecology.

Hypotheses to Explain Brain Size-Life History Correlations

The results presented in the previous section indicate that within primates, there are evolutionary associations between brain size and at least one life history variable—life span—that cannot be attributed to obvious confounds.

Hence, in this section, we evaluate the hypotheses that argue for direct brain size-life history associations (see table 10.1). Although each of these hypotheses deals chiefly with one aspect of life history, the strong correlations among life history variables (Harvey, Martin, and Clutton-Brock 1987; Charnov 1993; Charnov and Berrigan 1993) mean that any single hypothesis may turn out to have fairly general explanatory value. Conversely, these intercorrelations could ultimately make it difficult to distinguish among the hypotheses.

Although most of the hypotheses we review here have already been introduced in the literature, they have been underdeveloped with regard to their crucial assumptions and plausible evolutionary mechanisms. This situation has hampered attempts to understand how compatible these hypotheses are with modern life history theory and how general their predictions are (e.g., whether exceptions are allowed). Thus, besides testing predictions derived from the most plausible hypotheses, we devote considerable attention to exploring these fundamental issues.

Physiological Regulator Hypothesis

The physiological regulator hypothesis holds that brain size determines the overall regulation of physiological function: because larger-brained animals have more precise regulation of function, they are able to live longer (Sacher 1959, 1978; Hofman 1983; see also Friedenthal 1910; Mallouk 1975; Cutler 1976). According to this view, brain size determines longevity directly through physiological mechanisms, so that if there is selection on either brain size or longevity, the other trait necessarily changes.

The physiological regulator hypothesis is best understood as an extension of the now discredited rate-of-living theory (Rubner 1908; Pearl 1928; Stahl 1962; Lindstedt and Calder 1976, 1981; Boddington 1978). This theory holds that molecular damage is the inevitable by-product of energy consumption, and that a lower rate of energy consumption leads to a slower rate of molecular deterioration and therefore enhanced longevity. The basis of the rate-of-living theory was the general correlation between long life span and low metabolism (both of which are correlated with body size). However, in order to explain exceptional cases in which life span exceeds predictions based on metabolism (e.g., humans), brain size was incorporated via the physiological regulator hypothesis.

The rate-of-living theory no longer enjoys support because growing evidence indicates that metabolic rate is not directly related to either accumulated molecular damage or longevity. First, although metabolic by-products are indeed major sources of molecular damage, the key to predicting accumulated damage is the efficiency of mechanisms that mitigate that damage (Orr and Sohal 1994). For instance, aerobic respiration produces free oxygen radicals, or oxidants, which damage other molecules by dislodging their electrons. However, cells possess mechanisms to remove or repair such damage and specialized molecules (antioxidants) that bind to oxidants before they can do their damage. Species differ greatly in their mitigating mechanisms (e.g., Hart and Setlow 1974; Sohal, Sohal, and Brunk 1990; Sohal, Sohal, and Orr 1995; Ku, Brunk, and Sohal 1993; Austad 1997b), a fact that is inconsistent with the rate-of-living theory (cf. Sacher 1978). A second line of evidence indicating a dissociation between metabolism and life span is experiments in which reduced caloric intake is found to extend longevity; in these cases, metabolic rate is not reduced (reviewed by Masoro 1995). Finally, the rate-of-living theory is weakened by the fact that several long-lived taxa (e.g., bats, birds) have high metabolic rates (Lindstedt and Calder 1981; Austad and Fischer 1991).

Could the brain provide the key link between metabolism and life span, as the physiological regulator hypothesis claims? Probably not. Although progress has been made in identifying mechanisms for reducing or repairing molecular damage (reviewed by Austad 1997b; Finch and Tanzi 1997), there is no evidence that the efficiency of these mechanisms is in any way dependent on the size of the central nervous system. Moreover, it seems unlikely on theoretical grounds that there must automatically exist a tight relationship between brain size and the accumulation of molecular damage. We would expect, for instance, that novel enzymes for eliminating oxidized DNA would occur through specific genetic alterations rather than as correlates of increased overall brain size. In summary, the physiological regulator hypothesis is a mechanistic hypothesis without a plausible mechanism (see also Economos 1980b).

Growth Regulator Hypothesis

The growth regulator hypothesis holds that brain growth governs body growth: specifically, for a given body size, that a large-brained animal will be characterized by slow, prolonged somatic growth. The impetus for this hypothesis was the observation of cross-species correlations between brain size indices and life history variables (Sacher and Staffeldt 1974; Western and Ssemakula 1982; cf. Friedenthal 1910; Ricklefs 1979). In particular, Sacher and Staffeldt (1974) emphasized the relationship between prolonged gestation length and large neonatal brain size.

Like the physiological regulator hypothesis, the growth regulator hypothesis holds that brain size and life history are not distinct traits; rather, large brain size and prolonged somatic growth co-occur because, for physiological reasons, they must. There is no doubt that numerous mechanisms (e.g., the endocrine system) are in place to ensure that growth in each part of the body, including the brain, is complementary to that of other areas (harmony of growth: reviewed by Bogin 1999). Nevertheless, the tremendous diversity of brain and body growth trajectories across species indicates that natural selection can alter these mechanisms (Case 1978; Read and Harvey 1989). Moreover, there is no plausible physiological account of how larger brains are constrained to produce slower body growth. The only such proposal is Sacher and Staffeldt's (1974) "minimax" theory, and it is underpinned by assumptions that the brain is "the slowest growing organ in the mammal" and "the pacemaker for growth of all other somatic tissues, which are constrained to grow at its pace." Both of these claims are incorrect: in mammals, relative to adult body size, brain growth substantially outpaces body growth in the fetal and early postnatal periods (e.g., Count 1947; Deacon 1990; Bogin 1999). Thus, the argument that large brains are physiologically constrained to "drag out" body growth is unsupported (see also Read and Harvey 1989).

Neuronal Investment Hypothesis

The neuronal investment hypothesis holds that brain size limits longevity: because the functioning population of neurons diminishes throughout an organism's life, the only way to maintain adequate behavioral performance at an extended age is to invest initially in more neurons, and thus a larger brain. A crucial assumption of this hypothesis is that neurogenesis in mammals occurs only early in development. Allman (1995) proposed this hypothesis in light of evidence that neuronal damage, particularly in the cerebellum, accumulates in older animals and that this damage causes a decline in behavioral performance. To appreciate this idea, imagine the case of an old arboreal animal that is at high risk of falling from trees because it has a reduced population of cerebellar neurons. According to the neuronal investment hypothesis, if the animal's cerebellum initially had more neurons, more would be functioning for the animal in old age, and it would be less likely to fall.

Given the costs of growing and maintaining brain tissue (e.g., Holliday 1971; Aschoff, Gunther, and Kramer 1971; Armstrong 1985a; Aiello and Wheeler 1995), having to generate and maintain "extra neurons" to ensure that some will be available later in life would be highly inefficient; it would be better to invest in mechanisms to minimize or repair neuronal damage. Therefore, the neuronal investment hypothesis must assume that mechanisms to prevent neuronal damage do not exist, or are extremely limited in scope. A strong adaptationist would question this assumption because natural selection has a tremendous capacity to create mechanisms for repairing or avoiding cellular damage. For instance, a major cause of neuronal damage is ingested toxins (Allman 1995); selection for longevity might therefore act through the development of mechanisms to ensure that fewer toxins reach the brain, or, more simply, animals might change their diets. Despite this a priori argument, it is possible that constraints do exist such that in some contexts, neuronal damage is truly unavoidable. Allman (1995) implied that this might be the case for highly active inhibitory neurons, such as Purkinje cells of the cerebellum.

Fortunately, the validity of the unavoidable damage assumption can be tested empirically. If this assumption is correct, then neuron numbers should decrease throughout the life span. On the other hand, if the number of neurons does not begin declining until mid- to late adulthood, this finding would indicate that selection has acted to mitigate early damage. Because the Purkinje cells of the cerebellum are one of the few neuronal populations in which declines are undisputed (see Haug 1985; Allman 1995; Albert and Moss 1996; Scheibel 1996; Peters, Sethares, and Moss 1998), studies of this area are most relevant. Hall, Miller, and Corsellis (1975) and Torvik, Torp, and Lindboe (1986) both found that human Purkinje cells did not decline in number until about sixty years of age (see also Ellis 1920, chart 6). Thus, the assumption of unavoidable neuronal damage appears weak.

The neuronal investment hypothesis holds that the evolution of increased longevity must *often* be accompanied by increases in brain size. Conversely, evolutionary decreases in brain size must *often* be accompanied by evolutionary decreases in longevity. We stress the word "often" because it is conceivable that some animals will have brains that, for some adaptive reason (see the hypotheses below), are larger than the minimum size necessary to achieve their current life span (i.e., in this case, life span would be intrinsically limited by the deterioration of systems other than the brain). In this way, the neuronal investment hypothesis can be viewed as a "weak" or limited constraint hypothesis (as compared with the former two hypotheses, which argue for "strong" constraints). Because it allows exceptions, this hypothesis predicts a positive, but imperfect, evolutionary correlation between brain size and life span.

As already noted, there is some evidence for this brain size-life span linkage. Nevertheless, this prediction is readily derived from other hypotheses that have better-supported assumptions. Hence, we will focus here on the following prediction, which, as far as we know, is unique to the neuronal investment hypothesis: Because the cerebellum is apparently more sensitive to neuronal loss than other brain parts, cerebellum size and life span should be evolutionarily correlated, and the cerebellum's relationship to life span should be stronger than that of the whole brain or that of another large brain structure, such as the neocortex (Allman 1995).

We tested this prediction in three mammalian taxa: primates, bats, and insectivores.[3] Contrary to the prediction of the neuronal investment hypothesis, we found no evolutionary relationship between cerebellum size and life span in bats or insectivores (table 10.5). In primates, we did find evidence of a significant relationship, but, contrary to the prediction, it was not stronger than the relationship between neocortex and life span (cf. Allman, McLaughlin, and Hakeem 1993b; Hakeem et al. 1996). In summary, neither the chief assumption nor the unique prediction of this hypothesis is supported.

Maturational Constraints Hypothesis

In considering the evolution of life history and brain size in hominids, it has commonly been noted that prolonged juvenility may permit greater learning opportunities (e.g., Dobzhansky 1962; Mann 1972; Poirier and Smith 1974; Gould 1977; Lancaster and Lancaster 1983; Bogin 1999; Kaplan et al. 2000). Some researchers have therefore examined the hypothesis that large brains and slow life histories (especially prolonged juvenility) are evolutionarily associated across primates for the straightforward reason that large-brained animals need more time to learn adult skills than smaller-brained animals do (Janson and van Schaik 1993; Ross and Jones 1999). These and other researchers concluded, however, that this "needing-to-learn" hypothesis is weak on both empirical and theoretical grounds (Janson and van Schaik 1993; Ross and Jones 1999; Blurton Jones, Hawkes, and O'Connell 1999; N. G. Blurton Jones and F. W. Marlowe, unpub.; but see Kaplan et al. 2000). Nevertheless, we believe that a modified version of this hypothesis can withstand previously raised criticisms. We call this new version the maturational constraints hypothesis. Instead of focusing on the need for learning opportunities, the maturational constraints hypothesis emphasizes that some types of complex behavior can be supported only by mature nervous systems (cf. Altmann and Alberts 1987; Allman 1999; All-

Table 10.5 Relationships of life span residuals to whole brain residuals, neocortex residuals, and cerebellum residuals in five mammalian taxa

| | Whole brain | | | | | | Neocortex | | | | | | Cerebellum | | | | | |
| | All ICs | | | Old ICs | | | All ICs | | | Old ICs | | | All ICs | | | Old ICs | | |
	r	P	N	r	P	N	r	P	N	r	P	N	r	P	N	r	P	N
Primates	.31	.03	52	.36	.07	26	.17	.32	35	.48	.04	18	.16	.36	35	.46	.05	18
Insectivores	.11	.76	9	*	*	*	.22	.54	9	*	*	*	.06	.87	9	*	*	*
Bats	.10	.71	16	.08	.83	8	.27	.40	18	*	*	*	.45	.14	18	*	*	*
Carnivores	.35	.02	41	.31	.16	21	*	*	*	*	*	*	*	*	*	*	*	*
Odontocetes	.40	.12	11	*	*	*	*	*	*	*	*	*	*	*	*	*	*	*

Note: Asterisk denotes insufficient data available for analysis.

man and Hasenstaub 1999). Its claim is that an evolutionary increase in brain size often delays maturation.

The first assumption of this hypothesis is that a lengthy period of brain and behavioral maturation is unavoidable. In other words, no matter what environmental stimuli or nutritional resources are available, (some) complex behaviors, and the neural circuitry underlying them, cannot emerge immediately, but instead require long maturational periods. This proposition is supported by numerous findings that most skills emerge at approximately the same age across individuals of a given species (reviewed by Parker and McKinney 1999). The paradigmatic example is language: virtually all humans, in virtually all cultures, begin to understand and produce words at roughly one year of age (Pinker 1994).

Why maturational constraints exist is not entirely clear, but one reason seems to be that brain and behavioral development occurs in a stepwise fashion: because more complex patterns of behavior and neural connectivity are built on simpler ones, complex behavior cannot emerge immediately (e.g., Hebb 1949; Piaget 1980; Case 1992; Elman et al. 1996; Bjorklund 1997). Some support for this interpretation can be drawn from research on neural networks. In general, powerful networks grow or add nodes slowly, keeping pace with their learning, so that larger networks take longer to establish (reviewed by Quartz and Sejnowski 1997).

The second assumption of the maturational constraints hypothesis is that brain and behavioral maturation often takes so long that it actually affects life history. Most importantly, the duration of brain and behavioral maturation determines the onset of sexual maturity (the hypothesis is silent with regard to juvenile growth rates). If this assumption is correct, then animals should reach neuroanatomical and behavioral markers of maturity just prior to reaching adulthood (Janson and van Schaik 1993). Studies of corticospinal pathways, the fiber tracts that transmit efferent signals from the motor cortex to the extremities, support this claim: white matter density and conduction velocity increase until the age of three years in captive macaques (for which sexual maturity may occur as early as three and a half years: Melnick and Pearl 1987) and until the mid-teens in humans (Nezu et al. 1997; Olivier et al. 1997; Paus et al. 1999). In addition, in corresponding experimental tests of fine manual motor control, there are significant improvements until three years in the macaque and the early teens in humans (reviewed by Olivier et al. 1997).

The evidence regarding development in wild primates is more difficult to evaluate. Janson and van Schaik (1993) reviewed the ontogeny of foraging skill in wild primates and concluded that, although juveniles are usually less successful than adults, this fact is better attributed to differences in body size than to differences in experience or skill (see also N. G. Blurton Jones and F. W. Marlowe, unpub.). Nevertheless, this conclusion was based on only a few types of foraging in a few species (see Janson and van Schaik, 1993, p. 64), and it is possible that there are more difficult techniques that cannot be mastered until adulthood. Indeed, there is some evidence of late-developing skills that are apparently not limited by strength. For instance, mountain gorillas employ a variety of complex manual movements in leaf gathering, and adults (nine years or older) show a larger repertoire of these techniques than subadults (Byrne and Byrne 1993). Similarly, while most types of chimpanzee feeding tool use are established by five or six years, nut cracking with stone tools is not mastered until roughly ten years (Matsuzawa 1994; Boesch and Boesch-Achermann 2000; chimpanzee sexual maturity may occur as early as eleven years: see Goodall 1986). Kaplan et al. (2000) argue that in many hunter-gatherer societies, men and women do not become fully proficient in acquiring many types of crucial resources until their third decade, or even later. Thus, although more evidence is needed, the assumption that the length of brain and behavioral maturation delays adulthood is at least plausible.

The maturational constraints hypothesis is similar to the neuronal investment hypothesis in that it is most plausibly understood as arguing for a "weak" constraint. It predicts a positive, but imperfect, evolutionary correlation between brain size and the duration of brain and behavioral development. This correlation is imperfect because in some cases brain development might not be limiting the length of the juvenile period (i.e., some animals will have periods of development that are "longer than necessary" for their brain sizes).

The life history variable that captures the developmental period best is probably age at first reproduction, but we showed above that there is little evidence of an evolutionary association between age at first reproduction and brain size in primates. To further test this prediction, we examined whether age at first reproduction and brain size are evolutionarily correlated in four nonprimate taxa: insectivores, bats, carnivores, and odontocetes. Contrary to the maturational constraints hypothesis, brain size-age at first reproduction correlations were not significant in any of these nonprimate taxa (table 10.6). Nevertheless, in odontocetes, there was one obvious and significant outlier, and when it was removed, the relationship became significant ($n = 8$, $r = .78$, $p = .008$). The outlying contrast involved the

Table 10.6 Relationships of brain size residuals to age at first reproduction residuals and gestation length residuals in five mammalian taxa

| | Age at first reproduction | | | | | | Gestation length | | | | | |
| | All ICs | | | Old ICs | | | All ICs | | | Old ICs | | |
	r	P	N	r	P	N	r	P	N	r	P	N
Primates	.20	.18	45	.27	.21	23	.21	.16	44	.18	.43	22
Insectivores	.15	.57	16	*	*	*	-.23	.27	24	-.42	.16	12
Bats	.23	.22	29	.47	.07	15	.21	.11	58	.09	.64	29
Carnivores	.19	.21	44	.11	.59	22	.10	.41	72	.04	.83	37
Odontocetes	.36	.26	9	*	*	*	-.44	.59	11	*	*	*

Note: All results are based on independent contrasts. Asterisk denotes insufficient data available for analysis.

genus *Platanista,* a riverine dolphin that apparently evolved a slow life history without a corresponding brain enlargement, an occurrence that does not contradict this hypothesis.

Deriving other predictions from the maturational constraints hypothesis is difficult. Allman and Hasenstaub (1999) suggest that, although brain size should be correlated with age at first reproduction, it should be negatively correlated or uncorrelated with gestation length. Their idea is that a shorter gestation will allow the larger-brained organism to spend a greater proportion of its development outside the womb, where, compared with prenatal development, environmental stimulation is greater; this would in turn allow maturation to occur more rapidly. We tested for evolutionary correlations between brain size and gestation length and found no evidence for them (see table 10.6). Nevertheless, we do not believe that these results truly bear on the maturational constraints hypothesis. Our objection is that environmental stimulation, although necessary for brain and behavioral development, probably does not limit it. In fact, in young animals, high levels of stimulation in one modality often interrupt development in another one (overstimulation: reviewed by Bjorklund 1997). Mounting evidence also shows that there is substantial stimulation in the womb (e.g., DeCasper and Spence 1986). Furthermore, aspects of natural history, such as relative predation risk, apparently determine the relative timing of birth (i.e., precociality vs. altriciality) independently of general fast-slow life history trends (Read and Harvey 1989). Hence, predictions regarding gestation length and brain size are probably irrelevant to the maturational constraints hypothesis.

Ross and Jones (1999) attempted to test the needing-to-learn hypothesis by asking whether evolutionary changes in the length of juvenility are negatively correlated with evolutionary changes in percentage of foliage in the diet and positively correlated with evolutionary changes in group size. They suggested that these were reasonable predictions of the needing-to-learn hypothesis because both a larger group size and a nonfolivorous diet might require more complex behavior or learning, which should take longer to master. Ross and Jones (1999) did not find evidence of the predicted relationships. Whatever their implications for the needing-to-learn hypothesis, these tests certainly do not weaken the maturational constraints hypothesis. For one thing, although these relationships are often taken for granted, group size and folivory are variables that have never been demonstrated to correlate with behavioral complexity or learning (cf. Dunbar 1992; Deaner, Nunn, and van Schaik 2000). Even more crucial is the fact that the analyses of Ross and Jones (1999) controlled for differences in brain size. According to the maturational constraints hypothesis, brain size and behavioral demands are two sides of the same coin; thus, once brain size is controlled, juvenility *should not* be correlated with behavioral demands, even if they could be identified.

Cognitive Buffer Hypothesis

The cognitive buffer hypothesis holds that a large brain, and its concomitant behavioral flexibility and improved learning ability, serves to reduce extrinsic mortality (e.g., deaths due to food shortages or predation). Because larger-brained animals are better buffered against ecological dangers, they are likely to experience reduced mortality—a necessary condition for selection of slower life history (Allman, McLaughlin, and Hakeem 1993a; Hakeem et al. 1996; Allman and Hasenstaub 1999; cf. Kaplan et al. 2000).

The crucial underlying assumption of the cognitive buffer hypothesis is that larger brains substantially reduce extrinsic mortality. Although there are countless cases of wild animals employing unusual behavior to solve life-threatening problems, it is extremely difficult to test whether large-brained animals do this more frequently. Nevertheless, researchers have quantified the frequency of foraging innovations reported in the literature per given species and tested whether this measure is correlated with measures of brain size. Lefebvre et al. (1997, 1998; see also Timmermans et al. 2000) found significant positive cross-species correlations in birds, and S. M. Reader (unpub.) did likewise in primates. If feeding innovations do indeed lead to better survival, then these studies support the assumption that larger brains reduce mortality. In addition, cross-species comparisons of learning abilities, although contentious (e.g., Macphail 1987 and commentaries therein), have generally found that brain size (or related measures) predicts performance on learning tasks (Passingham 1975a; Riddell and Corl 1977; Rumbaugh, Savage-Rumbaugh, and Washburn 1996; R. O. Deaner et al., unpub.). Thus, although more work is needed to test this assumption, it is likely to hold.

Like the previous two hypotheses, the cognitive buffer hypothesis predicts positive, but imperfect, correlations between brain size and life history variables. Correlations will be imperfect because organisms might achieve slower life histories in other ways than through enhanced cognition, and, conversely, a larger brain could evolve without a corresponding increase in longevity. In other words, the cognitive buffer claims only that once larger brains have evolved, all else being equal, enhanced longevity becomes *more likely*.

The first prediction of the cognitive buffer hypothesis is that evolutionary changes in brain size and longevity should be positively correlated.

Several previous studies have found this association in primates (e.g., Harvey, Martin, and Clutton-Brock 1987; Austad and Fischer 1992; Allman, McLaughlin, and Hakeem 1993a), and above we showed that it holds even when conservative methods are employed to account for body size and phylogeny.

We can also test for a brain size-longevity relationship in nonprimate mammals. Earlier studies found little or no evidence for this relationship (Mace 1979; Economos 1980b; Gittleman 1986b; Read and Harvey 1989; Austad and Fischer 1991, 1992; cf. Sacher 1959, 1975; Mallouk 1975), but this could be because they did not consider the issue of phylogenetic independence. An evolutionary association between two traits could be obscured in an analysis taking species as independent data points because one (or more) major evolutionary event or grade shift (e.g., Martin 1990; Barton and Harvey 2000) at the base of the phylogenetic tree occurred contrary to the general association (Purvis and Webster 1999). Because the method of ICs calculates all changes within a phylogeny, it should not be unduly weighted by any particular event. Thus, we used ICs to test for an association between brain size and life span in carnivores, bats, insectivores, and odontocetes.

As shown in table 10.5, there was not a significant life span-brain size correlation within insectivores or bats. Within carnivores, there was a positive life span-brain size correlation in the "all contrasts" analysis, but not the "old contrasts" analysis. Within odontocetes, there was an apparent positive life span-brain size correlation, although it did not reach significance. Small sample size ($n = 11$ contrasts), partially due to several unresolved phylogenetic relationships, clearly limited statistical power. We thus decided to reconsider this relationship in an analysis taking species as independent data points. In this case, the relationship reached significance ($n = 14$, $r = .70$, $p = .005$), providing preliminary evidence of a life span-brain size linkage in odontocetes.

A second prediction can also be derived from the cognitive buffer hypothesis: if longevity and cognition are truly associated, then life span should be positively correlated with the size of the neocortex, even if it is not correlated with total brain size. This prediction derives from the fact that most structures implicated in higher-order cognition (e.g., planning, working memory) are located in the neocortex, which can therefore be considered superior to the whole brain as a cognitive assay (e.g., Sawaguchi and Kudo 1990; Dunbar 1992). Although some investigators argue that viewing any part of the brain as more "cognitive" than another is overly simplistic (e.g., Barton 1996; Barton and Dunbar 1997), to the extent that such a distinction is possible, the neocortex is clearly the best cognition candidate. We therefore tested for a neocortex-life span association within primates, insectivores, and bats. We found no evidence for this relationship in insectivores or bats, but in primates there was a significant relationship in the analysis of "old contrasts," although not in the analysis of "all contrasts" (table 10.5; cf. Allman, McLaughlin, and Hakeem 1993b; Hakeem et al. 1996). Overall, then, both predictions of the cognitive buffer hypothesis have received some support.

Brain Malnutrition Risk Hypothesis

The high energy costs of growing and maintaining the brain are often hypothesized to be linked to the evolution of brain size and life history (Martin 1981; Armstrong 1983, 1985a; Leonard and Robertson 1992; Aiello and Wheeler 1995; Martin 1996; Ross and Jones 1999; Kaplan et al. 2000). Nevertheless, evolutionary biologists have generally overlooked the implications of the growing brain's unique sensitivity to energy shortages (but see Nowicki, Peters, and Podos 1998). Here we present a brain malnutrition risk model showing that a large brain's high energy demands and extreme sensitivity to nutritional perturbations during development are most compatible with conservative brain and body growth trajectories. Hence, an evolutionary extension of the body growth period, and a corresponding decrease in the body growth rate, should increase the probability of an evolutionary increase in brain size.

This model is based on two assumptions about mammalian brain and body growth. First, if the energy needs of the developing brain are not met, even for a brief period, there is a high risk of long-lasting or even permanent brain damage and behavioral abnormalities (Shoemaker and Bloom 1977; Smart 1991; Levitsky and Strupp 1995). The timing of the energy shortage is crucial to predicting its consequences: if it occurs while the brain is growing, it will be far more debilitating than if it occurs after brain growth has been completed (Winick and Noble 1966; Shoemaker and Bloom 1977; Smart 1991; Levitsky and Strupp 1995). Because energy shortages during brain growth often lead to permanent or long-term incompetence, they are likely to severely curtail survival and reproductive success. In contrast, if body growth is stunted due to resource scarcity, the body retains a remarkable capacity to achieve the target adult size, provided resources later become abundant (catch-up growth: Prader, Tanner, and van Harnack 1963; Elias and Samonds 1977; Tanner 1986). Although the mammalian data showing the brain's relatively greater sensitivity to energy shortfalls come from only a few species, the phenomenon may occur generally (Schew and Ricklefs 1998).

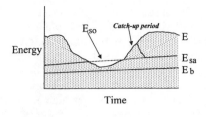

FIG. 10.3. Schematic representation of the brain malnutrition risks model. E represents the animal's potential energy intake; E_b represents the energy needed to maintain and grow the brain; E_{so} represents the energy needed to grow and maintain the soma optimally; E_{sa} represents the actual energy used to grow and maintain the soma. E fluctuates, but there is always enough energy to meet E_b. When E cannot also meet E_{so}, the soma is compromised. Later, when E is high, more energy is devoted to E_{sa}, so that the soma can return to its optimal growth trajectory. Note that E_b and E_{so} increase slowly, reflecting the demands of a growing animal.

The model's second assumption is that body growth trajectories lag behind brain growth trajectories in a similar fashion across mammals. In particular, the brain reaches its peak growth velocity and its adult size considerably before the body does (Count 1947; Holt et al. 1975; Deacon 1990). This relationship between brain and body growth trajectories probably reflects selection for a functional balance between behavioral capabilities and body size. In other words, an animal with a very small brain in an adult-sized body would probably lack the experience and motor control to effectively forage and move (see Mace 2000). Conversely, an animal possessing an adult-sized brain in an extremely small body would have great difficulty obtaining enough energy to maintain its large brain as its body grew.

Given these two assumptions, we can model brain and body growth as follows (fig. 10.3). At any given time during development, an animal has the potential to ingest a certain amount of energy, E, which fluctuates due to temporal variation in food abundance and the animal's improving foraging skills. In the allocation of E, first priority is given to the needs of the growing brain (E_b). The remainder of E is allocated to optimally growing and maintaining the soma (E_{so}). If there is not enough energy to maintain both optimal brain and optimal body growth trajectories, the brain is spared and the body is compromised (Stewart 1918; Shoemaker and Bloom 1977; Smart 1991; Peeling and Smart 1994). In this case, the actual energy devoted to the soma (E_{sa}) falls below the optimal amount (E_{so}). When E exceeds the joint needs of E_b and E_{so} (i.e., periods of superabundance), the animal limits its food consumption to maintain a functionally balanced relationship between brain and body size. If, however, the animal's body growth has been previ-

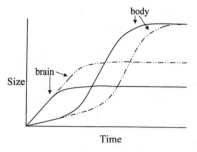

Fig. 10.4. Extension of brain growth and somatic growth could promote the evolution of a larger adult brain. In the ancestor (solid lines), brain growth occurs at a high rate and is finished quickly. Somatic growth first occurs slowly, but when brain growth is complete, it accelerates. In the larger-brained descendant (dashed lines), brain growth follows the ancestral trajectory, but continues at the high rate for a prolonged period, ultimately leading to a larger adult brain size. The descendant's somatic growth curve is also similar to the ancestor's, but the period of slow somatic growth is extended until brain growth terminates. Both the ancestor and its larger-brained descendant end up at the same adult body size, but the descendant reaches it later.

ously stunted, any surplus energy is devoted to the soma (E_{sa}) until body size has returned to its optimal trajectory.

The mean and variability of E are the main factors that select for the level of E_b during brain growth. In particular, natural selection is expected to set E_b only as high as allowed by the minimum reliable E, so that brain damage will be avoided even during periods of resource scarcity. E_{so}, in turn, will be set according to the requirement of maintaining a functional balance between brain and body size. So, although there may be some circumstances in which animals use E fully (e.g., food shortages, during catch-up growth), body growth will be generally limited by brain growth, not by total available energy. Nevertheless, once brain growth is complete, body growth may occur as rapidly as available resources permit, often resulting in growth spurts (Leigh 1996; Bogin 1999).

The evolutionary link between a slower, longer body growth trajectory and a larger adult brain size follows from ontogenetic consideration of how a descendant can achieve a larger adult brain size than its ancestor. Essentially, there are two potential evolutionary pathways by which this can occur. First, the brain may grow faster, allowing the descendant to reach a larger adult brain size without changing the ancestral duration of brain growth. Second, as illustrated in figure 10.4, the duration of brain growth may be prolonged, so that the ancestral brain growth rate eventually produces a larger brain (hypermorphosis: McKinney and McNamara 1991; McKinney 1998; see also Gould 1977). The first pathway—accelerated brain growth—

requires that more energy be allocated to E_b at any given time. Unless the minimum reliable E also increases in the descendant, which is generally unlikely (see below), this change will increase its risk of facing an energy crisis and suffering long-lasting brain and behavioral defects. Hence, if there is no evolutionary change in the somatic growth trajectory, an evolutionary increase in brain size is improbable. The second ontogenetic pathway for gaining a larger brain—extending the ancestral brain growth trajectory—is more likely because it is less risky. The risk of brain damage is low because the energy devoted to E_b at any given time remains the same as in the ancestor. Because of the functional balance requirement, an extended period of brain growth is possible only if there is an extended period of slow body growth. Therefore, on an evolutionary scale, a slower, extended body growth trajectory should increase the likelihood of an evolutionary increase in brain size.

We suggest that this argument applies throughout mammalian development, but that the degree of maternal dependence is an additional, crucial factor. In particular, during gestation and lactation (i.e., when all or most energy comes from the mother), E will tend to be high and subject to relatively less variability. This reflects the fact that adults tend to be large, efficient, and consistent foragers and may also have access to nutritional reserves. Thus, the mother serves as an energetic buffer for her offspring. Just after weaning, however, the offspring's potential energy intake will be low and variable, because it will have no foraging experience and its brain and body will not yet have achieved adult size. As the animal's brain size, body size, and experience all increase, E will increase and become less variable. In sum, the higher and more consistent energy supply during the period of maternal dependence can explain why brain growth is most rapid early in development, and why juvenile somatic growth is reduced relative to preweaning levels in animals with large brains (Janson and van Schaik 1993; Bogin 1999).[4]

The brain malnutrition risk model can be viewed as an extension of Janson and van Schaik's (1993) juvenile risks model, which holds that primate juvenility should be primarily viewed as a period of slow, conservative growth, serving to maximize survivorship to breeding age. However, unlike the juvenile risks model, the brain malnutrition risk model is able to explain why primates do not markedly increase their growth rates when food is abundant.

One objection to the brain malnutrition risk model can be derived from the "expensive tissue" hypothesis, which holds that the energetic costs of maintaining a larger brain can be paid by reducing the size of the gut, another metabolically expensive organ (Aiello and Wheeler 1995; but see

Hladik, Chivers, and Pasquet 1999). Specifically, one could argue that when E_b is raised, E_{so} could be lowered in a corresponding fashion while still achieving the ancestral body size. In other words, growing to and maintaining a certain body size may be cheaper for a larger-brained organism because a smaller gut consistently accompanies a larger brain (Foley 1995). This is an interesting idea, but it does not seriously weaken the model, the central point of which is not simply that if more energy is required to grow the brain, less must be available to grow the body. Instead, the model holds that body growth trajectories are usually determined by conservative brain growth trajectories, not by total E.

A second potential objection is that the evolution of a larger brain size is invariably accompanied by an increase in E or a reduction in E's variance. If so, then a larger-brained animal would be able to allocate more energy to both E_b and E_{so} and thus achieve a larger adult brain size without any change in life history. In this view, the proposed model becomes irrelevant because one of its assumptions—that E is unchanged by an increase in adult brain size—is met so rarely. Martin (1981, 1996) has argued similarly, noting that maternal basal metabolic rate (BMR), a measure that may be linked to E, is tightly linked to neonatal brain size. Although this neonatal brain size-maternal BMR relationship holds in primates (Martin 1996; R. A. Barton, unpub.), it does not generalize to nonprimates (Pagel and Harvey 1988a). Even more importantly, once the effects of body size have been controlled, BMR appears to be unrelated to adult brain size (McNab and Eisenberg 1989; Allman, McLaughlin, and Hakeem 1993a; Martin 1996; Barton 1999). Hence, there is little evidence that changes in adult brain size are achieved by altering E.

Therefore, we conclude that the brain malnutrition risk model is based on reasonable assumptions and is not susceptible to two potentially damaging criticisms. To further test its validity, we should derive and test additional predictions.

The first prediction of the model is that evolutionary changes in brain size should co-occur with evolutionary changes in the duration of brain growth; evolutionary changes in the rate of brain growth, by contrast, should be less common. Unfortunately, this prediction is presently difficult to test, despite the fact that patterns of brain and body growth have now been documented in approximately ten mammalian species (humans, chimpanzees, rhesus monkeys, dogs, cats, rats, mice, guinea pigs, rabbits, pigs: Holt et al. 1975; Dobbing and Sands 1979; see also Pereira and Leigh, chap. 7, this volume). The problem is that the model applies mainly to closely related species that share broadly similar body sizes and foraging patterns; in these cases, E is likely to be similar. The species in which brain and body growth patterns have been documented, however, are highly diverse in these characteristics and thus cannot offer reasonable tests. Nevertheless, one relevant comparison is clearly in agreement with the prediction: chimpanzees versus humans. These closely related species are similar in body size and BMR (Aiello and Wheeler 1995), but have dramatically different adult brain sizes. Crucially, they have highly similar early brain and body growth trajectories; the difference in adult brain size is due to the fact that humans greatly extend the most rapid period of brain growth and reach adult brain and body size much later (Count 1947; Passingham 1975b; Holt et al. 1975; McKinney and McNamara 1991).

A second prediction of the model is that if resources are abundant, a somatic growth spurt will occur shortly after the termination of brain growth. This prediction follows because, according to the model, the end of brain growth permits body growth to begin closely tracking available resources (E) (see also Leigh 1996). To our knowledge, information on both brain and body growth trajectories is available for only four taxa that show somatic growth spurts: macaques, mangabeys (*Cercocebus* spp.), chimpanzees, and humans.[5] Nevertheless, these limited data support the prediction. Rhesus macaques complete brain growth just after their first birthdays (Cheek 1975, fig. 1.7) and show somatic growth spurts as early as eighteen months (Leigh 1996, fig. 5). Mangabeys complete their brain growth between three and four years of age (Pereira and Leigh, chap. 7, this volume) and show growth spurts at the same age (Leigh 1996, fig. 6). Chimpanzees complete their brain growth between four and five years of age (Count 1947, fig. 4 and table 2), and their growth spurts begin at approximately five years (Leigh 1996, fig. 8). In well-fed humans, the adolescent growth spurt begins between ten and twelve years of age (Bogin 1999), and the vast majority of human brain growth is completed by eight or nine years, although there are slight increases until the late teens (Dobbing and Sands 1973; Dekaban and Sadowsky 1978).

Because its underlying model is reasonably well supported, the brain malnutrition risk hypothesis could be correct, and it is therefore worthwhile to consider predictions that follow directly from it. The main prediction is that there should be a modest positive correlation between evolutionary changes in brain size and evolutionary changes in the duration of somatic growth. Although there is relatively little cross-species data on growth durations, age at first reproduction is probably an excellent first approximation. Thus, the comparative tests performed above bear on the model: the fact that we found no robust correlations in any mammalian taxa fails to support the prediction.

Delayed Benefits Hypothesis

Modern life history theory holds that animals delay reproduction only if the costs of doing so are eventually offset by benefits (Williams 1957; Gadgil and Bossert 1970; Harvey, Read, and Promislow 1989; Stearns 1992; Charnov 1993). Brain investment is an excellent example of this trade-off (the immune system could be another). In particular, even if it is physiologically possible to quickly grow a large brain (see the maturational constraints hypothesis) and the energetic risks of rapid brain growth can be borne (see the brain malnutrition risk hypothesis), investing in a large brain still makes less sense for an animal with a fast life history than for an animal with a slow life history. The reasoning is as follows: A primary benefit of a large brain is that it allows an animal to develop experientially based skills or knowledge that eventually lead to fitness benefits. The development of these abilities, by definition, involves costs of time and risk. Hence, the delayed benefits of such abilities must exceed the costs of developing them. Life history is relevant because once the initial costs of learning are paid, the benefits may accrue indefinitely. Thus, long-lived animals have a greater opportunity than short-lived animals to take advantage of large brains and the learning they allow (cf. Dukas 1998; Kaplan et al. 2000). Therefore, an evolutionary increase in longevity should increase the likelihood of an evolutionary increase in brain size.

The crucial assumption of the delayed benefits hypothesis is that brain size is closely related to an animal's learning ability. As noted under the cognitive buffer hypothesis, this assumption has some support.

The first prediction of the delayed benefits hypothesis is that there should be a modest positive evolutionary correlation between brain size and life span. We have addressed this prediction under the neuronal investment and cognitive buffer hypotheses and found some evidence to support it, in both primates and nonprimates.

The second prediction of the delayed benefits hypothesis is that there should be a positive evolutionary correlation between neocortex size and life span. This prediction is again shared by the cognitive buffer hypothesis, and it likewise follows from the reasoning that the neocortex is especially implicated in learning and other higher-order cognitive processes. Although this prediction was supported in the analysis of old contrasts within primates, it was not supported at all among bats or insectivores. Thus, the main

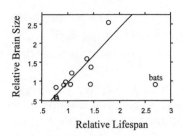

FIG. 10.5. The relationship between relative brain size and relative life span across eutherian mammals. Each point represents an "order average." The regression line is fit to all points save bats, which is a significant outlier.

assumption of the delayed benefits hypothesis is reasonable, and the comparative tests provide some support for its predictions.

Discussion
Did Brains and Life History Coevolve in Mammals?

Other than the brain size-life history correlation in primates, there is limited evidence for robust evolutionary relationships between life history variables and the size of the brain or brain structures. One might reasonably ask, then, "Is there truly a brain size-life history phenomenon?" We obviously do not yet know the answer to this question, but, for several reasons, we believe that the linkage is real.

First, although we found relatively few significant relationships, the correlation coefficients were, as predicted, overwhelmingly positive (e.g., a positive increase in brain size and a positive increase in longevity: see tables 10.5 and 10.6). The only exceptions to this pattern involved gestation length, a variable that is only weakly related to general fast-slow life history trends (see discussion of the maturational constraints hypothesis above; Read and Harvey 1989). It might be argued that the predominance of positive correlations reflects the fact that many of our analyses employed the same body mass estimates in calculating brain and life history residuals (i.e., the correlated errors problem). However, this criticism does not apply to the analyses involving primates, bats, and odontocetes.

Second, there appears to be a brain size-life span relationship across eutherian mammalian orders. Read and Harvey (1989) and Austad and Fischer (1992) did not report this relationship, but it is clear from plotting relative life span against relative brain size that the relationship is quite strong once one significant outlier—the bats—is removed (fig. 10.5; with bats: $r = .40$, $n = 12$, $p = .20$; without bats: $r = .86$, $n = 11$, $p = .0007$).[6] It could be argued that outliers should not be omitted, for they are the strongest refutation of a brain size-life history link. Although this argument applies to the "strong constraint" hypotheses, it does not pertain to any of the hypotheses that we deem plausible. The cognitive buffer hypothesis, for instance, is perfectly compatible with the possibility that some taxa evolve adaptations other than enhanced cognition to reduce their extrinsic mortality. Flying in bats is an excellent example.

Third, because our analyses strictly controlled for body size, they might have obscured or weakened genuine relationships between cognition (or other information-processing capabilities) and life history. The problem is that larger animals could systematically have greater cognitive capabilities than smaller ones (Deacon 1997), meaning that controlling for body size would be tantamount to controlling for cognition, the variable of interest (see Harvey and Pagel 1991). Of course, the reason that we (and others) controlled for body size is that the interrelationships among cognitive capabilities, brain size, and body size are unresolved, and the prudent approach at this stage is to ensure that neuroanatomical and life history relationships do not merely reflect body size. Nevertheless, if future studies (R. O. Deaner et al., unpub.) are able to establish the interrelationships among these variables, such studies will be likely to pave the way for stronger cognition and life history correlations.

Finally, it bears noting that we found at least some support for the brain size-life history linkage in primates, odontocetes, and carnivores, but not in bats or insectivores (tables 10.5 and 10.6). The former taxa are relatively large-brained, whereas the latter are relatively small-brained (Jerison 1973; Austad and Fischer 1992; Marino 1998). Similarly, within primates, the brain size-life history linkage appears to be far stronger in haplorhines than in the relatively small-brained strepsirrhines (Allman, McLaughlin, and Hakeem 1993a; Hakeem et al. 1996; Judge and Carey 2000; see also Kappeler 1996). Thus, it is possible that the forces that produce the brain size-life history linkage operate only in large-brained taxa.

The Competing Hypotheses Are Compatible

Correlations among life history variables are widely known and theoretically expected (Williams 1957; Gadgil and Bossert 1970; Harvey, Read, and Promislow 1989; Stearns 1992; Charnov 1993). Perhaps most basically, an animal with a long life expectancy can afford to delay its reproduction. Thus, despite the fact that each of the candidate hypotheses bears most directly on either maturation or longevity, any one of them could potentially explain virtually all associations between brain size and life history. For instance, if the cognitive buffer hypothesis holds, the following scenario might unfold: an evolutionary increase in brain size is followed by an evolutionary increase in life span, which is followed by an evolutionary increase in age at first reproduction. Therefore, although it is desirable to set up comparative tests so that the hypotheses may compete (i.e., make differing predictions about the importance of particular variables), it should be remembered that such tests are unlikely to be decisive (Harvey, Martin, and Clutton-Brock 1987). We especially urge caution when investigators employ variables that are likely to differ in their degree of error (i.e., maximum longevity vs. age at first reproduction).

Another important point is that, although the hypotheses might be viewed as competing to explain variation in brain size-life history associations, two or more explanations could be correct (see Kaplan et al. 2000). The maturational constraints and cognitive buffer hypotheses, for example, are complementary. According to the maturational constraints hypothesis, a lineage undergoing selection for increased brain size would be forced to postpone sexual maturation. The cognitive buffer hypothesis would then predict that the larger brain would promote the evolution of greater longevity. The brain malnutrition risk and delayed benefits hypotheses are similarly compatible. The former holds that only when the somatic growth period has been extended can animals afford to grow large brains with minimal risk, whereas the latter argues that only when life span is extended do the delayed benefits of large brains favor their evolution. In this way, the evolution of a slower life history could favor increased brain size from the point of view of both costs and benefits.

The fact that the plausible hypotheses are compatible should not deter attempts to falsify them individually. Clearly, it is unlikely that all of them are important, and it is certainly desirable to know which ones hold, and in which taxa. As we have noted, future comparative tests of their predictions, although important for investigating the generality of the brain size-life history phenomenon, are unlikely to reject any of the plausible hypotheses conclusively. We recommend, therefore, that investigators also pay attention to testing the assumptions unique to each hypothesis.

Summary and Conclusions

In the past decade, the link between brain evolution and life history evolution has generally been deemphasized (e.g., Read and Harvey 1989; Harvey, Read, and Promislow 1989; Austad and Fischer 1992; Kappeler 1996; but see Allman 1995; Hakeem et al. 1996). This can be attributed to two factors. First, there existed uncertainty as to whether the brain size-life history correlations would remain once important confounds—body size, phylogeny, socioecology—were removed. Here we have shown that in primates, there is evidence that these correlations are robust, at least in the case of life span. The second reason that the brain-size life history correlations were deemphasized was that there existed a shortage of sound theoretical ideas to explain them. In particular, "strong constraint" hypotheses were discussed most frequently, and, as we have noted, these hypotheses are very unlikely to hold. Even more importantly, these hypotheses were almost invariably viewed as candidates to explain life history variation in toto. As we have shown, however, there are four plausible, even compelling, hypotheses about brain size-life history associations that are fully compatible with modern life history theory. These hypotheses make a variety of predictions, and we have shown that some of them are supported. In the future, investigators should focus on testing the assumptions of the contending hypotheses, determining the generality of the brain size-life history connection, and exploring the theoretical implications of this linkage.

Acknowledgments

We thank Vicky Horton and Rachel Lutwick-Deaner for assistance in entering data, David Haring for providing data on the maximum life span of Duke University Primate Center animals, and Steve Austad for sharing unpublished maximum life span data. For comments on previous versions of this chapter, we thank John Allman, Atiya Hakeem, Debra Judge, Peter Kappeler, Steve Leigh, Michael Pereira, and two anonymous reviewers.

Notes

1. Sexual dimorphism is a potential complication in assigning species values for brain and body mass. Rather than treating males and females separately (Deaner and Nunn 1999; see also Prothero and Jürgens 1987), we calculated species values as the mean of the mean male value and the mean female value. For several species, however, the only available brain mass data came from specimens of unknown sex. A species value calculated in this way could be very misleading for sexually dimorphic species. However, for species with minimal sexual dimorphism (values between 0.9 and 1.1 for two-step ratio: Smith 1999), this should not be a serious issue, and hence for *Callimico goeldi* and *Eulemur rubriventer,* we calculated species averages from specimens of unknown sex. An additional concern in compiling the data set was whether to include *Homo.* We followed other workers in doing so, because there is no clear biological justification for omitting it, save that it would clearly support the hypothesized relationship (i.e., humans have longer lives and larger brains than any other primate species). Nevertheless, the relationship between life span and brain size shown in table 10.2 remains significant even with *Homo* excluded (based on different body size estimates: $r = .34, p = .011; n = 55$).

2. The converse method of performing this analysis is to use the Smith and Jungers (1997) data to calculate brain size residuals and the Stephan, Frahm, and Baron (1981; H. Stephan, H. D. Frahm, and G. Baron, unpub.) data to calculate life span residuals. We did this, and also repeated the analysis using the Smith and Jungers (1997) data to calculate both residual types. We again found that in both kinds of analyses, all three life history

variables were significantly correlated with brain size, although the relationships were, as expected, weaker when the residuals were calculated with different body size values.

3. In these and the subsequent analyses described in this section, we employed ICs as described above. The only difference was that, except in the case of bats, primates, and odontocete cetaceans, we were unable to employ different body mass estimates in calculating brain and life history residuals. Data on brain and body mass came from the following sources: bats: Baron, Stephan, and Frahm 1996; insectivores: Stephan, Baron, and Frahm 1991; carnivores: Gittleman 1986b; odontocetes: Marino 1998, Whitehead and Mann 2000. Data on neocortex and cerebellum volumes came from the following sources: primates: Stephan, Frahm, and Baron 1981; bats: Baron, Stephan, and Frahm 1996; insectivores: Stephan, Baron, and Frahm 1991. Data on maximum recorded life span came from the following sources: bats and insectivores: Austad and Fischer, unpub. (see S. N. Austad and K. E. Fischer 1991, 1992); carnivores: Gittleman 1986a, S. N. Austad and K. E. Fischer, unpub.; odontocetes: Whitehead and Mann 2000, Nowak 1991. Data on age at first reproduction came from the following sources: bats: Jones 1998; insectivores: Hayssen, van Tienhoven, and van Tienhoven 1993; carnivores: Gittleman 1986b; odontocetes: Whitehead and Mann 2000, Hayssen, van Tienhoven, and van Tienhoven 1993; Nowak 1991. Data on gestation length came from the following sources: bats: Jones 1998; insectivores: Eisenberg 1981, Parnham 1998; carnivores: Gittleman 1986a,b; odontocetes: Whitehead and Mann 2000, Marino 1997. Phylogenetic information came from the following sources: bats: Jones 1998; insectivores: R. Grenyer, pers. comm.; carnivores: Binida-Emonds, Gittleman, and Purvis 1999; odontocetes: Heyning 1997, Rice 1998, Le Duc, Perrin, and Dizon 1999.

4. We are, of course, explaining the model in very general terms here: other natural history characteristics will also be necessary for fully understanding brain and body growth trajectories. Pereira and Leigh (chap. 7, this volume) discuss several relevant examples, including the contrasting brain growth trajectories of squirrel monkeys and saddleback tamarins.

5. Pereira and Leigh (chap. 7, this volume, fig. 7.6) provide data on brain growth for baboons (*Papio* spp.); however, even an approximate point of growth termination is not yet clear, save that it is probably earlier than for mangabeys.

6. Order averages for relative brain size and relative life span were calculated from species values of "EQ" and "LQ" reported by S. N. Austad and K. E. Fischer (unpub.; see Austad and Fischer 1992, table 1). Bats are significant outliers in a Mahalanobis distance plot (calculated in JMP, SAS Institute, Cary, NC). Once bats are removed, primates become a significant outlier (i.e., Mahalanobis jackknife distance plot), but the relationship between relative brain size and relative life span remains significant when both bats and primates are excluded ($r = .76, n = 10, p = .01$). These analyses do not account for phylogeny or for the possible problem that body size error could affect both relative values of interest (see fig. 10.1), and hence must be considered preliminary.

11 Life History, Infant Care Strategies, and Brain Size in Primates

CAROLINE ROSS

Some of the most salient features of primate life histories are their relatively slow growth rates and late ages at maturity compared with mammals of similar body weight (Ross 1988; Charnov 1993). Previous work (reviewed below) has suggested a number of correlates of late maturation times and/or slow maturation rates in primates and other mammals. In particular, relative brain weight has been found to be an important predictor of age at first reproduction. Additionally, some previous work has indicated a link between infant care strategies and life history parameters in primates. In this chapter I look at the relationships between age at maturity, growth rates, brain weight, and infant care strategies in primates.

Delayed Maturation

Explanations of the relatively late maturation times found in primates have either been adaptive (i.e., postulating that there is an advantage to late maturation) or have suggested that late maturation arises as a result of some constraint or trade-off—in other words, that late maturation is not itself an adaptation, but arises as a consequence of other adaptations. In the latter case, late maturation is frequently suggested to be correlated with slow postnatal growth rates, either during infancy (between birth and weaning) or during the juvenile period (between weaning and maturity).

Adaptive explanations of delayed maturity usually concentrate on the advantages of a long juvenile period for animals that need to learn particular skills in order to survive and reproduce. Skills that might be learned during this time could include infant care and foraging. Experience of infant handling, for example, leads to better infant survival in primipara vervet monkeys (Fairbanks 1990); however, there is no clear association between previous exposure to infants and maternal behavior in *Macaca fascicularis*

(Timmermans and Vossen 1996). Janson and van Schaik (1993) suggest that, although experience may help to improve foraging efficiency, the main reason that juvenile primates are less able to forage than adults is because of their relatively small size.

Adaptive changes in age at maturity may also arise as an adaptive response to variation in mortality rates. Some models of life history evolution suggest that juvenile mortality rates are expected to correlate negatively with age at maturity (Charnov 1993), as animals with a high mortality rate will be selected to mature early so that they can reproduce before dying. This relationship between mortality rate and age at maturity is found for mammals (Purvis and Harvey 1995), although similar analyses for primates give ambiguous results, possibly because the mortality data are not of high quality (Ross and Jones 1999). The relationship between adult and juvenile mortality rates may also be predicted to have an influence on the evolution of age at maturity. For example, when juvenile mortality is high relative to adult mortality, delayed maturation may evolve (Stearns 1976; Pereira 1993b).

A number of studies also suggest that maturation times and/or rates may vary because of adaptive changes in or constraints on rates of growth—that is, that slow maturation may be due to slow growth rates. Maturation may be delayed if either infant growth rate (i.e., growth from birth to weaning) or juvenile growth rate (from weaning to maturity) is slowed for any reason.

Janson and van Schaik (1993) suggest that a slowing of juvenile growth rates, and hence a delay in maturation, may be adaptive in some circumstances. They note that juveniles are often found to have lower foraging success than adults in the same population, and that this low foraging success leads to juveniles having to spend more time foraging than do adults, as well as making them more susceptible to malnutrition or starvation when there are food shortages. They also present evidence suggesting that juveniles are more susceptible to predators than are adults. If the main reason for the foraging and predation risk problems faced by juvenile primates is their relatively small size, then these problems might be overcome by growing more rapidly to adult size. However, juvenile primates appear to grow more slowly than they are physiologically capable of doing, suggesting that there may be an adaptive advantage to slow growth. Janson and van Schaik suggest that juvenile primates grow slowly in order to cope with the foraging competition that comes with group living. Trade-offs between time spent foraging and predation avoidance are also discussed by Janson (chap. 5, this volume). The disadvantages of slow postnatal growth rates may be outweighed by the advantage of lowered mortality rates resulting from a reduction of predation risk. In Janson and van Schaik's model, folivorous diets are predicted to select for a high postnatal growth rate (because leaves are easy to find, so increased foraging for them has a smaller influence on mortality risk than does increased foraging for fruit or animal foods). There is some support for a link between diet and growth rates (Leigh 1994a), although another study did not find a clear link between maturation rates and diet (Ross and Jones 1999).

Another constraint on developmental rates may be the energetic costs of specific body organs. This possibility was first suggested by Sacher (1959) in his work on life span and brain size and has since been supported by a number of studies (these ideas are discussed further by Deaner, Barton, and van Schaik in chapter 10 of this volume, in which they also review work linking brain weight to life span in mammals). The brains of primates are large relative to those of other mammals of the same body weight, and the brain is an organ that is known to need a high energetic input (Martin 1983, 1996). Charnov and Berrigan (1993) suggest that large brain weight in primates may result in slow somatic growth and late maturation. More recent comparative analyses support this idea, as brain weight and age at first reproduction are strongly correlated in primates after controlling for body weight (e.g., Barton 1999). Slow growth rates in other taxa without large brains will need alternative explanations, as they may have other correlates (Rubenstein 1993; Pereira 1993a).

Ross and Jones (1999) used comparative analyses to investigate correlates of late maturation within the haplorhine primates. Their results show a correlation between relative brain weight and relative age at maturity, supporting the idea that maintaining a relatively large brain may have an energetic cost. However, it is unclear whether that energetic cost is due to the costs of supporting a large brain or the costs of brain growth (or both). Ross and Jones suggest that primates with relatively large brains may use a number of strategies to mitigate these costs and thus increase infant growth rate. These strategies may include social strategies (e.g., gaining high rank) that result in increased resources being available to mothers. Altmann and Samuels (1992) calculate that infant carrying in baboons may increase the energetic costs of infant care by 5% in the first month of life. Hence, mothers may also benefit from help with infant care, as this may give them increased foraging time and may also reduce the mother's energetic costs. Comparative studies support this idea, as species with high levels of nonmaternal care have higher infant growth rates, and higher birth rates, than do species in which levels of nonmaternal infant care are low (Ross and MacLarnon 1995; Mitani and Watts 1997; Ross and MacLarnon 2000). Although time spent in infancy may have an effect on age at maturity, it is clear that the length of the juvenile period is also likely to influence age at maturity. However, the above studies do not examine the link between age at maturity, juvenile growth rate, and brain size. That link is examined in this chapter.

Variation in Infant Care Strategies

The infant care strategies observed in mammals vary considerably both within and between orders. There is variation in (1) who cares for infants (mother, father, siblings, other kin or nonkin); (2) the age at which infants become independent; and (3) the type of contact between caregiver and infant. Although several primate species have some nonmaternal care (allocare), the majority of care in most primate species, as in other mammals, is carried out by the mother (Nicolson 1987; Wright 1997; Ross and MacLarnon 2000). The reasons for this variation have been discussed previously (Chism 2000; Ross and MacLarnon 2000). The length of time that offspring are cared for, which may be measured as age at weaning or some other benchmark of independence, also varies. Variation in these parameters has been investigated in many mammalian taxa, both allometrically and in relation to ecological adaptation (Case 1978; Stearns 1993; Derrickson 1992). Discussion of the factors influencing weaning age and infant postnatal growth in primates can be found in a number of papers (Lee and Bowman 1995; Ross and MacLarnon 1995; Lee 1996, 1999).

The third source of variation in infant care strategies is the type of contact between the infant and the mother, father, or other caregivers (all referred to as "caregivers"). The caregivers may remain in close contact with an infant for most of the time, or they may leave the infant for extended periods of time in order to forage. Infants that remain with their caregivers may do so either by following them as they move around or by being carried. Infants that are left for extended periods are usually hidden from predators, either in a specially constructed nest or den or in vegetation. As no primates have highly precocial young that can follow them from birth, primates show variation in five aspects of infant contact behavior: (1) whether infants are carried orally; (2) whether infants are carried clinging to fur; (3) whether infants are left in nests or other shelters; (4) whether infants are left in vegetation ("parking"); (5) whether allocare is present or absent. As animals could potentially show any combination of these behaviors, there are thirty-two potential infant care strategies. Of these, only nine are found in the primate species for which I could find data (tables 11.1 and 11.2).

Table 11.1 Infant care strategies observed in primates included
in the data set

Infant care strategy	Behaviors				
	Oral	Cling	Nest	Park	Allocare
1	+	−	+	+	−
2	+	−	−	+	−
3	+	+	+	+	−
4	+	+	−	+	−
5	+	−	−	+	−
6	+	+	−	−	−
7	−	+	−	+	+
8	−	+	−	−	+
9	−	+	−	−	−

Note: Species are classified as either showing or not showing each of five behaviors:
Oral = oral carrying of infants; Cling = carrying infants clinging to fur; Nest = leaving infants in
nests or other protected shelters; Park = leaving infants "parked" in vegetation; Allocare = infant
is carried or protected by an individual other than the mother for more than 5% of the total care
time. (+) = the behavior is seen; (−) = the behavior is absent.

Infant Care Strategies and Life Histories

Several previous papers have investigated correlations between infant care
strategies and life history patterns. Across mammals, the differences between those that have altricial and those that have precocial young have
been widely discussed (Eisenberg 1981; Martin and MacLarnon 1985,
1988, 1990; Martin 1990). Correlations between the developmental state of
the young and ecological and life history variables indicate that species
with poorly developed (altricial) young characteristically have small body
weights and have (relative to their body weight) short gestation periods,
large litter sizes, rapid maturation, and small brains that grow rapidly after
birth; they also have limited motor and thermoregulatory abilities at birth,
are nocturnal, and use nests. Species with more precocial young tend to have
the opposite set of characteristics (Martin 1990). Within the primates, differences between species with different infant care strategies have also been
looked at. For example, Kappeler (1998a) found links between low body
weight, the use of nests, and large litter size in strepsirrhine primates, although he did not find a link between growth rates (prenatal or postnatal)
and the use of infant shelters. Tilden and Oftedal (1997) found a link between the milk composition of prosimian primates and their infant care
strategies: species that leave their young for long periods of time have rich,

Table 11.2 Data used in this analysis

Species	1	2	3	4	5	6	7	8	9	10	11	12
Allenopithecus nigroviridis	9	—	—	3225	57	—	—	—	—	—	—	—
Alouatta caraya	8	11	8	4882	—	187	—	—	—	—	3.71	—
Alouatta palliata	8	—	12	5824	51.5	186	528	630	1100	1.36	3.58	1.35
Alouatta seniculus	8	8.4	7.1	5807	—	191.3	—	372	—	3.5	4.58	3.04
Aotus trivirgatus	8	91.6	3.8	724	17.2	133	97	75	360	3.5	2.38	1.81
Arctocebus calabarensis	5	—	—	254	—	134	25.2	105	160	2.14	1.12	.47
Ateles fusciceps	9	—	—	9163	108.8	226	—	486	—	6	4.86	2.91
Ateles geoffroyi	9	—	15	7669	104.9	225	426	821	2000	1.92	5.62	2.76
Ateles paniscus	9	—	18.2	8554	101.2	—	452.5	760	3790	4.39	5	—
Avahi laniger	9	—	2	875	10.6	—	—	—	—	—	—	—
Brachyteles arachnoides	9	—	20.9	8070	—	—	—	638	—	—	7.5	—
Cacajao calvus	—	—	17.4	3600	—	180	—	638	—	—	—	—
Callicebus moloch	8	85.1	3.3	1004	—	163.5	100	60	—	—	3	2.39
Callimico goeldii	8	70.5	7.2	582	11	155	50.6	70	215	2.95	1.32	.7
Callithrix argentata	8	—	5.5	353	—	—	70.2	—	—	3.4	1.67	—
Callithrix jacchus	8	46.7	11	287	8	148	56.7	90	128	2.98	1.5	.85
Cebuella pygmaea	7	57	2	79	—	137	31.5	90	70	.9	1.88	1.25
Cebus albifrons	8	—	25.1	2067	60.7	155	234	270	—	4.7	4.02	2.85
Cebus apella	8	—	12.9	2201	63.5	155	239.7	264	1000	2.87	5.5	—
Cebus capucinus	8	—	15.1	2578	71.9	162	230	516	1350	2.17	4	—
Cebus olivaceus	8	—	18.2	2520	—	—	—	731	—	—	6	—
Cercocebus albigena	8	—	15.8	6209	94.4	175	425	365	2170	5	4.08	2.61
Cercocebus galeritus	9	—	18.2	5473	88.8	171	—	—	—	4.04	6.5	—
Cercocebus torquatus	9	—	26.9	7420	110.5	171	—	—	—	—	4.67	—
Cercopithecus aethiops	8	19.6	19.5	3469	60.1	163	314	365	1170	4.2	5	3.55
Cercopithecus ascanius	—	—	28.8	2943	61.3	172	371	—	—	—	5	—
Cercopithecus campbelli	—	—	10.5	2200	—	—	—	365	—	—	—	—
Cercopithecus cephus	—	—	11.5	2805	—	170	340	—	—	2.87	5	—

continued

Table 11.2 continued

Species	1	2	3	4	5	6	7	8	9	10	11	12
Cercopithecus diana	—	—	30.2	4533	—	—	475	365	—	—	5.42	—
Cercopithecus lhoesti	—	—	17.4	4700	84.5	—	—	699	—	—	—	—
Cercopithecus mitis	—	—	18.6	4280	56.5	—	402	—	—	3.5	5.92	—
Cercopithecus mona	—	—	—	2500	62	—	284	—	—	—	—	—
Cercopithecus neglectus	8	0	4.8	4081	—	165	260	365	1640	3.2	4.67	3.22
Cercopithecus nictitans	—	—	16.6	4216	—	170	—	—	—	2.4	5	—
Cercopithecus pogonias	—	—	15.5	3021	—	170	340	—	—	2.57	5	—
Cheirogaleus major	1	—	1	403	7	70	—	70	—	—	—	—
Cheirogaleus medius	1	—	1	173	—	62	38	61	—	2.64	1.19	.86
Colobus badius	9	—	26.9	7421	80.2	—	—	790	—	—	4.08	—
Colobus guereza	8	—	12	8102	72	—	395.5	390	1600	3.6	4.75	—
Colobus polykomos	8	52	13	7662	81.4	—	597	219	1240	6	8.5	2.04
Daubentonia madagascariensis	1	—	1.5	2800	44	169	109	365	1535	3.81	3.5	—
Erythrocebus patas	8	10.3	28.2	6317	100	168	504.5	213	1950	6.78	3	1.96
Euoticus elegantulus	4	—	4	283	—	—	—	—	—	—	—	—
Galago alleni	1	—	6	262	—	135	31.2	—	—	1.82	1.04	—
Galago moholi	1	—	6	179	5	124	18.4	100	95	1.66	1	.39
Galago senegalensis	1	—	—	195	—	142	20.3	98	150	1.75	1.4	.74
Galagoides demidoff	1	—	10	63	3.3	110	9	—	—	1.07	.97	—
Galagoides zanzibaricus	1	—	3.5	132	—	125	16.8	—	—	1.73	1	—
Gorilla gorilla	9	—	7.1	82475	457.3	260	2122.9	1004	—	15.32	10.04	6.58
Hapalemur griseus	3	—	4.5	790	—	140	56.7	122	—	—	2.38	1.67
Homo sapiens	7	32.4	—	55000	1228	267	3375	731	10980	10.7	14	11.27
Hylobates agilis	9	—	4.4	5530	87.6	—	—	—	—	—	—	—
Hylobates concolor	9	—	7.6	5749	—	202	—	—	—	—	—	—
Hylobates hoolock	9	—	3.5	6500	—	—	—	—	—	—	—	—
Hylobates lar	9	0	3.3	5464	93.9	213	400	730	1070	.9	9.31	6.73
Hylobates moloch	9	—	3.5	5292	93.1	195	—	—	—	—	9	—
Hylobates syndactylus	8	50	4	10568	130	232	517	—	—	—	9	—

Indri indri	9	—	3.2	6250	38.3	160	300	365	—	—	—
Lagothrix lagotricha	9	—	33.1	5585	82.6	223	450	315	—	9.15	5
Lemur catta	9	—	17	2290	22.6	135	103	105	—	2.56	2.01
Leontopithecus rosalia	8	50.8	5	559	12.3	128	100	90	165	2.65	2.38
Lepilemur mustelinus	1	—	1	602	—	135	34.7	75	—	.76	1.88
Loris tardigradus	5	—	2	255.5	—	167	11.4	169	139	—	1.5
Macaca cyclopsis	9	—	—	6200	99.8	162	—	207	—	4.55	—
Macaca arctoides	9	4.1	27.5	8523	62.1	178	487	398	2300	5.86	3.84
Macaca fascicularis	9	0	53.7	3574	—	160	345	420	1700	6.27	3.86
Macaca fuscata	9	2.2	40.7	9100	84.7	173	496	365	2730	4.39	5.54
Macaca mulatta	9	—	18.2	5445	110	165	475	365	1454	2.6	4.5
Macaca nemestrina	9	—	47	5571	97.5	167	472	365	1417	3.53	3.92
Macaca nigra	9	0	34.7	4600	—	172	455	—	—	4.37	5.44
Macaca radiata	9	0	20.9	3700	—	162	404	365	2000	—	4
Macaca silenus	9	—	24.5	5000	—	180	407	365	—	—	4.9
Macaca sinica	9	13	18.2	3590	—	—	—	—	—	4.28	5
Macaca sylvanus	8	—	—	8283	—	165	—	—	2420	3.34	4.8
Macaca thibetana	8	—	17.4	12800	124	170	—	568	2400	—	5
Mandrillus leucophaeus	9	—	22.4	8450	—	173	—	—	—	6.53	4
Mandrillus sphinx	9	—	—	11350	1.8	175	613	350	3000	1.35	1
Microcebus murinus	1	—	1	72	37.3	60	12.3	40	—	1.1	4.38
Miopithecus talapoin	9	—	114.8	1120	—	162	175	180	450	—	1
Mirza coquereli	1	—	1	302	87	87	22.5	—	—	6.93	4.5
Nasalis larvatus	8	—	12	9593	12.8	166	525	213	2000	5.41	2.11
Nycticebus coucang	5	—	—	630	—	193	49.3	180	520	10.16	2.21
Otolemur crassicaudatus	3	—	6	1120	9.8	136	70.7	134	500	—	—
Otolemur garnettii	1	0	3	739	—	132	49.7	—	—	—	—
Pan paniscus	9	3.8	85.1	33200	—	—	—	1094	8500	6.71	1.58
Pan troglodytes	9	1.3	63.1	40300	380	235	1162	1702	8500	3.97	13
Papio cynocephalus	9	—	55	11532	164	173	1742	365	2500	5.1	5.5
Papio hamadryas	9	—	66.1	10404	142.6	170	854	—	3100	—	6.1

continued

Table 11.2 continued

Species	1	2	3	4	5	6	7	8	9	10	11	12
Papio papio	9	—	—	16166	192	187	—	—	—	—	—	—
Papio ursinus	9	—	34.7	14773	164.5	187	—	—	—	—	3.67	—
Perodicticus potto	5	—	2	935	13.5	194	51.2	150	—	4.73	2.03	1.09
Petterus fulvus	9	—	8.9	2428	—	119	74.7	135	—	9.75	2.16	1.46
Petterus macaco	9	—	10	2428	23.8	128	100	—	—	—	2.18	—
Petterus mongoz	9	0	3	1890	32.2	128	55	152	—	2.66	2.52	1.75
Pithecia monachus	8	—	2.9	2170	—	—	121	—	—	—	—	—
Pithecia pithecia	9	0	2.7	1604	—	164	—	—	—	—	2.08	—
Pongo pygmaeus	8	0	2	37078	301.7	250	1735	1277	11000	7.26	9.68	7.88
Presbytis cristata	8	33.2	32.4	5856	66	200	500	396	—	—	3.42	1.78
Presbytis entellus	8	—	30.2	10280	—	—	380	—	2100	6.34	—	—
Presbytis obscura	8	—	10.2	6530	60.8	165	—	—	—	—	—	—
Presbytis phayrei	8	—	—	10500	—	200	360	274	—	—	—	—
Presbytis vetulus	8	—	7.9	5797	—	140	107	180	1100	2.7	3.5	—
Propithecus verreauxi	9	—	5	3183	26.7	210	—	—	—	—	—	2.62
Pygathrix nemaeus	8	—	15	8180	—	149	—	90	—	—	—	—
Saguinus fuscicollis	8	86.7	5.6	350	—	145	79.8	—	—	—	2.33	1.68
Saguinus labiatus	8	69.3	4.2	520	—	—	87	70	—	—	—	—
Saguinus midas	8	—	4.7	558	—	168	80	84	—	3.68	2	—
Saguinus nigricollis	8	—	6.8	350	—	171	87	50	175	3.5	—	—
Saguinus oedipus	8	68.5	7.9	425	9.6	178	83.6	168	130	1.78	1.89	1.29
Saimiri sciureus	8	30	38	699	23	—	95.2	80	418	.82	2.5	1.57
Tarsius bancanus	2	—	2	109	—	180	24.4	68	—	—	2.52	1.82
Tarsius spectrum	1	—	3	220	—	—	28.5	82	—	—	1.42	—
Tarsius syrichta	2	—	6	120	3.3	—	26.2	—	—	—	—	—
Theropithecus gelada	8	—	10	11427	119.7	170	553	547	3900	6.12	4	2.3
Varecia variegata	1	—	2.8	2700	31	102	175	90	1328	28.93	1.95	1.42

Note: 1 = infant care strategy (see table 11.1 for definitions); 2 = proportion of allocare; 3 = social group size (of breeding group); 4 = female body weight (g); 5 = female brain weight (g); 6 = gestation length (days); 7 = litter weight at birth (g); 8 = age at weaning (days); 9 = weight at weaning (g); 10 = growth rate of litter from birth to weaning (g/day); 11 = female age at first reproduction (years); 12 = length of juvenile period (years).

energy-dense milk compared with species that carry their young continuously and suckle them more frequently.

In haplorhine primates, allocare has been found in species with high infant growth rates and high birth rates (Ross and MacLarnon 1995; Mitani and Watts 1997). These studies suggest that mothers that have help with infant care are able to save resources that they would otherwise be investing in infant carrying and protection and invest them in growing infants rapidly to independence. There is also evidence to suggest that primate mothers that carry infants clinging to their fur have different life history patterns than those that leave their infants in nests or parked. After body weight and phylogenetic influences are taken into account, primates that carry infants for extended periods have later ages at weaning, and later ages at first reproduction, than do other species (Ross 2001). Again, this difference in life history pattern is probably best explained as a result of the energy constraints imposed on mothers that carry their infants with them. The later age at first reproduction also leads to species that carry clinging infants having a lower reproductive rate (as measured by the intrinsic rate of population increase) than parkers. However, precociality of infants is not correlated with nesting or parking behavior (Ross 2001).

In this analysis, I further explore the relationships between life history, brain weight, infant and juvenile postnatal growth rates, and social strategies. In particular, I address the following questions:

1. How do life history patterns and brain weight vary with the nine infant care strategies identified in extant primates?

2. Can the link between brain weight and age at maturity be explained by brain weight being correlated with either infant or juvenile growth rate, when other parameters are controlled for? Such a correlation might be expected if the slow growth rate of primates (compared with other mammals) is explained by their relatively large brain weights, as suggested by previous work (Charnov and Berrigan 1993; Ross and Jones 1999).

3. Do mothers with large-brained infants need more help than mothers with relatively small-brained infants? Ross and MacLarnon (1995) suggest that the relatively large brain weight of human infants may explain why humans grow relatively slowly during infancy despite having relatively high levels of allomaternal care. If this is the case, one might expect that after body weight and postnatal growth rates are controlled, brain weight might be positively correlated with levels of allomaternal care. Similarly, brain weight might be expected to have a negative relationship to infant growth rates after both body weight and infant care strategy are controlled.

Methods

Database

The data used in this analysis and their definitions are given in table 11.2. Body weight and life history data are taken from a data set collated from the literature by Ann MacLarnon, Bob Martin, Ben Rudder, and Caroline Ross. Data on infant care strategies are taken from compilations by Ross and MacLarnon (2000) and Ross (2001). Proportions of allocare are given as percentages of carrying time contributed by nonmothers (Ross and Mac-Larnon 1995).

Care was measured as the amount of time the infant was carried by caregivers. Species were classified as having allocare (i.e., placed in care groups 7 or 8) if they had more than 5% allocare (Ross and MacLarnon 1995). Quantitative data of this sort were found for thirty-one haplorhine species, but very little quantitative information on the amount of nonmaternal care infants received could be found for strepsirrhine species. Analyses that included the parameter "proportion of allocare" were therefore carried out for haplorhine species only. Species were also classified as having allocare or no allocare by estimating the amount of allocare from qualitative reports, and this classification was also done for strepsirrhine species.

Comparative Analyses

Because the life history variables examined in this analysis are known to be correlated with body weight (Harvey, Martin, and Clutton-Brock 1987; Ross 1988; Lee 1999), the effects of body weight were controlled in all cases. When multiple regressions were used, this was done by including female body weight in the analyses. For other analyses, body weight was removed by taking the residual values (or relative values) from the best-fit least-squares regression line.

All life history and body weight data were log-transformed before analysis, and data were analyzed and allometric methods carried out using least-squares regression. Least-squares regression underestimates the value of the slope of the best-fit line when there is error in the measurement of the X variable (as there will be in this data set) (Harvey and Mace 1982), but was preferred to major axis regression so that the results could be directly compared with those of other studies (Charnov 1993; Purvis and Harvey 1995).

Most analyses were carried out using a comparative method based on Felsenstein's (1985) method of independent contrasts. This method, "Comparative Analysis by Independent Contrasts" (CAIC), was used to control for the phylogenetic bias that results when species are treated as independent data points (Harvey and Pagel 1991). Because there have been criticisms of the use of CAIC analyses (Martin 1996; Ricklefs 1996; MacLarnon 1999), the analyses were also carried out using species values; however, these results are reported only when they differ from those found using independent contrasts or when analysis using CAIC could not be carried out.

All analyses using CAIC were carried out as detailed by Purvis and Rambaut (1995), using the primate phylogeny, including branch lengths, produced by Purvis (1995). Contrasts were analyzed using least-squares regression through the origin. One potential problem in using this method on life history data is that a large amount of variance occurs at the species level (Ross 1988) and at least some of this variance is due to random error. As there are more contrasts at the top levels of the phylogeny, these contrasts can have undue influence on the results (Purvis and Harvey 1995). Hence, the youngest third of the contrasts were removed before the analyses were carried out, and only the older contrasts were used (Purvis and Harvey 1995; Ross and Jones 1999).

Analyses of variation in infant care strategy could not be carried out using CAIC, as CAIC does not deal well with categorical data when there are more than two possible categories (Purvis and Rambaut 1995). Hence, differences in brain weight and life history between infant care strategies were tested by using an ANOVA on the residual values of these variables. A single ANOVA was carried out for each life history variable across all infant care strategies, and significant results are reported (table 11.3).

Results
Analysis of Infant Care Strategies
Species with different infant care strategies show some significant differences in brain weight and life history patterns (table 11.3). Species with infant care strategies 1, 8, and 9 have relatively short gestation periods when compared with other care groups, whereas those with care strategies 2 and 5 have relatively long gestation periods. However, these differences in gestation length do not lead to similar differences in litter weight or prenatal growth rates. Whereas the short gestations of care groups 1 and 9 lead to relatively low litter weights, those of care group 8 are associated with high litter weights. The long gestations of group 5, but not those of group 2, are associated with a very low relative fetal growth rate, and hence with a low litter weight relative to body weight. Infant growth rates are relatively high in group 3 (although this does not lead to an early maturation). In groups 2 and 7 the juvenile period is long, leading to a late age at first reproduction, whereas in group 5 the length of the juvenile period is relatively short.

Table 11.3 Comparison of life history parameters for species with different infant care strategies

Infant care strategy	2. Carry orally, park	3. Carry orally and clinging to fur, park, nest	5. Carry on fur, park	7. Carry on fur, park, allocare	8. Carry on fur, allocare	9. Carry on fur, no allocare
1. Carry orally, park, nest	gest (1 < 2) litwt (1 < 2) juv (1 < 2) AR (1 < 2)		gest (1 < 5) LFGR (5 < 1)	gest (1 < 7) litwt (1 < 7) juv (1 < 7) AR (1 < 7) brnwt (1 < 7)	gest (1 < 8) litwt (1 < 8) wnwt (8 < 1)	gest (1 < 9)
2. Carry orally, park		gest (3 < 2) LPNGR (2 < 3) AR (3 < 2)	litwt (5 < 2) LFGR (5 < 2) juv (5 < 2) AR (5 < 2)	brnwt (2 < 7)	gest (8 < 2) juv (8 < 2) AR (8 < 2)	gest (9 < 2) juv (9 < 2) AR (9 < 2)
3. Carry orally and clinging to fur, park, nest				litwt (3 < 7)	LPNGR (8 < 3)	LPNGR (9 < 3)
5. Carry on fur, park				litwt (5 < 7) LFGR (5 < 7) AR (5 < 7) brnwt (5 < 7)	gest (8 < 5) litwt (5 < 8) LFGR (5 < 8) juv (5 < 8) AR (5 < 8)	gest (9 < 5) litwt (5 < 9) LFGR (5 < 9) juv (5 < 9) AR (5 < 9)
7. Carry on fur, park, allocare					AR (8 < 7) brnwt (8 < 7)	litwt (9 < 7) AR (9 < 7)
8. Carry on fur, allocare						litwt (9 < 8)

Note: For definitions of infant care strategies, see table 11.1; for further information on the variables tested, see table 11.2. Comparisons were made using ANOVA for the following variables (all using residual values to remove the influence of female body weight): gestation length (gest), litter weight at birth (litwt), litter fetal growth rate (LFGR), age at weaning (wnage), weight at weaning (wnwt), infant growth rate of litter from birth to weaning (LPNGR), length of the juvenile period (juv), age at first reproduction (AR), and female brain weight (brnwt). Juvenile growth rate was not tested owing to the small sample size. Only one species, *Euoticus elegantulus*, was classified as having infant care strategy 4, and only *Colobus verus* was classified as having infant care strategy 6; as there is very little life history data available for these species, strategies 4 and 6 are not included in these analyses. An ANOVA was carried out for each parameter, with each infant care strategy being treated as a separate category. Significant differences ($P < .05$) are given for each pair of infant care categories (Fisher test).

Brain Weight, Maturation Rates, Growth Rates, and Group Size

The relationship between brain weight and age at first reproduction is significant after controlling for body weight ($t = 3.377$; d.f. $= 2,32$; $p = .007$). This link between brain weight and maturation time is also found with the length of the juvenile period ($t = -2.57$; d.f. $= 2,23$; $p = .014$), but not with age at weaning ($t = -0.147$; d.f. $= 2,21$; $p = .88$). Brain weight is positively linked to group size after controlling for body weight ($t = -2.138$; d.f. $= 2,35$; $p = .04$).

When growth rates are considered, litter infant postnatal growth rate is not significantly linked to brain weight after controlling for body weight ($t = -0.060$; d.f. $= 2,26$; $p = .95$). However, the juvenile growth rate is significantly linked to brain weight after controlling for body weight ($t = -0.060$; d.f. $= 2,26$; $p = .95$). Multiple regressions also indicate that both the length of the juvenile period and the juvenile growth rate are significantly correlated with brain weight, whereas age at weaning and infant growth rate are not (table 11.4).

Brain Weight, Life History Parameters, and Proportion of Allocare

The relationship between relative brain size and proportion of allocare is negative (fig. 11.1). Multiple regressions show that after controlling for body

Table 11.4 Results of multiple regression analysis on haplorhine species

N	r	p	Multiple regression statistics x	p (sign of coefficient)
Growth rates and brain size				
23	0.93	0.0001	W	0.0001 (+)
			IGR	0.7524
			JGR	0.0460 (−)
Age at weaning, age at maturity, and brain size				
27	0.91	0.0001	W	.0001 (+)
			wnage	0.8921
			juv	0.0138 (+)
Juvenile period, group size, allocare, and brain size				
10	95	0.0001	W	0.0001 (+)
			juv	0.0264 (+)
			group size	0.0735 (+)
			% allocare	0.5145

Note: Results of multiple regression analysis on haplorhine species using CAIC-generated independent contrast values, where female brain weight is the dependent variable and the independent variables are as follows: W = female body weight; IGR = growth rate from birth to weaning; JGR = juvenile growth rate from weaning to female age at first reproduction; wnage = age at weaning; juv = length of juvenile period; group size; % allocare = proportion of allocare.

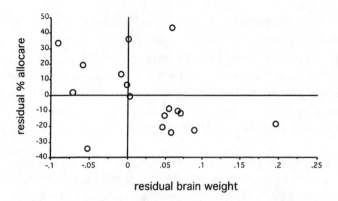

FIG. 11.1. The relationship between relative brain weight
and the proportion of allocare across primates. Analysis of
independent contrasts revealed that evolutionary increases
in relative brain weight were negatively correlated with the
proportion of allocare.

weight, litter infant postnatal growth rate is positively, but not significantly,
linked to proportion of allocare ($t = 2.073$; d.f. $= 2,12$; $p = .06$). The results
indicate that female brain weight has a positive relationship to length of the
juvenile period, female body weight, and group size, but does not have a
significant positive relationship to proportion of allocare (table 11.4).

Discussion

The results of the analysis of variance indicate that different infant care
strategies are associated with different patterns of life history variation.
However, the confounding influence of phylogeny makes it hard to interpret
these results. Many species in the strepsirrhine suborder use infant care
strategy 1 (oral carrying with nesting and parking), and this pattern is associated with a short gestation and small litter weight. This finding supports previous work by Kappeler (1998a) and Ross (2001), which showed
that nesting species tend to have low litter weights in comparison with other
primates. In contrast, species that park infants but do not nest (care groups
2 and 5) have relatively long gestations. These two groups include the tarsiers and lorisines, both of which also share the trait of having single infants.
It is unclear whether species with short gestations nest because they have
relatively small infants or whether the use of a nest allows a short gestation,
as less well developed young can be easily guarded and protected in nests.
However, animals of care group 5 (all non-nesting lorisine species) also have

relatively small infants, suggesting that nesting may be more closely associated with the birth of multiple offspring than with small offspring.

It is also notable that groups with high proportions of allocare (care groups 7 and 8) generally have relatively high litter weights compared with other groups. In the callitrichids, these high litter weights are produced by twinning; it could be the production of two infants, rather than a high litter weight, that is linked to high levels of allocare. Dunbar (1995a) notes that the high energetic demands of twin production combined with postpartum estrus in the callitrichids mean that is unlikely that a female could carry and suckle two infants while simultaneously gestating the next litter. He suggests that the evolution of paternal care in this group preceded the evolution of twinning. High levels of allocare are also found in related New World monkey species that do not twin (in particular, Goeldi's monkey), which also suggests that high allocare levels arose before the evolution of twinning (Pook 1978; Ross 1991).

Differences in rates of postnatal growth and maturation are also associated with the different infant care strategies, but again, the confounding influence of phylogeny makes it difficult to interpret these results. Two infant care groups (2 and 7) have relatively late maturation, but both are groups with very few species included, and hence the relevance of this finding is unclear. Group 2 includes two tarsier species (which do not have relatively large brains when compared with the other infant care groups), and group 7 consists of *Cebuella pygmaea* and *Homo sapiens* (which do have relatively large brains when compared with the other groups). Although these two groups share infant parking behavior, they also differ, as group 2 does not show significant amounts of allocare, whereas group 7 includes species with high levels of allocare.

Some insight as to why these two groups may have relatively slow maturation rates may be found in the multiple regression analyses, which also investigate links between late maturation times and other life history variables. The results of the multiple regression support previous studies using CAIC (e.g., Barton 1999; Ross and Jones 1999), which suggest that large brain size is linked to large body size, late maturation age (as measured by age at first reproduction), and large group size. The link between large brain size and a late age at first reproduction is due to species with larger brains having a longer juvenile period, rather than an extended period between birth and weaning. Ross and Jones (1999) suggest that this link is probably a result of the energetic costs of supporting and growing a large brain, which prevent large-brained species from growing rapidly to adulthood. This suggestion is supported by the results shown here, as juvenile growth rates (but not fetal or infant growth rates) are correlated with brain size after controlling for the other variables.

Ross and Jones (1999) also suggest that a mother might allow her infants to mature more rapidly by increasing their growth rate between birth and weaning, thus compensating for the slow growth rates between weaning age and maturity imposed by the energetic costs of a large brain. Since infants that receive high levels of allocare grow more rapidly between birth and weaning than those that receive lower levels of allocare (Ross and Mac-Larnon 2000; this study), one way of increasing infant growth rates might be for mothers to solicit allocare. There is no link, however, between litter infant postnatal growth rate and brain size. The multiple regression indicates that there is a negative relationship between relative brain size (i.e., relative to body size) and proportion of allocare, and that this relationship is lost completely when juvenile period and group size are also held constant. Hence, contrary to the predictions made above, the amount of nonmaternal care received by infants does not have a positive relationship to brain size; rather, species with high levels of allocare actually have smaller relative brain sizes that those with low levels of allocare.

Given the link between brain size and energy expenditure, the lack of a positive correlation between levels of allocare and brain size suggests that mothers of large-brained infants cannot compensate for the slow maturation rates of their infants by growing them rapidly to weaning age. This suggestion is supported by the lack of a link between age at weaning and adult brain size. If brain size does act as an energetic constraint on maturation rates, it must be primarily constraining growth rates after weaning—that is, once the offspring is nutritionally independent from its mother. This assumption, in turn, suggests that the advantages of rapid infant growth and early weaning that are linked to high levels of allocare (Mitani and Watts 1997; Ross and MacLarnon 2000) are direct advantages for the mother, as she may thereby increase her reproductive rate. However, her offspring does not benefit from a significant increase its own reproductive rate, as it is constrained by its brain size to have a certain juvenile period, and hence a certain age at maturity. Similarly, Ross and MacLarnon (2000) found no evidence that infants directly benefit from nonmaternal care, as there was no clear link between the amount of allocare received and an infant's "need" for early weaning (e.g., because of increased risk of mortality to unweaned infants). Rapid weaning of infants may, however, have an indirect advantage for offspring, as their inclusive fitness may be increased by their mother's increased reproductive output.

The reason why some primate species are apparently constrained to have slow growth rates between weaning and reproduction is not clear from these analyses. As noted by Pereira and Leigh (chap. 7, this volume), there is considerable variation in the brain growth patterns seen in primates. Primates are unusual among mammals in that rapid brain growth may continue after birth; however, this rapid phase of brain growth is generally completed at around weaning age (Deacon 1990; Wood 2000). However, some brain growth may occur after weaning, and the robust relationship between the length of the juvenile period and brain size may indicate that the energetic constraints of absolute brain size are limiting the somatic growth rates of some species. Further analyses of the rates of increase in both overall body weight and brain weight are needed to confirm this, as it is by no means clear that the energetic demands of brain tissue are sufficient to slow overall growth rates. In addition, the constraints of seasonality and juvenile diet (cf. Janson and van Schaik 1993) need to be examined in relationship to patterns of growth between weaning and age at first reproduction if the reasons for delayed maturation in some species are to be completely understood.

Summary and Conclusions

Infant care strategies are linked to the life history characteristics of primate species, and these links include differences in rates of postnatal growth and maturation. However, the confounding influence of phylogeny makes it difficult to interpret these results. Analyses of the links between brain size, age at maturity, and maturation rate show clearer results. The results presented indicate that when body size has been accounted for, juvenile growth rates are slow, and the juvenile period is long, in species with relatively large brains. This finding suggests that the commonly found positive correlation between brain size and age at maturity is primarily due to species with larger brains having a slow juvenile growth rate and a long juvenile period. It is unclear from these results whether brain size and/or brain growth during the juvenile period act as energetic constraints preventing rapid growth in juveniles. The growth rates of infants and their ages of weaning are not good predictors of primate brain size, and rapid growth during infancy does not result in an early age at maturity relative to brain size. Mothers that are helped by other caregivers to care for their infants may increase their offspring's growth rates, but this does not result in their offspring reaching maturity at an earlier age than other species (of the same body and brain weight) that do not have allocare. Although infants receiving allocare may be weaned at a relatively early age, they are then constrained by their brain size to have a certain juvenile growth rate, and hence a certain age at maturity. Similarly, there is no relationship between amount of allocare received and relative brain size in primates. If increased levels of allocare cannot enable offspring to mature rapidly or to have large brains, this suggests that it is the mother that gains the main reproductive advantage when allocare is given.

Acknowledgments

I am grateful to Peter Kappeler for inviting me to speak at the 2nd Göttinger Freilandtage and must thank him and all those who worked so hard to make this meeting a success. I am also grateful to those who funded this conference and my attendance at it. This chapter has benefited from comments made by Peter Kappeler, Michael Pereira, Deborah Curtis, and two anonymous reviewers.

12 Why Are Apes So Smart?

Robin I. M. Dunbar

Brain size is a major life history variable; indeed, some have argued that it is the primary life history variable that is responsible for driving all other life history variables (Harvey and Bennett 1983; Harvey, Martin, and Clutton-Brock 1987). This argument, in turn, raises the question of what drives brain size evolution. Conventional biological wisdom assumes that, ultimately, there is some ecological factor underpinning the evolution of life history variables (see, for example, Charnov 1993; Charnov and Berrigan 1993). However, the correlations between ecological variables, group size, brain size, and life history variables raise a fundamentally important question: is the evolution of life history variables (including brain size) driven directly by ecological considerations (with variables such as group size being a by-product of brain size), or is it driven indirectly via socio-demographic variables?

This distinction is important because on the latter view (the so-called "social brain" hypothesis), brain size is a constraint on species' generating appropriate social solutions to the ecological problems they face, whereas on the former view group size is a mere by-product of evolving extended life histories as a solution to an ecological problem. In both cases, the driving ecological problem is assumed to be the same—namely, predation (see Dunbar 1988; Charnov 1993)—but the mechanisms involved are different. It is important to appreciate, however, that the social brain hypothesis is not necessarily incompatible with Charnov's life history model: it is incompatible only with a version of Charnov's model that assumes adjustments in birth rate to be the *only* mechanism that mammals use for countering predation risk. The alternative version assumes that life history variables have the form they do because brain growth occurs at a slow, constant rate in mammals (Harvey and Bennett 1983; Pagel and Harvey 1988a); hence, in order to evolve the large brains necessary to maintain the social coherence of large groups, it is necessary to continue brain growth for proportionately longer (and this necessitates an extended life history).

In this chapter, I first review briefly the evidence for the social origins of differences in primate brain size, and argue that attempts to provide alternative explanations have largely failed to adduce convincing evidence for these hypotheses. I then draw attention to the fact that the hominoids appear to lie on a separate grade from the simians in this respect, a fact that begs explanation. This observation raises two questions: what cognitive mechanisms are involved in this grade shift, and why did hominoids come to evolve them? My aim here is to raise an issue that seems to have been overlooked. I don't pretend to have answers to either of these two questions, but I will try to sketch in what seem to me to be the critical issues and likely possibilities.

The Social Brain Hypothesis

It is now widely accepted that the principal pressure selecting for increased brain size in primates was the need to weld together large social groups (Byrne and Whiten 1988; Dunbar 1992; Barton 1996; Barton and Dunbar 1997; Dunbar 1998), probably as a defense against predators. The principal evidence for this claim is shown in figure 12.1. The robustness of the relationship between brain size and group size is evidenced by the fact that it has survived repeated tests using different statistical methods and by the fact that it predicts social group sizes in an independent set of species (Dunbar 1995b). The importance of the social dimension as the explanatory variable underlying this relationship is emphasized by the fact that, within the primates, brain size (or, more properly, neocortex size) correlates with the species-typical size of grooming cliques (equivalent to alliances: Kudo and Dunbar 2001), the frequency of social play (Lewis 2000), and the ability to exploit subtle social strategies (Byrne 1995b; Pawlowski, Lowen, and Dunbar 1998). In addition, Joffe (1997) was able to show that the relative size of the nonvisual areas of the primate neocortex (that is, the neocortex anterior to the primary visual area, $V1$) correlates best with the length of the juvenile period (the time between weaning and puberty or first reproduction), even though the volume of the brain as a whole correlates best with the duration of more conventional life history variables, such as the duration of direct parental investment (i.e., gestation plus lactation). This finding implies that the software programming associated with socialization is crucial in the complex of processes associated with enlarged brain size in primates. (Note

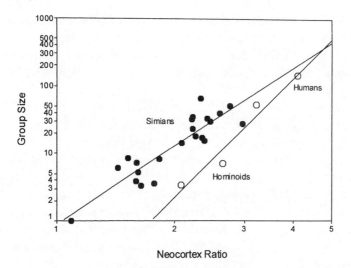

FIG. 12.1. Mean taxon group size plotted against neocortex
ratio (neocortex volume divided by the volume of the rest of
the brain). The ape group sizes have been updated using the
data of Bean (1998) for gorillas (mean group size 11.7, $N = 14$
populations) and of Lan (2001) for gibbons (mean group size
3.8, $N = 21$ populations). Remaining data points are as given
by Dunbar (1992).

that the enlarged brains of primates are really associated with an enlarged
neocortex, rather than a general enlargement of all parts of the brain.)

Although a number of alternative hypotheses have been advanced to
explain large brain size in primates (these include several different ecological hypotheses and at least one suggestion based on a developmental
constraint: for review, see Dunbar 1992, 1998), no convincing empirical evidence has ever been offered for any nonsocial explanation. More importantly, direct tests between competing social and ecological hypotheses
come down unequivocally in favor of the social hypothesis (Dunbar 1992).
Although Deaner, Nunn, and van Schaik (2000) have questioned the claim
that these tests are unequivocal in their support for the social hypothesis, their analysis (1) does not offer any more convincing support for the
ecological hypotheses, (2) is confounded by body size scaling effects with
respect to the ecological variables it uses, and (3) suffers from a rather dis-
turbing (and apparently unnoticed) methodological problem (namely, the
fact that analyses based on neocortex ratios and residuals from a common

regression line yielded different results, despite the fact that these two procedures are in fact mathematically exactly equivalent). Nor do they offer any explanation as to why, if an ecological explanation is to be preferred, so many social variables should correlate so strongly with relative neocortex size in primates (see above).

Byrne (1995b and elsewhere) has, of course, argued that an alternative explanation (e.g., in terms of food processing) may be necessary for gorilla brain size. However, conventional procedure in science does not normally admit the claim that a single counterexample is a valid test of a statistical hypothesis: that gorillas have defaulted to an alternative ecological mode in terms of brain function does not contradict the claim that brain size evolution in primates *as a whole* has been driven by social considerations. In any case, as I shall show below, the claim that gorillas are exceptional in this respect is questionable: in actual fact, they have exactly the social group size that would be predicted for their neocortex size.

In sum, the occasional claims to the contrary notwithstanding, no serious challenges to the social hypothesis for brain size evolution in primates have been marshaled during the past decade. Such claims as have been advanced have invariably been based either on arguing special cases (e.g., the extractive foraging hypothesis for *Cebus* and *Gorilla:* Parker and Gibson 1977; Byrne 1995b) or on empirically unsubstantiated assertions. Indeed, if anything, there has been increasing evidence over time that the same relationship between social group size and neocortex size holds in taxa other than the primates (including, to date, advanced insectivores, carnivores, bats, and cetaceans: see Marino 1996; Dunbar and Bever 1998; Tschudin 1999; R. A. Barton, unpub.). The only taxa for which such a relationship has not been demonstrated are the (neurologically) primitive insectivores (Dunbar and Bever 1998) and, rather surprisingly given the preceding results, the prosimians (Barton 1996) — although the evidence for the latter case remains open to serious question and awaits more detailed analysis (see Dunbar 1998).

It is important to be clear that the social brain hypothesis as outlined above does not seek to claim that ecology plays no role in the process of brain size evolution. To the contrary: ecology is the driving force behind the whole process. Rather, the issue is whether ecology alone is an adequate explanation for brain size evolution in primates (with the social correlates identified above being merely by-products of the possibilities opened up by having large brains) or whether brain size is a response to the social difficulties created by species' attempts to solve their ecological problems. The difficulty has, if anything, been in trying to tease apart the independent effects of these two biologically correlated factors (a point that is amply evidenced by the sometimes surprisingly confused discussions of this issue). If nothing else, biological parsimony would enjoin us to be suspicious of any claim that sought to interpret social phenomena as a mere by-product of having a large brain size: living in social groups incurs very significant costs that are not idly borne merely because brain size allows large social groups—a point that has been patently clear ever since van Schaik's (1983) lucid analysis of the problem (see also Dunbar 1988). Moreover, the massive costs incurred by evolving (and maintaining) large brains (Aiello and Wheeler 1995) make any purely ecological claims inherently questionable, since they require us to assume that primate ecological strategies in general (as opposed to particular cases, such as those considered by Parker and Gibson [1977] and Byrne [1995b]) are significantly different in their cognitive demands from those of all other mammalian taxa—a suggestion for which no convincing evidence has ever been seriously advanced. Indeed, it is conspicuous and quite remarkable that those who argue in favor of a purely ecological origin for primate brain size never address this crucial issue.

The same problem attends any attempt to claim that large brain size is simply a by-product of differences in life history patterns among primate species. Life histories (and other developmental arguments: see Martin 1981; Armstrong 1985b) are unquestionably constraints on brain size evolution. But it is one thing to claim that life history is a constraint on evolving large brains, and an entirely different matter to claim that brain size evolution is driven by life history evolution and nothing else. To do so ignores the costs of maintaining large brains and thus fails the conventional requirements of any well-formed evolutionary explanation by being a mere just-so story. The only biologically plausible way of interpreting this apparently confusing set of intercorrelations is to argue that life history processes are a constraint on brain growth that has to be overcome if a species is to be able to evolve the size of group needed to cope with the particular ecological challenge(s) it faces. Species that cannot overcome that constraint must inevitably go extinct or be confined to specialized ecological niches where the need for a larger group size does not arise.

More recently, it has been possible to extend the analyses of the social brain hypothesis to narrow down the focus of the relationship in neuro-anatomical terms. Joffe and Dunbar (1997), for example, showed that social group size in anthropoids correlates significantly better with the relative volume of nonstriate neocortex (i.e., neocortex volume anterior to the primary visual area, V1) than it does with the neocortex as a whole or with the relative volume of the primary visual area. In addition, recently published data

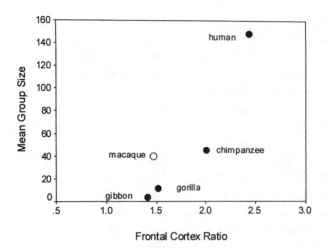

Fig. 12.2. Mean species group size plotted against relative frontal cortex volume. Frontal cortex volume (based on data from Semendeferi et al. 1997) is given as a ratio of frontal lobe divided by hindbrain volume (based on data from Stephan, Frahm, and Baron 1981). The hindbrain is taken to be total brain volume excluding the neocortex (see Dunbar 1992). Because total brain volumes in these two sources differ (Stephen, Frahm, and Baron's values were standardized to mean species body weight; Semendeferi et al.'s are absolute raw volumes), hindbrain volumes have been corrected by the ratio of total brain sizes given by the two papers. For sources of group size data, see table 12.1.

on frontal neocortex volumes for a small selection of anthropoids (principally great apes) exhibit a very consistent linear relationship between the volume of the frontal cortex and social group size, with a clear grade difference between the great ape species and the only simian (a macaque) in the sample (fig. 12.2). With or without the data point for modern humans, the macaque has a significantly higher residual from the common reduced major axis regression line than the four hominoid data points, all of which have negative residuals (with humans: $z = 5.238$, $P < .0001$; without humans: $z = 14.2$, $P \ll .0001$).

The latter finding draws attention to a curious feature that has largely escaped comment in all previous discussions of the data shown in figure 12.1; namely, the fact that even though the apes fit the general primate pattern, they are, as a group, nonetheless displaced significantly to the right on the graph that plots social group size against relative neocortex size (differences between simian and hominoid regression equations: intercept: $t = 3.04$, $P < .01$; slope: $t = 0.54$, $P > .5$; d.f. $= 26$ in each case).

As it happens, this grade difference helps to explain an apparent anomaly that many have noticed; namely, the position of the orangutan in these analyses. Zilles and Rehkämper (1988) provide data on orangutan brain volume components, from which we can estimate a neocortex ratio of $C_R = 2.47$. Interpolating this ratio into the ape equation from figure 12.1 yields a predicted group size of 7.6. In other words, contrary to almost everyone's assumptions hitherto, the neocortex-predicted group size for orangutans is considerably smaller than that for chimpanzees (approximately 55) and only slightly larger than that for gorillas (approximately 7). This estimate appears to be well in line with the statements made in almost every field study of wild orangutans to the effect that stable local communities on the order of 5–15 individuals can be readily identified in almost all study populations. The point, perhaps, is that even though there is a scaling relationship between neocortex size and total brain size in primates (see, for example, Aiello and Dunbar 1993; Finlay and Darlington 1995), we should beware of assuming that large brains necessarily imply large neocortices: both gorilla and orangutan are characterized by relatively small neocortices relative to body size. Instead, both species are characterized by cerebella that, as a proportion of total brain volume, are about 1.5 times the size of those of chimpanzees and humans, possibly reflecting the need to manipulate a large body mass in an arboreal habitat.

Why should hominoids differ so strikingly from simians in the relationship between group size and relative neocortex size? In trying to answer this question, two other questions need to be addressed. The first is a purely mechanistic one as to why apes need more computing power to sustain a given group size than monkeys do (what do they do that is cognitively different?). The second is the evolutionary question as to why the ape lineage as a whole set off on what seems like a completely separate evolutionary trajectory at what must have been a very early stage in its evolution.

What Do Apes Do with Their Brains?

Irrespective of what apes actually do with their brains, some interesting and unexpected features emerge from a consideration of the neuroanatomical data. Joffe and Dunbar (1997) showed that if the volume of the striate cortex (primary visual cortex, area V1) is plotted against the volume of the rest of the neocortex (nonstriate cortex), the linear relationship between the two appears to level off at about the brain size of great apes (fig. 12.3). It is

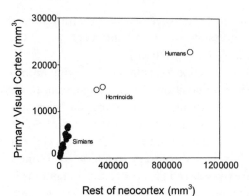

FIG. 12.3. Volume of the primary visual cortex (area V1) plotted against the rest of the neocortex for individual anthropoid species (each species represents a different genus). Humans are the data point on the extreme right. (Adapted from Joffe and Dunbar 1997.)

not immediately clear why this should be the case (especially given the more or less linear relationship between striate and nonstriate areas among the monkeys).

However, even with this limited degree of resolution, the effect is very striking: once past the brain size of apes, the amount of striate cortex increases at a much slower rate than does the amount of nonstriate cortex. As a result, the absolute amount of spare cortical material available for higher processing functions (the residual between that observed and that predicted by the simian data in figure 12.3) increases exponentially (fig. 12.4). If nothing else, then, the computing power available for managing social relationships (and anything else aside from primary visual processing) increases quite dramatically once past the brain/body size of the extant monkeys.

An obvious interpretation of this relationship is that the acuity of visual processing approaches an asymptotic value as the volume of cortex assigned

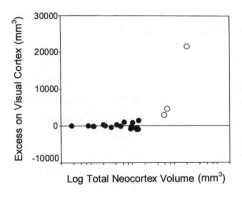

FIG. 12.4. Relative volume of spare neocortical capacity in different anthropoid primate species (solid circles = simians; open circles = hominoids). The plotted values are the residuals between observed nonstriate neocortex volume and that predicted from striate neocortex volume by the relationship shown in figure 12.3 for simian primates (monkeys). Humans are the data point on the extreme right.

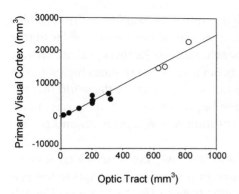

FIG. 12.5. Volume of the primary visual cortex (area V1) plotted against the volume of the optic tract for anthropoid primates (solid circles = simians; open circles = hominoids; humans are the data point on the extreme right). (Data from Stephan, Frahm, and Baron 1981.)

to it increases. In other words, the gain in acuity of vision as processing capacity increases is marginal after a point that appears to coincide with the brain volumes of the extant great apes. This suggestion is reinforced by one other anatomical observation. The efficiency of visual processing ultimately depends on the number and cross-sectional area of the neurons bringing visual signals from the retina to the primary visual cortex. This is a simple fact of physical and physiological constraints. This relationship is reflected in the fact that the size of the primary visual cortex in primates correlates virtually linearly with the volume of the optic tract (fig. 12.5), suggesting that the cortical volume devoted to primary visual processing is directly determined by the volume of input from the retina. (Although Stephan, Frahm, and Baron [1981] do not provide data on the cross-sectional area of the optic tract, the fact that its volume correlates so closely with the volume of the primary visual cortex is in itself enough to establish the point.)

Thus, while the primary visual cortex is directly linked to input from the peripheral visual system, the nonstriate cortex appears not to be so constrained, suggesting that, despite the considerable size of the secondary (area V2) and tertiary (frontal lobe) visual processing areas, the nonstriate cortex has functions other than the analysis of visual patterns. This does not imply that it plays no role in the analysis of visual input; rather, the point is that its role changes, perhaps reflecting a shift from the analysis of visual patterns to an analysis of the social meaning of those visual patterns (see also Frith 1996).

Beyond this, however, it is difficult to see exactly what apes do that is different. This is not because there is no difference between apes and monkeys, but rather because we know almost nothing about their cognitive and behavioral differences in the social domain (see Tomasello and Call 1997). There are some hints that point at possibilities, however. One of the core features of human cognitive development that has been extensively studied over the past decade is theory of mind (a rather awkward term that identifies the ability to understand that another individual has a mind similar to that one experiences for oneself). The benchmark test for theory of mind is now universally accepted to be the false belief task, which requires the subject to understand that another individual holds a belief about the world that the subject supposes to be false. Children begin to pass this test at around four years of age, with a rather rapid transition between incorrect and correct responses.

Although developmental psychologists have shown considerable interest in these aspects of social cognition, only two direct attempts to test primates using a false belief task have ever been carried out. One (O'Connell 1995) obtained (weak) positive results suggesting that chimpanzees were about as good as children on the threshold of acquiring theory of mind; the other (Call and Tomasello 1999) yielded negative results, again with chimpanzees. However, a more sophisticated experimental design by Hare et al. (2000) has since provided convincing evidence that even though chimpanzees may fail to understand human minds, they do understand other *chimpanzees'* knowledge states, suggesting that they might well pass false belief tasks that have a more chimpocentric design. Although no monkeys have been tested on these kinds of tasks, there is more or less universal agreement that simians do not possess theory of mind, even though they are highly competent in the social domain (Cheney and Seyfarth 1990; Byrne 1995b).

The fact that apes might just be able to pass theory of mind tests is, however, unimpressive by the standards of normal human adults. False belief tasks represent second-order intentionality (having a belief about someone else's belief), and there is now clear evidence that normal human adults can successfully pass fourth-order tasks (having a belief about someone else's belief about a third person's belief about a fourth person's belief), although fifth-order tasks seem to be beyond their grasp (Kinderman, Dunbar, and Bentall 1998). Although this remains the only study to date that has attempted to explore the natural history of advanced social cognition in normal human adults, it nonetheless emphasizes the scale of the contrast between human and ape social cognition. More importantly, it appears to mirror rather closely the distribution of spare neural capacity reflected in figure 12.5. Indeed, taking the implied levels of intentionality suggested by the data (one for macaques, two for chimpanzees, and four for humans), there is a relationship between level of intentionality achieved and both frontal cortex ratio (fig. 12.6) and absolute volume of the frontal cortex. Despite the small sample, the linearity of the results is impressive. More intriguingly, perhaps, the fact that absolute frontal cortex volume appears to

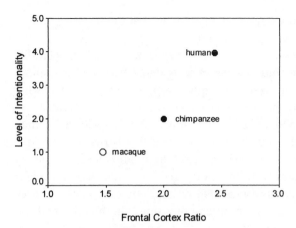

FIG. 12.6. Levels of intentionality achieved by macaques, chimpanzees, and humans plotted against frontal cortex ratio (the ratio of frontal lobe volume to the volume of the noncortical brain).

be just as good as relative frontal cortex volume at predicting group size reinforces earlier claims that absolute frontal cortex size may in fact be critical in determining a species' social "computing power" (see Dunbar 1992; Byrne 1995b).

An Evolutionary Puzzle

The real puzzle in all of this is just why and when this grade shift in cognitive ability came about. The story outlined in the preceding section does not in itself necessarily imply a grade shift of any kind. The patterns shown in figures 12.3 and 12.6 would seem to suggest that these contrasts reflect simple allometric scaling relationships in brain part size. Ape social cognition is simply what emerges when you expand the standard monkey brain up to the size typical of great apes. However, this is clearly not the whole story, because figure 12.1 suggests that gibbons lie on the same group size-neocortex size grade as the great apes. If simple scaling relationships were all there was to it, then gibbons would nestle neatly in among the monkeys (which is where, cortically speaking, they belong). That they do not implies that something peculiar happened to the ape lineage at a very early stage in its evolution (or, at least, sometime after the divergence of the ape and Old World monkey lineages some 28 million years ago, but before the divergence of the gibbon and great ape lineages about 18 million years ago).

This grade shift on the part of the ape lineage is clearly demonstrated by the fact that the New World and Old World monkeys are indistinguishable in figure 12.1: they clearly fall on the same grade, despite being more distantly related to each other than the Old World monkeys are to the apes. Moreover, since we know that the New World monkeys diverged from the Old World anthropoids sometime before the split between the apes and Old

World monkeys, parsimony enjoins us to assume that the simian grade in figure 12.1 is in fact the ancestral condition.

This assumption leaves us with a decided puzzle: what could conceivably account for the ape grade shift at such an early date? Just what is it about ape social systems that requires so much more computing power? And what possible factors could have forced the evolution of such requirements?

There seems to me to be only one plausible difference between simian and hominoid behavior that might fit the bill, and that is the fact that dispersed social systems are characteristic of the hominoids and completely absent from all the monkeys (with the questionable exception of the spider monkeys and the perhaps more plausible example of some callitrichids). Dispersed social systems (in which, at any one time, the members of the community are divided among a number of foraging parties) are very dramatically characteristic of the chimpanzees, orangutan, and humans. This kind of social system must place a significant load on the information processing capacity associated with managing the coherence of social groups compared with that experienced by monkeys in their more compact groups. The task is not just a memory problem (remembering who is who in the group), but one of maintaining an updated matrix of social knowledge that can be used to facilitate social relationships. Time out of sight is time during which very significant changes in social relationships can occur, as when individuals form alliances or fall out with each other. Gelada baboons provide an instructive contrast in this respect: they appear to have a dispersed form of social grouping (bands) whose size is significantly larger than the group size predicted from the species' (estimated) neocortex size (Dunbar 1995b). In contrast to the remarkably cohesive and well-integrated communities of chimpanzees and humans, gelada bands exhibit very weak, unstable relationships between their constituent units (reproductive units) (Dunbar 1984; Kawai et al. 1983).

We can gain some idea of the magnitude of the problem faced by hominoids by comparing the observed group sizes for different ape species with those that would be predicted by the simian relationship between group size and neocortex ratio. Dunbar (1993) gives the regression equation for the simians as

$$\log_{10} N = -0.221 + 4.135 \log_{10} C_R$$

where N is the species mean group size and C_R is the species' neocortex ratio (neocortex volume divided by the volume of the rest of the brain). Using this equation to predict group sizes for the ape species yields the values shown in table 12.1. As a group, the hominoids have an observed group size

Table 12.1 Observed and predicted values for social group size for ape genera

Genus	$C_R{}^a$	Mean group size		Ratio
		Observed[b]	Predicted[c]	Predicted/ observed
Gibbon	2.08	3.8	12.4	3.263
Gorilla	2.65	11.7	33.8	2.889
Chimpanzee	3.22	45.2	75.7	1.675
Human	4.10	148.4	205.5	1.385

Sources: Dunbar 1992, 1993
[a]Neocortex ratio (volume of neocortex divided by volume of rest of brain).
[b]Observed mean group sizes for 21 gibbon populations by Lan (2001); for 14 gorilla and 13 chimpanzee populations by Bean (1998), and for clans of 9 human hunter-gatherer societies by Dunbar (1993).
[c]Predicted from neocortex ratio by the simian equation (see text).

that is about half what would be expected of simians with the same neocortex sizes. One interpretation of this finding is that it represents the information overload costs of dispersed social groups.

If we accept the suggestion that dispersed social systems place a significant additional information processing load on a species, we are left to wonder why dispersed social systems should be so characteristic of the ape lineage. The only plausible answer that I can see is a dietary one. As Wrangham (1980b) noted a long time ago, the great apes (at least) differ from the monkeys in that they cannot digest the tannins that plants use to defend unripe fruit. Being a large-bodied ripe fruit feeder probably has two important consequences. First, it is necessary to range farther afield in order to be sure of finding enough trees in this fruiting stage within the home range; second, since patch sizes are relatively smaller, it is invariably necessary to forage in smaller groups. Hence, in order to maintain a large community size, the community is forced to disperse.

Some additional evidence pointing in the same direction is provided by the fact that terrestrial Old World monkeys seem to lie on a grade intermediate between the arboreal (New and Old World) monkeys and the apes, at least in terms of the relationship between grooming clique size and relative neocortex size (Kudo and Dunbar 2001). Kudo and Dunbar interpret this finding as suggesting that the higher predation risk associated with terrestrial habitats may have been responsible for triggering the shift toward cognitively more expensive social strategies in order to permit groups to be more cohesively bonded so as to allow them to function as more effective defenses against predators. Thus the high predation risk characteristic of terrestrial habitats and the enforced dispersal of social groups might have been the key factors that gave rise to the grade shift suggested by figure 12.1.

This suggestion has at least one interesting implication; namely, that the early ape lineages may have been more terrestrial than their simian cousins. The fossil evidence would support such a claim: the Miocene apes, at least, occupied the full range of habitats used by modern simians, including both arboreal and terrestrial quadrupedal niches (Fleagle 1999).

Summary and Conclusions

I began by reviewing the evidence in support of the claim that the evolution of large brain size within the primate order was driven by the cognitive demands of living in complexly organized social systems (the social brain hypothesis). Evidence to support alternative functional explanations (principally, the demands of different ecological strategies) is, at best, sparse; despite this, ecological hypotheses continue to be advocated with surprising enthusiasm. Many commentators have argued for an explanation based on a developmental or life history constraint; however, such an argument confuses functional with ontogenetic explanations. Such constraints are not alternative explanations for brain evolution, but rather form part of the functional complex. New analyses suggest that the frontal neocortex plays a more prominent role in social cognition than the anterior lobes.

I then went on to point out that hominoids appear to lie on a separate grade from simian primates with respect to the group size that is supported by a given neocortex volume. Advanced social cognitive abilities such as theory of mind are considered likely candidates for the kinds of processes that might underpin these taxonomic differences. The limited data that are available suggest that there may be a linear relationship between frontal neocortex volume and the extent to which such abilities are developed in individual taxa. The question of why hominoids apparently took off in this new evolutionary direction remains to be explained. Although it is possible that these abilities are simply a consequence of allometric scaling effects, the fact that gibbons lie on the hominoid line rather than with the simians suggests that a more disjunct process is involved. One possibility is that dispersed social groupings combined with increased terrestrialization (and the associated higher predation risks) may have been responsible for their evolution.

PART FOUR

Where Do We Go from Here?

13 Primate Life Histories and Future Research

STEPHEN C. STEARNS, MICHAEL E. PEREIRA, AND PETER M. KAPPELER

Most primates offer poor model systems for researching life history evolution. Their long lives require years to collect data on development and reproductive success, and many kinds of experiments are unfeasible or unethical to conduct using primates. Primates tend to be found in places that harbor threats to researchers' welfare, and many are difficult to observe because they are nocturnal, live in dense forest, or both. Even with more tractable primate subjects, samples large enough for satisfying statistical analyses are generally difficult and sometimes impossible to obtain.

Clearly, one must have good reasons to study primate life histories and behavioral ecology. In this concluding chapter, we review some of the most important ones. We discuss some outstanding issues in life history research per se, asking whether primate work might contribute to their resolution, and we suggest some areas for future integrative research.

Three major aspects of life history theory call specifically for research on primate life histories. First, our understanding of life history evolution relies substantially on comparative study. Given that Primates comprises a large, diverse, and well-described order, extending the overall program of research on mammalian life histories without appreciable attention to primates would be nothing short of folly. Second, many aspects of primate life histories are of general theoretical interest due to their uniqueness or extreme nature among the mammals and other vertebrates. The preceding chapters specifically address several such traits, including delayed maturation (Purvis et al., chap. 2), postponement of somatic growth (Pereira and Leigh, chap. 7), very large brains (Deaner, Barton, and van Schaik, chap. 10), and large semi-closed social groups (Janson, chap. 5) strongly overlapping two and sometimes even three generations (Hawkes, O'Connell, and Blurton Jones, chap. 9). Every chapter explores important correlates or consequences of primates' unusual features. Third, research on nonhuman primates offers important opportunities for understanding human evolution (e.g., Hawkes, O'Connell, and Blurton Jones, chap. 9, this volume).

Primates and Research on Life Histories

Although primates exhibit great diversity in their life histories (Harvey and Clutton-Brock 1985; Ross 1998), they also share particular combinations of traits setting them far enough apart from other mammalian orders to require specific evolutionary explanations. The formulation and refinement of these explanations informs the study of life history evolution in general, as the preceding chapters demonstrate. Certain new developments and questions in the quest to understand life history evolution, however, are unlikely to receive important contributions from primatology. We present examples of such studies below as well to help young scientists think clearly about careers in research and whether they value work with particular organisms over work with particular questions, or vice versa.

The Evolution of Juvenility

Primates are exceptionally well qualified to illuminate the principles underlying protracted development because they are so heavily concentrated at one end of the continua of absolute and relative lengths of juvenility in animals (Stearns 1992; Pereira 1993b). As Pereira and Leigh (chap. 7, this volume) emphasize, primates exhibit not only distinctive maturational delays shared by few other mammals (Promislow and Harvey 1990; Janson and van Schaik 1993), but also pronounced long-term shifts in growth rate following parental care, unrelated to seasonality or food deprivation, not known in other vertebrates (Case 1978; Martin and McLarnon 1988; Lee, Majluf, and Gordon 1991; Pereira and Fairbanks 1993b; Teather and Weatherhead 1994; cf. Pereira et al. 1999).

Variation in the existence and timing of prereproductive growth rate suppressions and spurts within and among primates surely derives in part from divergent intensities of intrasexual selection (Jarman 1983; Stamps 1993; Alberts and Altmann 1995; Leigh 1995; Pereira and Leigh, chap. 7, this volume). But much remains to be learned about this factor as well as the relative importance of other factors, such as adult size, patterns of brain growth, species-typical allometry, predation risk, interannual patterns of food supply, mechanisms of rank acquisition, and patterns of alliance formation (Pereira 1995; in this volume: Deaner, Barton, and van Schaik, chap. 10; Ganzhorn et al., chap. 6; Godfrey et al., chap. 8; Janson, chap. 5; Lee and Kappeler, chap. 3; Pereira and Leigh, chap. 7). As socioecological data accumulate, life history effects on variation in the antipredator and foraging strategies of juvenile primates should elucidate the functional significance of growth schedules in primates, other mammals, and other vertebrates (cf. Janson, chap. 5, this volume; Alberts and Altmann, chap. 4, this volume).

Brain Size and Cognition

Primates' unusually large brains in relation to body size and impressive cognitive abilities, discussed by Deaner, Barton, and van Schaik (chap. 10) and by Dunbar (chap. 12) in this volume, constitute another suite of traits requiring special explanation and providing additional foci for investigations of general interest (e.g., Parker, Langer, and McKinney 2000). Important progress has been made in analyzing patterns of brain growth and adult size variation through the application of modern comparative methods (e.g., Barton 1996; Barton and Harvey 2000; Nunn and Barton 2001). However, the information available on species-typical brain and brain part sizes is still insufficient for powerful analyses (see appendix, this volume), and considerable debate continues over the relative importance of various factors affecting brain evolution (Parker, Langer, and McKinney 2000; in this volume: Deaner, Barton, and van Schaik, chap. 10; Ross, chap. 11; Dunbar, chap. 12).

Primates' large brains and great cognitive capacities have evolved in an unusual mammalian social context. Unlike the vast majority of mammals, most primates form cohesive, semi-closed social groups in which adult males and females associate permanently. Moreover, social groups vary widely in size and composition across primate species. This variation, combined with that in relative brain size and in the relative sizes of different brain parts, provides rich grounds for analyses of coevolution between brain and behavior in particular and between life histories and socioecology in general.

Because primate anatomy, reproductive physiology, ecology, and sociobiology are easily among the best described among mammalian orders, comparative research on these and other topics can be achieved with primates that as yet remains impossible with most other taxa. Ultimately, however, such work should be complemented by efforts with counterpart taxa having impressive cognitive abilities and social complexities of their own but, in most cases, faster life histories (e.g., squirrels, mongooses, canids, hyenas, corvids, parrots, and a variety of communally breeding birds). Whether primates' unusual combinations of traits relate importantly to their predominantly arboreal lifestyle is another example of a question remaining for future comparative work to explore (Kappeler, Pereira, and van Schaik, chap. 1, this volume; van Schaik and Deaner 2002).

Reinventing Life History Theory with Complex Dynamics and Spatial Structure

A life history is an evolutionary solution to a set of ecological challenges developed in the context of intrinsic trade-offs and manifested as adjustments of vital rates (Alberts and Altmann, chap. 4, this volume). The classic explanation of life history evolution uses optimality models (Roff 1992; Stearns 1992). While such an explanation functions reasonably well, it also suffers from logical shortcomings. Its assumption of constant birth and death rates, for example, implies simple population dynamics (e.g., no frequency or density dependence) and is known to be invalid. The classic model also assumes well-mixed populations in homogeneous environments, which is equally unrealistic. The first outstanding issue in life history evolution, then, is a theoretical one: what life histories do models predict that allow for frequency and density dependence and realistic population dynamics in an explicitly spatial framework?

Some answers come from the extension of classic methods to simple spatial frameworks—source-sink environments—without explicit dynamics (Houston and McNamara 1992; McNamara and Houston 1996; Kawecki and Stearns 1993). The new prediction is that when most reproduction occurs in one habitat (source) and little in another (sink), the population adapts to the source and remains fairly maladapted to the sink. Whereas longitudinal research on primate populations across varied portions of species' geographic ranges is an important objective for primatology and conservation (Pereira and Leigh, chap. 7, this volume), it is hard to imagine either retrospective or prospective primate studies helping to evaluate the sink-source hypothesis definitively.

Another new approach, called adaptive dynamics (Dieckmann 1997), allows exploration of the consequences of complex population dynamics for life history evolution in a spatially explicit context. It is starting to yield results that differ from classic predictions. For example, when selection favors semelparity or other life histories weighted toward early reproduction, complex population dynamics, including chaos, may result (Ferrière and Clobert 1992). More work is needed to see how reliably the new model predicts results sufficiently and meaningfully different from classic expectations to be detected in realistic experiments using bacteria or insects. In any case, such efforts are unlikely ever to be complemented by work with primates.

What Causes Trade-offs?

Trade-offs are an essential component of life history theory. They represent the intrinsic constraints on multi-trait responses to selection revealed when changes in one trait increase fitness while changes in others decrease fitness. We do not yet understand the genetic and physiological causes of trade-offs; that is, we do not understand the developmental mechanisms of change and failure to change under selection (Schlichting and Pigliucci 1998). As yet, trade-offs are black boxes within theories that are far more explicit about mechanisms among organisms and their populations than among and within organisms' cells.

Can dissection of the complex connections between genotype and phenotype reveal clear mechanisms underlying trade-offs? Trade-offs may result from conflicts among whole-organism functions over whole-genome patterns of gene expression. Such conflicts can now be measured in straightforward steps for any species whose genome has been sequenced and for which, therefore, a whole-genome microarray can be produced to measure mRNA levels in tissue samples. That has been done for one primate, *Homo sapiens,* and the possibility of doing it for another, the chimpanzee, is under discussion. The lessons learned from functional genomics, however, depend not on the species chosen, but rather on the experimental designs from which data emerge. Two aspects of design are particularly important here: a level of replication that yields ample statistical power, and the possibility of performing evolutionary experiments to test preliminary interpretations directly. Progress on this front, then, will continue to be more rapid with fruit flies, nematode worms, and laboratory mice than with primates, in which sample sizes are inevitably small and evolutionary experimentation is largely out of reach.

Population-Level and Community-Level Consequences of Variation in Life History and Social Behavior

We know from theory that the dynamics of heterogeneous populations are more stable than those of homogeneous populations (e.g., Doebeli and Koella 1994; Doebeli 1995; Doebeli and Ruxton 1998; Doebeli and de Jong 1999). Furthermore, we know from experiments that individual behavioral variation has important consequences for population dynamics and the stability of community-level interactions (Werner and Anholt 1996; Peacor and Werner 2001; Wellborn, Skelly, and Werner 1996). The heterogeneity within populations that stabilizes their dynamics comprises variation in both behavioral and life history traits. Can population dynamics be understood to result from the coordinated evolution of life histories and behavior? Can the evolution of life histories and behavior be understood, at least in part, as an indirect consequence of population dynamics? Answers to these fascinating questions are yet unknown.

If these questions were answered, the next ones could be posed: Can the structure and stability of communities be understood as consequences of species interactions (where the species' population dynamics are determined by life histories and behavior)? Does the answer to that question depend on the amount of variation in behavioral and life history traits within species?

Will such an audacious act of reduction become a concrete research program? The ideal model system would allow researchers to control the amounts of genetic variation and phenotypic plasticity affecting behavioral and life history traits in interacting species. Insofar as such conditions would not be easy to arrange for experimental communities of bacteria and nematodes, or of algae and zooplankton, they are almost certainly unattainable using primates. Nonetheless, there is a role here for primate fieldwork. Such work can quantify behavioral and life history variation among populations and species (Lee and Kappeler, chap. 3. this volume) by monitoring identified individuals throughout their lives (Alberts and Altmann, chap. 4, this volume; Pereira and Leigh, this volume). *That* is impossible to achieve with algae and zooplankton!

Primate Life Histories and Conservation Biology

A related topic of real importance is the integration of life history models into theoretical and practical efforts toward establishing feasible conservation strategies. The vast majority of primate species is at risk of going extinct in the near future (Myers et al. 2000; Jernvall and Wright 1998) and quite unlikely to persist without human intervention and management. A detailed understanding of primates' basic life history schedules, along with their consequences for genetic structure and population dynamics (e.g., Altmann et al. 1997; Hapke, Zinner, and Zischler 2001), constitutes a basic prerequisite for effective management of wild populations and captive propagation projects. Our best efforts to achieve conservation objectives will also require knowing whether less variable primates have population dynamics inherently less stable than those of species more variable and plastic in behavior and life history (Alberts and Altmann, chap. 4, this volume). In addition, we will need to identify spatial structures that would maximize each species' chances of persistence.

Whether our closest living relatives actually do persist will depend crucially on the economics and politics of human resource consumption and reproduction. Still, biological insights into life histories, behavior, population dynamics, and community ecology can help to make clear the ways in which, and degrees to which, humans must adjust their behavior to maximize the chances of attaining their conservation goals (Cowlishaw and Dunbar 2000).

Nonhuman Primates and Human Evolution

Issues in human life history evolution and behavioral ecology can be illuminated by comparative data from primates. Here we discuss a few examples of such research.

Oocytic Atresia and Menopause

Hawkes, O'Connell, and Blurton Jones (chap. 9, this volume) detail their grandmother hypothesis relating to human menopause. They argue that the timing of last reproduction has remained unchanged since the time of our common ancestor with chimpanzees, whereas the adaptiveness of post-menopausal longevity is quite another matter. We take this opportunity to illustrate how cytological work on primates contributes to the understanding of life history evolution. This effort underscores how life history traits (e.g., cessation of reproduction) are influenced by adaptations at "selection arenas" inside individual entities that are difficult or impossible to recognize through ecological or other "skin-out" biology alone.

By the third month following conception, ovaries containing about seven million oocytes have developed in human embryos. By parturition, that number has fallen to about one million, and by the onset of menses it is less than 500. Such oocyte attrition, called oocytic atresia in the medical literature, is shared by most mammals and continues until all oocytes are lost by the end of the reproductive life span (e.g., humans, rhesus macaques, and bonobos: Finch and Sapolsky 1999).

Jansen and de Boer (1998) and Krakauer and Mira (1999) provide an adaptive explanation for this attrition. Because mitochondria reproduce asexually and pass regularly through population bottlenecks, they accumulate deleterious mutations rapidly (Muller's ratchet: Muller 1964), so inevitably all would eventually be crippled. This problem could be circumvented if oocytes containing mitochondria with deleterious mutations were selectively purged or lost. The cells with defective mitochondria might issue signals used by maternal tissue to neglect or kill specifically those oocytes.

This hypothesis makes three predictions: First, the percentage of defective mitochondria isolated from embryonic ovaries should decline toward birth. If such a trend were demonstrated, however, it alone would not establish that the function of selective attrition is ridding females of oocytes with damaged mitochondria; factors invoking the pattern might simply correlate with the presence of defective mitochondria. Second, whatever factor(s) invokes the predicted pattern of selective oocyte loss should correlate reliably with the presence of damaged mitochondria. The nature of the putative factor is unknown. Finally, the number of mitochondria allocated to new oocytes should be one, or very low. Jansen and de Boer's (1998) review of microphotographs of primordial oocytes, in fact, suggests that the number is less than ten. If many mitochondria entered primordial oocytes, selection would be difficult because most oocytes would receive some defective mitochondria, and biochemical factors associated with cells containing more or fewer would be relatively indistinct. Estimating the number of mitochondria that enter primordial oocytes thus becomes a focal objective.

Whereas several adaptive explanations have been proposed for menopause (reviews in Gosden et al. 1998; Leidy 1999; cf. Kaplan et al. 2000; Hawkes, O'Connell, and Blurton Jones, chap. 9, this volume), Gosden et al. (1998) emphasize that menopause may be a straightforward consequence of oocytic atresia. Variation spanning differences of years in menopausal onset derive from very small variation in numbers of oocytes created and lost before birth. The rate of oocytic atresia does correlate with the age of "menopause" in mice (J. L. Tilly, pers. comm.), suggesting that mammalian menopause occurs when a female runs out of oocytes. The cytology provides useful corroboration of Hawkes et al.'s counterpart evolutionary analysis, offering a perspective from the cellular level that also indicates that no explicitly adaptive explanation of menopause per se is needed.

Choosing Mates for Disease Resistance

Perspectives on mate choice in humans developed independently along evolutionary and medical tracks before merging recently. The evolutionary track began with Hamilton and Zuk's (1982) suggestion that birds select mates on the basis of honest signals indicating genetic resistance to parasites and pathogens. If variation in genetic quality can be detected in external phenotypes, then it pays to select high-quality mates. (Experimental confirmation in stalk-eyed flies revealed bizarre underlying mechanisms: Wilkinson, Presgraves, and Crymes 1998.) Wedekind (1994) extended this idea to other levels: mothers, or ova, could differentially favor particular sperm before, during, or after zygote formation, for example, to increase proportions of heterozygous offspring (see also Birkhead and Møller 1993; Eberhard 1996; Pereira, Clutton-Brock, and Kappeler 2000).

As with the oocytic attrition underlying menopause, we encounter here a "selection arena," or a selection process inside an entity that is itself a unit of selection at a higher level. The nested process is an adaptation of the greater entity's—in this case a human's—reproductive physiology (Stearns 1987). Selection on variation in reproductive performance at the higher level has shaped the selection process at the lower, internal level. As we will see next, selection arenas of considerable external consequence clearly do operate within humans.

While the evolutionary ideas were being developed, Carol Ober and her colleagues were studying the Hutterite communities of South Dakota. The Hutterites live in small endogamous communities, relatively inbred, with large sibships, a communal lifestyle, and the smallest number of 5-locus HLA haplotypes that are known for any couples. For some Hutterite women, recurrent spontaneous abortions are a major difficulty. Ober et al. (1992) found that women with particularly high rates of abortions had husbands with similar HLA loci. It thus appeared that their reproductive tracts were rejecting embryos that were likely to die of infectious disease during childhood. Ober et al. (1997) then asked whether potential mates generally avoided partners with similar immune genes. They found significantly fewer marital matches of HLA haplotypes than would have occurred had couples united without regard to haplotype. Hutterites, then, somehow generally avoid mating with partners of the same HLA haplotype. The mechanism used to identify potential partners' immune genes is probably olfactory, at least in part (Jacob et al. 2002).

Thus, inbred humans can detect genetic variation in potential for disease resistance among potential mates and select mates partly on the basis of that information. Both inbred and seminatural populations of mice also do so; in this case, the signals also appear to be olfactory (Yamasaki et al. 1990). In outbred human populations, in which the probability of choosing a mate with the same MHC haplotype is very low, other factors affecting pairing are probably more important. It would be valuable for theoretical and conservation purposes to determine which, if any, nonhuman primate societies exhibit patterns of mating or reproduction that correlate with MHC haplotype (e.g., Sauermann et al. 2001) and whether degree of inbreeding influences the correlation.

Where Do Human Life Histories Fit among Primates?

Nonhuman primate life history variation provides the most immediate and meaningful context for research on adaptations in human life histories. Only solid data on central tendencies and degrees of plasticity for many traits across a wide spectrum of species (cf. appendix, this volume) permit us to distinguish plesiomorphic and derived life history traits in humans.

Huge brains and postmenopausal longevity are well-known human idiosyncrasies that have generated much debate and research over the past few decades. But it is our sharing of so many basic life history traits with our closest living relatives that enables research on our peculiar traits to inform interpretations of evolutionary origins in humans (Hawkes, O'Connell, and Blurton Jones, chap. 9, this volume).

The nature and origin of morphological sex differences in humans, for example, have long been a domain of particular interest to anthropologists (e.g., Martin 1980; Krishtalka, Stucky, and Beard 1990; Plavcan and van Schaik 1997). Hall (1982) and Bailey and Katch (1981) reviewed human sexual dimorphism. Adult males are 5–10% taller than their female counterparts in all societies. Women have wider hips; men have broader shoulders. Men are about 15% fat and 52% muscle by mass, on average, whereas women are about 27% fat and 40% muscle.

Explanations of sexual dimorphism offered for humans and other primates include ecological specialization of the sexes (e.g., men hunt, women forage) and polygyny, causing intrasexual selection via male-male competition for mates (Leutenegger and Cheverud 1982; Shine 1989; Kappeler 1990; Martin, Willner, and Dettling 1994; Mitani, Gros-Louis, and Richards 1996). Despite much comparative work on this issue, no field study of a primate has yet estimated the fractions of observed variation that are attributable to ecological specialization and sexual selection, respectively (cf. Plavcan, van Schaik, and Kappeler 1995). Such an estimate would require the partitioning of lifetime reproductive success into components using the kind of partial regression analysis developed by Lande and Arnold (1983). Perhaps it could be done in the Amboseli baboon study described by Alberts and Altmann (chap. 4, this volume) or the Tai chimpanzee study recently summarized by Boesch and Boesch-Achermann (2000), in which the most detailed longitudinal observations on individuals exist.

Functional links between sexual dimorphism and life history traits, particularly patterns of growth, have only begun to be explored (Allal 1998; Bogin 1999; Lancaster et al. 2000; Hawkes, O'Connell, and Blurton Jones, chap. 9, this volume; Pereira and Leigh, chap. 7, this volume). The differential existence of growth spurts and the relative timing of spurts and sexual maturation are important matters for socioecological research. Female primates clearly benefit from achieving adult stature and body fat proportions before initiating reproduction. But why do some juvenile females suppress growth for months or years before growing rapidly later (Pereira and Leigh, chap. 7, this volume)? And why do some male primates somatically conceal their adult status for their first years of fertility before suddenly completing growth to adult size and proportions? While they may do so to avoid aggression from adult males while sneaking copulations, thereby advancing their age at first reproduction, there are many other possibilities (see above).

Alternative explanations for characteristic male and female growth schedules, and the consequences of the patterns seen in each species (e.g., Pereira 1995), all call for investigation across the spectrum of primate social systems. Again, expanding the comparative database is essential to providing context for research on humans, and unequivocal answers are also more approachable in chimpanzees and other nonhuman primates than in humans, in which it is difficult, at best, to get unfiltered data on sexual behavior and actual paternity patterns.

Summary and Conclusions

We began this concluding chapter by taking a step back and asking whether, and for which questions, primates are good subjects for studies of life history evolution. Looking forward, we argue that work on primate life histories is important to comparative analyses of mammals and essential to garnering insight on the evolution of very large brains, juvenility, and advanced cognitive abilities. A detailed understanding of the diversity and plasticity of life history traits is also indispensable for research-based attempts to develop protocols for sustainable conservation efforts with endangered primates.

Of the three basic open issues in life history evolution identified here— reinventing the theory with complex dynamics and spatial structure, understanding the mechanisms that produce trade-offs, and relating population dynamics to behavioral and life history variation among individuals—work on primates will most likely contribute primarily to the third.

There are many cases in which the comparative perspective can lend insight into adaptive features of human life histories. The contribution of putative purging of defective oocytes to the timing of menopause and the genetic and physiological bases of human mate choice provide compelling examples, which illuminate relations between internal selection arenas and external selection dynamics.

Finally, using sexual dimorphism as an example, we indicate how comparative behavioral data from nonhuman primates can generate promising hypotheses about the origin and function of life history traits in human and nonhuman primates. Taken together, these examples make clear that primate life histories yet hold many fascinating secrets, and that research interrelating them to age-, sex-, and species-typical patterns of behavior awaits contributions by a new generation of innovative organismal biologists.

Acknowledgments

S. C. S. thanks Peter Kappeler for inviting him to Göttingen to speak in the symposium that gave birth to this book. He also thanks former students and colleagues, who have done path-breaking work on primates and on evolutionary and behavioral ecology cited here, for intellectual companionship and inspiration: Susan Alberts, Nadine Allal, Christophe and Hedwige Boesch, Michael Doebeli, Pascal Gagneux, Tad Kawecki, and Jacob Koella. M. E. P. and P. M. K. thank Sonia Cavigelli, Dario Maestripieri, and especially David Watts for comments on early drafts of this chapter.

Appendix
A Primate Life History Database

Family	Genus	Species	BM	BW	CC	AFR	GL	IBI	LS	NNW	LW	LGR	WA	WW	DJ
Cheirogaleidae	Allocebus	trichotis	83												
Cheirogaleidae	Cheirogaleus	major	356	7			71			~18.1			70		
			362												
			403												
Cheirogaleidae	Cheirogaleus	adipicaudatus													
Cheirogaleidae	Cheirogaleus	crossleyi													
Cheirogaleidae	Cheirogaleus	sibreei													
Cheirogaleidae	Cheirogaleus	ravus													
Cheirogaleidae	Cheirogaleus	minusculus													
Cheirogaleidae	Cheirogaleus	medius	139			1.19	62	12	2	12	38	2.64	61		0.86
			172			1.4									
Cheirogaleidae	Microcebus	berthae	30												
Cheirogaleidae	Microcebus	griseorufus	63												
Cheirogaleidae	Microcebus	murinus	63	1.8		1	60	12	2	4.6	12.3	1.35	40		0.73
			72							7					
Cheirogaleidae	Microcebus	myoxinus	49												
Cheirogaleidae	Microcebus	ravelobensis	56												
Cheirogaleidae	Microcebus	rufus	42												
Cheirogaleidae	Microcebus	sambiranensis	44												
Cheirogaleidae	Microcebus	tavaratra	61												
Cheirogaleidae	Mirza	coquereli	297			1	87	12	2	14	22.5				
			302			1.92				~17.5					
			326												
Cheirogaleidae	Phaner	furcifer	328												
Daubentoniidae	Daubentonia	madagascariensis	2,490	44		3.5	164	20	1	102	109	3.81	170	1,535	2.04
			2,572				169			147			365		
			2,800												
Lepilemuridae	Lepilemur	dorsalis	~500												

Family	Genus	species											
Lepilemuridae	*Lepilemur*	*edwardsi*	934										
Lepilemuridae	*Lepilemur*	*leucopus*	594										
Lepilemuridae	*Lepilemur*	*microdon*											
Lepilemuridae	*Lepilemur*	*mustelinus*	~777 602		1.83 1.88	135			~27	34.7	2.65	75 76 120	1.3
Lepilemuridae	*Lepilemur*	*ruficaudatus*	779	7.4	1.63								
Lepilemuridae	*Lepilemur*	*septentrionalis*	~750										
Lemuridae	*Eulemur*	*albocollaris*	2,150										
Lemuridae	*Eulemur*	*collaris*											
Lemuridae	*Eulemur*	*coronatus*	1,080 1,177			125		1	48 ~70				
Lemuridae	*Eulemur*	*fulvus*	2,206 2,250 2,428	25.6	2.16 2.66 2.83	120	24	1	67.8 76 ~83.3	74.7	9.75	135 183	1.46
Lemuridae	*Eulemur*	*macaco*	1,760 1,793 2,428 2,510	24.6	1.66 2.18 2.58	129	12	1	50 74	100		135	
Lemuridae	*Eulemur*	*mongoz*	1,481 1,560 1,890	23.8	2.5 2.52	129		1	57 59.8	55		152	1.75
Lemuridae	*Eulemur*	*rubriventer*	1,940 1,956			123			82 89				
Lemuridae	*Hapalemur*	*aureus*	1,390				12						
Lemuridae	*Hapalemur*	*griseus*	670 709 790	14.6	2.38 2.73 2.92	140 145	11	1	~45.2 54	56.7		120 122 154	1.67
Lemuridae	*Hapalemur*	*griseus alaotrensis*	~1,450										
Lemuridae	*Hapalemur*	*simus*	1,300										

continued

Family	Genus	Species	BM	BW	CC	AFR	GL	IBI	LS	NNW	LW	LGR	WA	WW	DJ
Lemuridae	Lemur	catta	2,210	22.6	23.4	2.01	135	14	1	65	103	9.15	105		1.35
			2,290			2.25				85			179		
						2.63				~88.2					
Lemuridae	Varecia	variegata	3,100	102.1	31.2	1.42	102	12	2	78.4	90	1.95	89	2,893	
			3,520			1.92	175			~87.2			90		
			3,600			2.72				100			146		
													150		
Indriidae	Avahi	laniger	875	10.6	9.8	2.58		12							
			1,050												
			1,316												
			1,320												
Indriidae	Avahi	occidentalis	777												
Indriidae	Avahi	unicolor													
Indriidae	Propithecus	tattersalli	3,500		30.5	4.5				98			153		
			3,590												
Indriidae	Propithecus	verreauxi	2,950	26.7	29.7	3.5	140			103.2	107		180		2.62
			3,183			4.57							183		
			3,620												
Indriidae	Propithecus	coquereli	3,244			4.2	141	12	1	103			182		
			4,280												
Indriidae	Propithecus	diadema	6,260		40.0	4	178	25		~135			183		
						4.51							363		
Indriidae	Indri	indri	6,250	38.3			159	30			300		363		
			6,840										365		
Lorisidae	Arctocebus	aureus	~210												
Lorisidae	Arctocebus	calabarensis	254			1.12	135	6		32	25.2	2.14	105	160	0.47
			298			1.17				37			115		
			306												
Lorisidae	Loris	grandis	269												
Lorisidae	Loris	lydekkerianus	193												
Lorisidae	Loris	malabaricus						1							

Family	Genus	species													
Lorisidae	Loris	nordicus	255												
Lorisidae	Loris	nycticeboides													
Lorisidae	Loris	tardigradus				1.5	166	6		10 ~11	11.4	0.76	170	139	0.58
Lorisidae	Nycticebus	bengalensis	1,020	12.8											
Lorisidae	Nycticebus	coucang	626 630			2.11	193 195	12	1	50.8 52	49.3	5.41	180 182	520	1.09
Lorisidae	Nycticebus	intermedius	798												
Lorisidae	Nycticebus	javanicus	511												
Lorisidae	Nycticebus	menagensis	~307												
Lorisidae	Nycticebus	pygmaeus					185	2		20 ~23 ~17.8					
Lorisidae	Nycticebus	intermedius	~1,100												
Lorisidae	Perodicticus	edwardsi	1,210												
Lorisidae	Perodicticus	ibeanus	836												
Lorisidae	Perodicticus	potto	935	13.5		2.03 2.08	195	12	1	33 38.9	51.2	4.73	150 151		1.09
Lorisidae	Pseudopotto	martini													
Galagidae	Euoticus	elegantulus	261 283 293												
Galagidae	Euoticus	pallidus	~300												
Galagidae	Galago	alleni	262 269			0.83 1.04	135			24	31.2	1.82			
Galagidae	Galago	gabonensis	212												
Galagidae	Galago	gallarum	173												
Galagidae	Galago	matschiei	179												
Galagidae	Galago	moholi	188	5	3.6	0.71 1	123	6	1	11.8 15	18.4	1.66	84 100	95	0.39

continued

Family	Genus	Species	BM	BW	CC	AFR	GL	IBI	LS	NNW	LW	LGR	WA	WW	DJ
Galagidae	Galago	senegalensis	195 199			1.4	142	6		19 19.5	20.3	1.75	98	150	0.74
Galagidae	Galagoides	demidoff	60 63	3.3		0.97 1	110	12	2	8 ~9.9	9	1.07			
Galagidae	Galagoides	granti													
Galagidae	Galagoides	orinus													
Galagidae	Galagoides	rondoensis													
Galagidae	Galagoides	thomasi	130												
Galagidae	Galagoides	udzungwensis													
Galagidae	Galagoides	zanzibaricus	132 137			1	126	6		~14.1 15	16.8	1.73			
Galagidae	Otolemur	crassicaudatus	1,110 1,120			2.17 2.21	135	12	2	~43.2 46	70.7	10.16	135	500	1.47
Galagidae	Otolemur	garnettii	721 734 739	9.8		1.58	132	7	1	49	49.7				
Tarsiidae	Tarsius	bancanus	109 117 107			2.52	178	8	1	~23.6 24	24.4	0.82	79		1.82
Tarsiidae	Tarsius	dianae							1						
Tarsiidae	Tarsius	pumilis							1						
Tarsiidae	Tarsius	spectrum	108 220			1.42		5	1	~23.1	28.5		68		
Tarsiidae	Tarsius	syrichta	120	3.3			180		1		26.2		82		
Cebidae	Aotus	azarai	1,230					12	1				231		
Cebidae	Aotus	hershkovitzi													
Cebidae	Aotus	lemurinus	874				133	12	1	98			75		
Cebidae	Aotus	miconax													
Cebidae	Aotus	nancymaae	780												
Cebidae	Aotus	nigriceps	1,040												

Family	Genus	Species													
Cebidae	*Aotus*	*trivirgatus*	724 736 698	17.2	16.1	2.38 2.42	133		1	94	97	3.5	75 179	360	1.81
Cebidae	*Aotus*	*vociferans*													
Cebidae	*Cebus*	*albifrons*	2,067 2,290	60.7	56.8	4.02	155	18	1	227.7 234	234	4.7	269 270 274		2.85
Cebidae	*Cebus*	*apella*	2,201 2,520	63.5	63.1	5.5 5.78	154	22	1	197.1 232	239.7	2.87	261 264 416	1,000	
Cebidae	*Cebus*	*capucinus*	2,540 2,578	71.9		4	162	26	1	~230 239 250	230	2.17	510 517	1,350	
Cebidae	*Cebus*	*kaapori*													
Cebidae	*Cebus*	*libidinosus*													
Cebidae	*Cebus*	*nigritus*													
Cebidae	*Cebus*	*olivaceus*	2,520			6							730		
Cebidae	*Cebus*	*xanthosternos*													
Cebidae	*Saimiri*	*boliviensis*	711 680												
Cebidae	*Saimiri*	*oerstedii*													
Cebidae	*Saimiri*	*sciureus*	662 699	23	23.2	2.5	170	9	1	106.4 146	95.2	1.78	168 183 240	418	
Cebidae	*Saimiri*	*ustus*	799 650												
Cebidae	*Saimiri*	*vanzolinii*													
Callitrichidae	*Callimico*	*goeldii*	355 ~582	11		1.32	151 155	9	1	48 53.3	50.6	2.75	65 70	215	0.7
Callitrichidae	*Cebuella*	*pygmaea*	79 122			1.88	137		2	12 14.1	31.5	0.9	90	70	1.25
Callitrichidae	*Callithrix*	*aurita*	429												
Callitrichidae	*Callithrix*	*flaviceps*	~406						2						

continued

Family	Genus	Species	BM	BW	CC	AFR	GL	IBI	LS	NNW	LW	LGR	WA	WW	DJ
Callitrichidae	*Callithrix*	*geoffroyi*	~359												
Callitrichidae	*Callithrix*	*jacchus*	287 381	8	7.7	1.37 1.5	148		2	27 30.23	56.7	2.98	60 77 90	128	0.85
Callitrichidae	*Callithrix*	*kuhlii*	~375												
Callitrichidae	*Callithrix*	*penicillata*	307						2						
Callitrichidae	*Leontopithecus*	*caissara*	572												
Callitrichidae	*Leontopithecus*	*chrysomelas*	535												
Callitrichidae	*Leontopithecus*	*chrysopygus*													
Callitrichidae	*Leontopithecus*	*rosalia*	559 ~598	12.3		2.38 2.42	129	6	2	50 62.4	100	2.56	90	165	1.78
Callitrichidae	*Saguinus*	*bicolor*	430												
Callitrichidae	*Saguinus*	*fuscicollis*	350 358		8.2	2.33 2.5	148	12	2	38.8 40	79.8		90		1.68
Callitrichidae	*Saguinus*	*geoffroyi*	502						2	48.1					
Callitrichidae	*Saguinus*	*graellsi*													
Callitrichidae	*Saguinus*	*imperator*	475							46.8					
Callitrichidae	*Saguinus*	*inustus*	803?												
Callitrichidae	*Saguinus*	*labiatus*	520 529				145	10	2	38 44 44	87				
Callitrichidae	*Saguinus*	*leucopus*	490												
Callitrichidae	*Saguinus*	*martinsi*													
Callitrichidae	*Saguinus*	*midas*	558 575			2		7	2	40	80		70, 71		
Callitrichidae	*Saguinus*	*mystax*	539						2	46.9					
Callitrichidae	*Saguinus*	*niger*													
Callitrichidae	*Saguinus*	*nigricollis*	350 484		8.9	2.33			1	43.9	87	3.68	77 83 84	175	

Callitrichidae	Saguinus	oedipus	404 425	9.6	1.89	168		2	42.1 44	83.6	3.5	50	130	1.29
Callitrichidae	Saguinus	tripartitus												
Callitrichidae	Mico	argentatus	353 360		1.67		7	2	36	70.2	3.4			
Callitrichidae	Mico	acariensis												
Callitrichidae	Mico	leucippe												
Callitrichidae	Mico	intermedius						2						
Callitrichidae	Mico	chrysoleucus												
Callitrichidae	Mico	emiliae	330											
Callitrichidae	Mico	humeralifer	380						~34.5					
Callitrichidae	Mico	humilis												
Callitrichidae	Mico	mauesi	398											
Callitrichidae	Mico	melanurus												
Callitrichidae	Mico	nigriceps	390					2						
Callitrichidae	Mico	marcai												
Callitrichidae	Mico	manicorensis												
Callitrichidae	Mico	saterei												
Pitheciidae	Cacajao	calvus	2,880 3,600			180						638		
Pitheciidae	Cacajao	melanocephalus	2,710											
Pitheciidae	Callicebus	baptista												
Pitheciidae	Callicebus	barbarabrownae												
Pitheciidae	Callicebus	brunneus	805											
Pitheciidae	Callicebus	cinerascens												
Pitheciidae	Callicebus	coimbrai												
Pitheciidae	Callicebus	cupreus	1,120			130	13	1	74			203		
Pitheciidae	Callicebus	donacophilus	909											
Pitheciidae	Callicebus	hoffmansi	1,030											
Pitheciidae	Callicebus	medemi												
Pitheciidae	Callicebus	melanochir												
Pitheciidae	Callicebus	modestus												

continued

Family	Genus	Species	BM	BW	CC	AFR	GL	IBI	LS	NNW	LW	LGR	WA	WW	DJ
Pitheciidae	Callicebus	moloch	956			3	164			74.4	100		60		2.39
			1,004												
Pitheciidae	Callicebus	nigrifrons													
Pitheciidae	Callicebus	oenanthe													
Pitheciidae	Callicebus	olallae													
Pitheciidae	Callicebus	ornatus													
Pitheciidae	Callicebus	pallescens	1,380												
Pitheciidae	Callicebus	personatus	1,210												
Pitheciidae	Callicebus	torquatus	2,490												
Pitheciidae	Chiropotes	albinasus	2,580												
Pitheciidae	Chiropotes	satanas	3,100												
Pitheciidae	Pithecia	aequatorialis	~2,250												
Pitheciidae	Pithecia	albicans	2,070												
Pitheciidae	Pithecia	irrorata	2,110	32.2						120	121	2.66			
Pitheciidae	Pithecia	monachus	2,170							121					
Pitheciidae	Pithecia	pithecia	1,580			2.08	164								
			1,604												
Atelidae	Alouatta	belzebul	5,520												
Atelidae	Alouatta	caraya	4,330		47.9	3.71	187			262			325		
			4,882												
Atelidae	Alouatta	coibensis													
Atelidae	Alouatta	guariba													
Atelidae	Alouatta	nigerrima													
Atelidae	Alouatta	palliata	4,020	51.5	55.4	3.58	186	20	1	320	528	1.36	325	1,100	1.35
			5,350			3.99							495		
			5,824										630		
			6,600												
Atelidae	Alouatta	pigra	6,430							~480					

Family	Genus	species													
Atelidae	*Alouatta*	*sara*	4,670; 5,210; 5,807; 6,020												
Atelidae	*Alouatta*	*seniculus*	7,850; 9,330			4.58	191			295		3.5	372		3.04
Atelidae	*Ateles*	*belzebuth*													
Atelidae	*Ateles*	*chamek*													
Atelidae	*Ateles*	*fusciceps*	9,160; 9,163	108.8		4.86	226					6	486		2.91
Atelidae	*Ateles*	*geoffroyi*	7,290; 7,669	104.9	106.4	5.62; 6.39	225; 229	37	1	426; ~512	426	1.92	750; 821	2,000	2.76
Atelidae	*Ateles*	*hybridus*													
Atelidae	*Ateles*	*marginatus*													
Atelidae	*Ateles*	*paniscus*	8,440; 8,554	101.2		5	230	24	1	425; 480	452.5	4.39	760; 810	3,790	
Atelidae	*Brachyteles*	*arachnoides*	8,070			7.5	233	34	1				638; 855		
Atelidae	*Brachyteles*	*hypoxanthus*													
Atelidae	*Lagothrix*	*cana*													
Atelidae	*Lagothrix*	*lagothricha*	5,585; 7,020	82.6		5; 7.58	223; 225	24	1	432; 447	450		315; 507		3.53
Atelidae	*Lagothrix*	*lugens*													
Atelidae	*Lagothrix*	*poeppigii*													
Atelidae	*Oreonax*	*flavicauda*	~10,000												
Cercopithecidae	*Allenopithecus*	*nigroviridis*	3,180; 3,225	57						242					
Cercopithecidae	*Miopithecus*	*talapoin*	1,120; 2,000	37.3		4.38	162; 165	12	1	~177.5; 180	175	1.1	180; 195	450	3.44
Cercopithecidae	*Erythrocebus*	*patas*	6,317; 6,500	100		3	167	12	1	468; 625	504.5	6.78	213; 255	1,950	1.96
Cercopithecidae	*Chlorocebus*	*aethiops*	2,980	59.2		4.75							219		

continued

Family	Genus	Species	BM	BW	CC	AFR	GL	IBI	LS	NNW	LW	LGR	WA	WW	DJ
Cercopithecidae	Cercopithecus	aethiops	2,980 3,100 3,469	60.1		5 3.5	163	12	1	335.9 430	314	4.2	201 365	1,170	3.55
Cercopithecidae	Cercopithecus	ascanius	2,920 2,943	61.3		5	172			~371	371				
Cercopithecidae	Cercopithecus	campbelli	2,200 2,700										365		
Cercopithecidae	Cercopithecus	cephus	2,805 2,880			5	170			339	340	2.87			
Cercopithecidae	Cercopithecus	diana	3,900 4,533			5.33 5.42		12	1	460 479	475		365		
Cercopithecidae	Cercopithecus	dryas													
Cercopithecidae	Cercopithecus	erythrogaster	2,400												
Cercopithecidae	Cercopithecus	erythrotis	~2,900												
Cercopithecidae	Cercopithecus	hamlyni	3,360												
Cercopithecidae	Cercopithecus	lhoesti	3,450 4,700	84.5											
Cercopithecidae	Cercopithecus	mitis	3,930 4,250 4,280 4,910	56.5		5.72 5.83		24	1	~337 398 ~402 495	402	3.5	692 699		
Cercopithecidae	Cercopithecus	mona	2,500	62						282	284				
Cercopithecidae	Cercopithecus	neglectus	3,550 4,081 4,130			4.67	165 168	12	1	260 450	260	3.2	365 420	1,640	3.22
Cercopithecidae	Cercopithecus	nictitans	4,216 4,260				170			406		2.4			
Cercopithecidae	Cercopithecus	petaurista	2,900												
Cercopithecidae	Cercopithecus	pogonias	2,900 3,021			5	170		1	339	340	2.57			

Family	Genus	Species													
Cercopithecidae	Cercopithecus	preussi	~4,500												
Cercopithecidae	Cercopithecus	salongo													
Cercopithecidae	Cercopithecus	sclateri	~2,500												
Cercopithecidae	Cercopithecus	solatus	3,920												
Cercopithecidae	Cercopithecus	wolfi	2,870												
Cercopithecidae	Macaca	arctoides	8,400 8,523	99.8		4.75	178	18 19	1 1	~435 489 497	487	4.55	393 398	2,300	
Cercopithecidae	Macaca	assamensis	6,900												
Cercopithecidae	Macaca	brunnescens													
Cercopithecidae	Macaca	cyclops	4,940 6,200			3.84 4.25	162	15	1	398 402			206		
Cercopithecidae	Macaca	fascicularis	3,574 3,590	62.1	62.5	3.9	160 167	13	1	326.1 375	345	5.86	330 420 427	1,700	2.27
Cercopithecidae	Macaca	fuscata	8,030 9,100			5.54	173	24	1	503 526.8	496	6.27	365 540	2,730	4.07
Cercopithecidae	Macaca	hecki													
Cercopithecidae	Macaca	maura	6,050			5		22	1	389 ~390			500		
Cercopithecidae	Macaca	mulatta	5,370 5,445 6,500 7,970 8,800	84.7	81.3	3 4.19 4.5	165 164	12	1	466.3 473	475	4.39	192 321 365	1,454	3.05
Cercopithecidae	Macaca	nemestrina	4,900 5,571 6,500	110	96.2	3.92	167 171	14	1	443.7 473	472	2.6	234 365	1416.5	2.46
Cercopithecidae	Macaca	nigra	4,600 5,470	97.5		5.44 5.5	170	18	1	457 ~461	455	3.53			
Cercopithecidae	Macaca	nigriscens													
Cercopithecidae	Macaca	ochreata	2,600												

continued

Family	Genus	Species	BM	BW	CC	AFR	GL	IBI	LS	NNW	LW	LGR	WA	WW	DJ
Cercopithecidae	Macaca	*radiata*	3,700 3,850			4	162	12	1	388 407	404	4.37	365	2,000	2.56
Cercopithecidae	Macaca	*sinica*	3,200 3,590									5			
Cercopithecidae	Macaca	*silenus*	5,000			4.92	180 182	17	1	407 438	407		365		3.41
Cercopithecidae	Macaca	*sylvanus*	8,280 8,283 ~11,000			4.75 4.8	165	22	1	450		4.28	210	2,420	
Cercopithecidae	Macaca	*thibetana*	9,500 12,800				170	24	1	500 ~550		3.34	561 568	2,400	
Cercopithecidae	Macaca	*tonkeana*	9,000												
Cercopithecidae	Mandrillus	*leucophaeus*	8,450 ~12,500	124		5	173			722					
Cercopithecidae	Mandrillus	*sphinx*	11,350 12,900			4	175 220	17	1	613 890	613	6.53	348 350	3,000	2.56
Cercopithecidae	Cercocebus	*albigena*	6,209 5,660	94.4		4.08	175				425	5	365	2,170	2.61
Cercopithecidae	Cercocebus	*agilis*	6,200												
Cercopithecidae	Cercocebus	*atys*	5,260 5,473			6.5	171			530					
Cercopithecidae	Cercocebus	*galeritus*	5,500 7,420	88.8								4.04			
Cercopithecidae	Cercocebus	*torquatus*		110.5		4.67	171								
Cercopithecidae	Lophocebus	*albigena*	6,020				186	33	1	500			210		
Cercopithecidae	Lophocebus	*aterrimus*	5,760												
Cercopithecidae	Papio	*anubis*	11,700 13,300 14,700		158.9	4.5	180	25	1	915 950			584 600		

Family	Genus	species												
Cercopithecidae	*Papio*	*cynocephalus*	9,750 11,000 11,532 12,300 13,600	164	145.5	5.5	173 175	23	1	710 803	854	5.1	365 450 456	2,500	4.03
Cercopithecidae	*Papio*	*hamadryas*	9,900 10,404 11,400	142.6		6.1	170	24	1	695 1,000			561	3,100	
Cercopithecidae	*Papio*	*papio*	12,100 16,166	192			187			603.6					
Cercopithecidae	*Papio*	*ursinus*	14,773 14,800	164.5		3.67	187	29	1	600 850					
Cercopithecidae	*Theropithecus*	*gelada*	11,427 11,700	119.7		4	170 180	24	1	465	553	6.12	540 547	3,900	2.3
Cercopithecidae	*Procolobus*	*verus*	4,200		64.8	4.08									
Cercopithecidae	*Procolobus*	*badius*	7,421 8,210 8,250	80.2									522 790.4		
Cercopithecidae	*Colobus*	*angolensis*	7,570												
Cercopithecidae	*Colobus*	*guereza*	7,900 8,102 9,200	72	69.0	4.75	170	20	1	445 549	395.5	3.6	330 390 394	1,600	
Cercopithecidae	*Colobus*	*polykomos*	7,662 8,300	81.4		8.5	170	22	1	400 ~600	597	6	215 219	1,240	
Cercopithecidae	*Colobus*	*satanas*	7,420												
Cercopithecidae	*Colobus*	*vellerosus*	6,900												
Cercopithecidae	*Semnopithecus*	*entellus*	6,910 9,890 10,280 10,900 14,800		98.2	3.42 3.9	184 200	17	1	500	500	6.34	249 396 416	2,100	
Cercopithecidae	*Semnopithecus*	*johnii*													

continued

Family	Genus	Species	BM	BW	CC	AFR	GL	IBI	LS	NNW	LW	LGR	WA	WW	DJ
Cercopithecidae	Semnopithecus	vetulus	5,797 5,900		56.6	4	200	32	1	360 ~447	360	2.7	219 270 274	1,100	
Cercopithecidae	Trachypithecus	auratus	5,760	66	54.5	4									
Cercopithecidae	Trachypithecus	cristata	5,856										365		
Cercopithecidae	Trachypithecus	francoisi	7,300 7,350		66.9	4				457			394		
Cercopithecidae	Trachypithecus	geei	9,500												
Cercopithecidae	Trachypithecus	johnii	11,200												
Cercopithecidae	Trachypithecus	obscura	6,260 6,530	60.8	60.8	4				341 380	380		365		
Cercopithecidae	Trachypithecus	phayrei	6,300 10,500				165	15	1				305		
Cercopithecidae	Trachypithecus	pileatus	9,860												
Cercopithecidae	Presbytis	comata	6,710												
Cercopithecidae	Presbytis	frontata	5,670												
Cercopithecidae	Presbytis	hosei	5,630												
Cercopithecidae	Presbytis	melalophos	6,470												
Cercopithecidae	Presbytis	potenziani	6,400												
Cercopithecidae	Presbytis	rubicunda	6,170												
Cercopithecidae	Presbytis	thomasi	6,690												
Cercopithecidae	Nasalis	larvatus	9,593 9,820	87	82.8	4.5	166	18	1	450	525	6.93	210 213 281	2,000	
Cercopithecidae	Simias	concolor	6,800												
Cercopithecidae	Pygathrix	nemaeus	8,180 8,440				210			463					
Cercopithecidae	Rhinopithecus	avunculus	8,000 8,500												

Family	Genus	Species													
Cercopithecidae	*Rhinopithecus*	*bieti*	9,960												427
Cercopithecidae	*Rhinopithecus*	*brelichi*	11,600												
Cercopithecidae	*Rhinopithecus*	*roxellana*	5,530; 5,820												
Hylobatidae	*Hylobates*	*agilis*	5,749	87.6											
Hylobatidae	*Hylobates*	*concolor*	7,620			202									
Hylobatidae	*Hylobates*	*gabriellae*	6,500												
Hylobatidae	*Hylobates*	*hoolock*	6,880												
Hylobatidae	*Hylobates*	*klossii*	5,920												
Hylobatidae	*Hylobates*	*lar*	5,340; 5,465	93.9; 102.2	6.69; 9.31	205; 213	30	1	400; 407	400	0.9	548; 730; 735	1,070	6.73	
Hylobatidae	*Hylobates*	*leucogenys*	7,320												
Hylobatidae	*Hylobates*	*moloch*	5,292; 6,250	93.1		195			500						
Hylobatidae	*Hylobates*	*muelleri*	5,350	130											
Hylobatidae	*Hylobates*	*pileatus*	5,440												
Hylobatidae	*Hylobates*	*syndactylus*	10,568; 10,700	113.5	5.18; 8.92; 9	232; 234	50	1	513; 552	517		639			
Pongidae	*Pongo*	*pygmaeus*	35,700; 35,800; 37,078	301.7; 374.5	7; 9.68; 14.3	250; 244	72	1	1,728; 1,653	1,735	7.26	720; 1,277; 1,825	11,000	7.88	
Pongidae	*Pongo*	*abeli*	35,600												
Pongidae	*Gorilla*	*gorilla*	71,000; 71,500; 80,000; 82,475; 97,500	457.3; 455.6	4; 10.0; 10.3	260; 285	47	1	1,996; 2,110	2,122.9	15.32	900; 1,004; 1,278		6.58	
Pongidae	*Pan*	*paniscus*	33,200			240	48	1	1,400; 1,494	1,162	6.71	1,080; 1,094	8,500		

continued

Family	Genus	Species	BM	BW	CC	AFR	GL	IBI	LS	NNW	LW	LGR	WA	WW	DJ
Pongidae	*Pan*	*troglodytes*	30,000	380	371.1	13	235	60	1	1,750	1,742	3.97	1,680	8,500	8.36
			33,700			13.9	240			1,814			1,702		
			40,300										1,825		
			40,400												
			41,600												
			45,800												
Hominidae	*Homo*	*sapiens*	42,200	1,228		14	267	36	1	2,900	3,375	10.7	730	10,980	11.27
			45,800			15	270			3,334			930		
			49,700												
			52,500												
			53,600												
			56,300												
			62,100												
			73,200												

Note: For each currently recognized primate species, principal life history data are presented. Note the large number of empty cells and the variation in some variables used in the literature. Details are available on request from the corresponding compiling authors:

Smith, R., and W. Jungers. 1997. Body mass in comparative primatology. *J. Hum. Evol.* 32:523–59.

Godfrey, L., K. Samonds, W. Jungers, and M. Sutherland. 2001. Teeth, brains, and primate life histories. *Am. J. Phys. Anthropol.* 114:192–214.

Smith, R., and S. Leigh. 1998. Sexual dimorphism in primate neonatal body mass. *J. Hum. Evol.* 34:173–201.

C. Ross, chap. 11, this volume.

P. Lee and P. M. Kappeler, chap. 3, this volume.

BM (Body mass): mean body mass (g) of wild adult females; ~ indicates mixed or unknown sex sample

BW (Brain weight): mean brain weight (g) of adult females

CC (Cranial capacity): mean adult female cranial capacity (cc)

AFR (Age at first reproduction): mean age (years) of first female reproduction

GL (Gestation length): mean gestation length (days)

IBI (Interbirth interval): mean interbirth interval (months)

LS (Litter size): modal litter size

NNW (Neonatal weight): mean body mass (g) of neonate females; ~ indicates mixed or unknown sex sample

LW (Litter weight): litter size * birth weight

LGR (Litter growth rate): mean growth rate (g/days) of litter between birth and weaning

WA (Weaning age): mean age at weaning (days)

WW (Weaning weight): mean body mass (g) at weaning

DJ (Juvenility): duration of juvenile period (years)

Contributors

Paul-Michael Agapow
Department of Biology
Imperial College
Silwood Park
Ascot
Berkshire SL5 7PY
U.K.

Susan C. Alberts
Department of Biology
Duke University
Box 90338
Durham, NC 27708
U.S.A.

Jeanne Altmann
Department of Ecology &
Evolutionary Biology
Guyot 401
Princeton University
Princeton, NJ 08544-1003
U.S.A.

Robert A. Barton
Department of Anthropology
University of Durham DH1 3HN
U.K.

Nicholas G. Blurton Jones
Departments of Anthropology,
Psychiatry, and
Graduate School of Education
University of California
Los Angeles, CA 90024
U.S.A.

Robert O. Deaner
Department of Neurobiology

Duke University Medical Center
325 Bryan Research Building
Durham NC 27710
U.S.A.

Robin I. M. Dunbar
School of Biological Sciences
University of Liverpool
Liverpool L69 3GS
U.K.

Jörg U. Ganzhorn
Institut für Zoologie
Universität Hamburg
Martin-Luther-King Platz 3
20146 Hamburg
Germany

Laurie R. Godfrey
Department of Anthropology
University of Massachusetts
240 Hicks Way
Amherst, MA 01003-9278
U.S.A.

Kristen Hawkes
270 E. 1400 E., rm. 102
University of Utah
Salt Lake City, UT 84112-0060
U.S.A.

Nick J. B. Isaac
Department of Biology
Imperial College
Silwood Park
Ascot
Berkshire SL5 7PY
U.K.

Charles H. Janson
  Department of Ecology and
  Evolution
  Stony Brook University
  Stony Brook, NY 11794-5245
  U.S.A.

Kate E. Jones
  Department of Biology
  Imperial College
  Silwood Park
  Ascot
  Berkshire SL5 7PY
  U.K.

William L. Jungers
  Department of Anatomical Sciences
  Stony Brook University
  Stony Brook, NY 11794-8081
  U.S.A.

Peter M. Kappeler
  Abteilung
  Verhaltensforschung/Ökologie
  Deutsches Primatenzentrum
  Kellnerweg 4
  37077 Göttingen
  Germany

Susanne Klaus
  Abteilung Biochemie und
  Physiologie der Ernährung
  Deutsches Institut für
  Ernährungsforschung
  Arthur-Scheunert-Allee 114–116
  14558 Bergholz-Rehbrücke
  Germany

Phyllis C. Lee
  Department of Biological
  Anthropology
  University of Cambridge
  Downing Street
  Cambridge CB2 3DZ
  U.K.

Steven R. Leigh
  Department of Anthropology
  University of Illinois
  Urbana, IL 61801
  U.S.A.

Robert D. Martin
  The Field Museum
  1400 S. Lake Shore Drive

Chicago, IL 60605-2496
  U.S.A.

James F. O'Connell
  270 E. 1400 E., rm. 102
  University of Utah
  Salt Lake City, UT 84112-0060
  U.S.A.

Sylvia Ortmann
  Abteilung Biochemie und
  Physiologie der Ernährung
  Deutsches Institut für
  Ernährungsforschung
  Arthur-Scheunert-Allee 114–116
  14558 Bergholz-Rehbrücke
  Germany

Michael E. Pereira
  The Latin School of Chicago
  59 West North Boulevard
  Chicago, IL 60610-1492
  U.S.A.

Andy Purvis
  Department of Biology
  Imperial College
  Silwood Park
  Ascot
  Berkshire SL5 7PY
  U.K.

Caroline Ross
  School of Life Sciences
  University of Surrey Roehampton
  West Hill
  London SW15 3SN
  U.K.

Karen E. Samonds
  Department of Anatomical Sciences
  Stony Brook University
  Stony Brook, NY 11794-8081
  U.S.A.

Jutta Schmid
  Department of Experimental
  Ecology
  University of Ulm
  Albert-Einstein-Allee 11
  89069 Ulm
  Germany

Stephen C. Stearns
  Department of Ecology and
  Evolutionary Biology
  Box 208106
  Yale University
  165 Prospect Street
  New Haven, CT 06520-8106
  U.S.A.

Michael R. Sutherland
  Statistical Consulting Center
  Lederle Graduate Research Tower
  University of Massachusetts
  Amherst, MA 01003-4535
  U.S.A.

Carel P. van Schaik
  Department of Biological
  Anthropology & Anatomy
  Duke University
  Box 90383
  Durham NC 27708-0383
  U.S.A.

Andrea J. Webster
  Department of Biology
  Imperial College
  Silwood Park
  Ascot
  Berkshire SL5 7PY
  U.K.

References

Abbott, D. H. 1989. Social suppression of reproduction in primates. In *Comparative socioecology: The behavioral ecology of humans and other animals,* edited by V. Standen and R. A. Foley, 285–304. Oxford: Blackwell.

Abrahams, M., and A. Sutterlin. 1999. The foraging and antipredator behavior of growth-enhanced transgenic Atlantic salmon. *Anim. Behav.* 58:933–42.

Abrams, P. A. 1993. Why predation rate should not be proportional to predator density. *Ecology* 74:726–33.

Abrams, P. A., O. Leiman, S. Nylin, and C. Wiklund. 1996. The effects of flexible growth rates on optimal sizes and developmental times in a seasonal environment. *Am. Nat.* 147:381–95.

Agapow, P.-M., and N. J. B. Isaac. 2002. MacroCAIC: Revealing correlates of species richness by comparative analysis. *Diversity and Distributions* 8:41–43.

Aiello, L. C., and R. I. M. Dunbar. 1993. Neocortex size, group size and the evolution of language. *Curr. Anthropol.* 34:184–93.

Aiello, L. C., and P. W. Wheeler. 1995. The expensive tissue hypothesis: The brain and the digestive system in human and primate evolution. *Curr. Anthropol.* 36:199–221.

Albert, M. S., and M. B. Moss. 1996. Neuropsychology of aging: Findings in humans and monkeys. In *Handbook of the biology of aging,* edited by E. L. Schneider and J. W. Rowe, 217–33. New York: Academic Press.

Alberts, S. C. 1992. Maturation and dispersal in male baboons. Ph.D. thesis, University of Chicago.

Alberts, S. C., and J. Altmann. 1995. Preparation and activation: Determinants of age at reproductive maturity in male baboons. *Behav. Ecol. Sociobiol.* 36:397–406.

Alexander, R. D. 1974. The evolution of social behavior. *Annu. Rev. Ecol. Syst.* 5:325–83.

Allal, N. 1998. The evolution of human life history. Diploma thesis, University of Basel, Switzerland.

Allison, P. D. 1995. *Survival analysis using the SAS system: A practical guide.* Cary, NC: SAS Institute.

Allman, J. M. 1995. Brain and life span in catarrhine primates. In *Delaying the onset of late-life dysfunction,* edited by R. N. Butler and J. A. Brody, 221–41. New York: Springer.

——. *Evolving brains.* 1999. New York: Scientific American Library.

Allman, J. M., and A. Hasenstaub. 1999. Brains, maturation times, and parenting. *Neurobiol. Aging* 20:447–54.

Allman, J. M., T. McLaughlin, and A. Hakeem. 1993a. Brain weight and life-span in primate species. *Proc. Natl. Acad. Sci. USA* 90:118–22.

——. 1993b. Brain structures and life-span in primate species. *Proc. Natl. Acad. Sci. USA* 90:3559–63.

Altmann, J. 1979. Age cohorts as paternal sibships. *Behav. Ecol. Sociobiol.* 5:161–164.

Altmann, J. 1980. *Baboon mothers and infants.* Cambridge, MA: Harvard University Press.

Altmann, J., and S. C. Alberts. 1987. Body mass and growth rates in a wild primate population. *Oecologia* 72:15–20.

Altmann, J., S. C. Alberts, S. A. Haines, J. Dubach, P. Muruthi, T. Coote, E. Geffen, D. J. Cheesman, R. S. Mututua, S. N. Saiyalel, R. K. Wayne, R. C. Lacy, and M. W. Bruford. 1996. Behavior predicts genetic structure in a wild primate group. *Proc. Natl. Acad. Sci., USA* 93:5797–5801.

Altmann, J., S. A. Altmann, and G. Hausfater. 1978. Primate infant's effects on mother's future reproduction. *Science* 201:1028–30.

Altmann, J., S. A. Altmann, G. Hausfater, and S. A. McCuskey. 1977. Life history of yellow baboons: Physical development, reproductive parameters, and infant mortality. *Primates* 18:315–30.

Altmann, J., and A. Samuels. 1992. Costs of maternal-care: Infant carrying in baboons. *Behav. Ecol. Sociobiol.* 29:391–98.

Altmann, J., D. Schoeller, S. A. Altmann, P. Murithi, and R. M. Sapolsky. 1993. Body size and fatness of free-living baboons reflect food availability and activity levels. *Am. J. Primatol.* 30:149–61.

Altmann, S. A. 1991. Diets of yearling female primates *(Papio cynocephalus)* predict lifetime fitness. *Proc. Natl. Acad. Sci. USA* 88:420–23.

——. 1998. *Foraging for survival: Yearling baboons in Africa.* Chicago: University of Chicago Press.

Alvarez, H. P. 2000. Grandmother hypothesis and primate life histories. *Am. J. Phys. Anthropol.* 133:435–50.

Anderson, C. M. 1986. Predation and primate evolution. *Primates* 27:15–39.

Ardito, G. 1975. Checklist of gestation ages for primates. *J. Hum. Evol.* 5:213–22.

Arenz, C., and D. Leger. 2000. Antipredator vigilance of juvenile and adult thirteen-lined ground squirrels and the role of nutritional need. *Anim. Behav.* 59:535–41.

Armbruster, P., and R. Lande. 1993. A population viability analysis for African elephant *(Loxodonta africana):* How big should reserves be? *Conserv. Biol.* 7:602–10.

Armstrong, E. 1983. Relative brain size and metabolism in mammals. *Science* 220:1302–4.

——. 1985a. Allometric considerations of the adult mammalian brain, with special emphasis on primates. In *Size and scaling in primate biology,* edited by W. L. Jungers, 115–46. New York: Plenum Press.

——. 1985b. Relative brain size in monkeys and prosimians. *Am. J. Phys. Anthropol.* 66:263–73.

Arnold, K., and I. Owens. 1999. Cooperative breeding in birds: The role of ecology. *Behav. Ecol.* 10:465–71.

Arnold, S. J. 1992. Constraints on phenotypic evolution. *Am. Nat.* 140:S85–107.

Arnold, S. J., and M. J. Wade. 1984a. On the measurement of natural and sexual selection: Theory. *Evolution* 38:709–19.

———. 1984b. On the measurement of natural and sexual selection: Applications. *Evolution* 38:720–34.

Aschoff, J., B. Gunther, and K. Kramer. 1971. *Energiehaushalt und Temperaturregulation.* München: Urban and Schwarzenberg.

Asfaw, B., T. White, O. Lovejoy, B. Latimer, S. Simpson, and G. Suwa. 1999. *Australopithecus garhi:* A new species of early hominid from Ethiopia. *Science* 284: 629–35.

Aureli, F., R. Cozzolino, C. Cordischi, and S. Scucchi. 1992. Kin-oriented redirection among Japanese macaques: An expression of a revenge system? *Anim. Behav.* 44:283–91.

Austad, S. N. 1997a. Postreproductive survival. In *Between Zeus and the salmon,* edited by K. W. Wachter and C. E. Finch, 161–74. Washington, D.C.: National Academy Press.

———. 1997b. *Why we age: What science is discovering about the body's journey through life.* New York: John Wiley and Sons.

Austad, S. N., and K. E. Fischer. 1991. Mammalian aging, metabolism, and ecology: Evidence from the bats and marsupials. *J. Gerontol.* 46:B47–53.

———. 1992. Primate longevity: Its place in the mammalian scheme. *Am. J. Primatol.* 28:251–61.

Bailey, S. M., and V. L. Katch. 1981. The effects of body size on sexual dimorphism in fatness, volume and muscularity. *Hum. Biol.* 53:337–49.

Bailey, W. G. 1987. *Human longevity from antiquity to the modern lab: A selected, annotated bibliography.* New York: Greenwood Press.

Baron, G., H. Stephan, and H. D. Frahm. 1996. *Comparative neurobiology in Chiroptera. I. Macromorphology, brain structures, tables and atlases.* Basel: Birkhäuser Verlag.

Barraclough, T. G., P. H. Harvey, and S. Nee. 1995. Sexual selection and taxonomic diversity in passerine birds. *Proc. R. Soc. Lond.* B 259:211–15.

Barton, R. A. 1996. Neocortex size and behavioural ecology in primates. *Proc. R. Soc. Lond.* B 263:173–77.

———. 1999. The evolutionary ecology of the primate brain. In *Comparative primate socioecology,* edited by P. C. Lee, 167–94. Cambridge: Cambridge University Press.

Barton, R. A., R. W. Byrne, and A. Whiten. 1996. Ecology, feeding competition and social structure in baboons. *Behav. Ecol. Sociobiol.* 38:321–29.

Barton, R. A., and R. I. M. Dunbar. 1997. Evolution of the social brain. In *Machiavellian intelligence II: Extensions and evaluations,* edited by A. Whiten and R. W. Byrne, 240–63. Cambridge: Cambridge University Press.

Barton, R. A., and P. H. Harvey. 2000. Mosaic evolution of brain structure in mammals. *Nature* 405:1055–58.

Barton, R. A., A. Purvis, and P. H. Harvey. 1995. Evolutionary radiation of visual and olfactory brain systems in primates, bats and insectivores. *Phil. Trans. R. Soc. Lond.* B 348:381–92.

Barton, R. A., and A. Whiten. 1994. Reducing complex diets to simple rules: Food selection by olive baboons. *Behav. Ecol. Sociobiol.* 35:283–93.

Basolo, A. L. 1990. Female preference predates the evolution of the sword in swordtail fish. *Science* 250:808–10.

Bateson, P. P. G. 1988. The active role of behavior in evolution. In *Evolutionary processes and metaphors,* edited by M. W. Ho and S. W. Fox, 191–207. London: Wiley and Sons.

Bean, A. 1998. The ecology of sex differences in great ape foraging behaviour and hunter-gatherer subsistence behaviour. Ph.D. thesis, University of Cambridge.

Begon, M., J. L. Harper, and C. R. Townsend. 1996. *Ecology.* Oxford: Blackwell.

Bekoff, M., T. Daniels, and J. Gittleman. 1984. Life history patterns and the comparative social ecology of carnivores. *Annu. Rev. Ecol. Syst.* 15:191–232.

Bekoff, M., J. Diamond, and J. B. Mitton. 1981. Life-history patterns and sociality in canids: Body size, reproduction, and behavior. *Oecologia* 50:386–90.

Bellomo, R. V. 1994. Methods of determining early hominid behavioral activities associated with the controlled use of fire at FxJj 20 Main, Koobi Fora, Kenya. *J. Hum. Evol.* 27:173–95.

Belyaev, D. K. 1969. Domestication of animals. *Science J.* 5:47–52.

Belyaev, D. K., A. O. Ruvinsky, and L. N. Trut. 1981. Inherited activation-inactivation of the "star" gene in foxes. *J. Hered.* 72:264–74.

Benton, T. G., and A. Grant. 1996. How to keep fit in the real world: Elasticity analyses and selection pressures on life histories in a variable environment. *Am. Nat.* 147:115–39.

———. 1999. Elasticity analysis as an important tool in evolutionary and population ecology. *Trends Ecol. Evol.* 14:467–71.

Benton, T. G., A. Grant, and T. H. Clutton-Brock. 1995. Does environmental stochasticity matter? Analysis of red deer life-histories on Rum. *Evol. Ecol.* 9: 559–74.

Bercovitch, F. B. 1986. Male rank and reproductive activity in savanna baboons. *Int. J. Primatol.* 7:533–50.

Bercovitch, F. B., and J. D. Berard. 1993. Life history costs and consequences of rapid reproductive maturation in female rhesus macaques. *Behav. Ecol. Sociobiol.* 32:103–9.

Bercovitch, F. B., and P. Nürnberg. 1996. Socioendocrine and morphological correlates of paternity in rhesus macaques. *J. Reprod. Fert.* 107:59–68.

Bercovitch, F. B., and S. C. Strum. 1993. Dominance rank, resource availability and reproductive maturation in female savanna baboons. *Behav. Ecol. Sociobiol.* 33:313–18.

Bermudez de Castro, J. M., and M. E. Nicolas. 1997. Paleodemography of the Atapuerca-Sima de los Huesos Middle Pleistocene hominid sample. *J. Hum. Evol.* 33:333–55.

Berven, K. A., and D. E. Gill. 1983. Interpreting geographic variation in life history traits. *Am. Zool.* 23:85–97.

Beynon, A. D., and B. A. Wood. 1987. Patterns and rates of enamel growth in the molar teeth of early hominids. *Nature* 326:493–96.

Bieber, C. 1998. Population dynamics, sexual activity, and reproductive failure in the fat dormouse *(Myoxus glis). J. Zool.* 244:223–29.

Binida-Emonds, O. R. P., J. L. Gittleman, and A. Purvis. 1999. Building large trees by combining phylogenetic information: A complete phylogeny of the extant Carnivora (Mammalia). *Biol. Rev. Camb. Phil. Soc.* 74:143–75.

Birkhead, T. R., and A. P. Møller. 1993. Female control of paternity. *Trends Ecol. Evol.* 8:100–104.

Bjorklund, D. F. 1997. The role of immaturity in human development. *Psychol. Bull.* 122:153–69.

Blueweiss, L., H. Fox, V. Kudzma, D. Nakashima, R. Peters, and S. Sams. 1978. Relationships between body size and some life-history parameters. *Oecologia* 37:257–72.

Blumenschine, R. J. 1991. Hominid carnivory and foraging strategies, and the socioeconomic function of early archaeological sites. *Phil. Trans. R. Soc. Lond.* B 334:211–21.

Blumenschine, R. J., and C. W. Marean. 1993. A carnivore's view of archaeological bone assemblages. In *From bones to behavior: Ethnoarchaeological and experimental contributions to the interpretation of faunal remains,* edited by J. Hudson, 273–300. Occasional Paper 21, Center for Archaeological Investigations. Carbondale: Southern Illinois University.

Blurton Jones, N. G., K. Hawkes, and J. F. O'Connell. 1989. Studying costs of children in two foraging societies: Implications for schedules of reproduction. In *Comparative socioecology: The behavioral ecology of humans and other animals,* edited by V. Standen and R. A. Foley, 365–90. London: Blackwell.

———. 1996. The global process and local ecology: How should we explain differences between the Hadza and !Kung? In *Cultural diversity among twentieth century foragers: An African perspective,* edited by S. Kent, 159–87. Cambridge: Cambridge University Press.

———. 1999. Some current ideas about the evolution of human life history. In *Comparative primate socioecology,* edited by P. C. Lee, 140–66. Cambridge: Cambridge University Press.

———. 2002. The antiquity of post-reproductive life: Are there modern impacts on hunter-gatherer post-reproductive life spans? A careful look at the demography of the Hadza, Ache and !Kung with an evaluation of the claims that the reported numbers of aged individuals arise from errors in the demography and/or modern influences on longevity. *Hum. Nat.* In press.

Blurton Jones, N. G., L. C. Smith, J. F. O'Connell, K. Hawkes, and C. Kamazura. 1992. Demography of the Hadza, an increasing and high density population of savanna foragers. *Am. J. Phys. Anthropol.* 89:159–81.

Bocquet-Appel, J.-P., and C. Masset. 1982. Farewell to Paleodemography. *J. Hum. Evol.* 11:321–33.

Boddington, M. J. 1978. An absolute metabolic scope for activity. *J. Theor. Biol.* 75:443–49.

Boesch, C., and H. Boesch-Achermann. 2000. *The chimpanzees of the Tai forest: Behavioural ecology and evolution.* Oxford: Oxford University Press.

Bogin, B. 1999. *Patterns of human growth.* Cambridge: Cambridge University Press.

Bogin, B., and B. H. Smith. 1996. Evolution of the human life cycle. *Am. J. Hum. Biol.* 8:703–16.

Boinski, S. 1987. Birth synchrony in squirrel monkeys *(Saimiri oerstedi):* A strategy to reduce neonatal predation. *Behav. Ecol. Sociobiol.* 21:393–400.

Boinski, S., and C. A. Chapman. 1995. Predation on primates: Where are we and what next? *Evol. Anthropol.* 4:1–3.

Boinski, S., and D. M. Fragaszy. 1989. The ontogeny of foraging in squirrel monkeys, *Saimiri oerstedi. Anim. Behav.* 37:415–28.

Boinski, S., A. Treves, and C. A. Chapman. 2000. A critical evaluation of the influence of predators on primates: Effects on group movement. In *On the move: How and why animals travel in groups,* edited by S. Boinski and P. Garber, 43–72. Chicago: University of Chicago Press.

Borries, C., K. Launhardt, C. Epplen, J. T. Epplen, and P. Winkler. 1999. DNA analyses support the hypothesis that infanticide is adaptive in langur monkeys. *Proc. R. Soc. Lond.* B 266:901–4.

Boyce, M. S. 1979. Seasonality and patterns of natural selection for life histories. *Am. Nat.* 114:569–83.

———. 1988. Evolution of life histories: Theory and patterns from mammals. In *Evolution of life histories of mammals,* edited by M. S. Boyce, 3–30. New Haven, CT: Yale University Press.

Boyce, M. S., and D. J. Daley. 1980. Population tracking of fluctuating environments and natural selection from tracking ability. *Am. Nat.* 115:480–91.

Brain, C. K. 1988. New information from Swartkrans Cave of relevance to "robust" Australopithecines. In *Evolutionary history of the "robust" Australopithecines,* edited by F. Grine, 311–16. Hawthorne, NY: Aldine de Gruyter.

Brandner, F., J. S. Keith, and P. Trayhurn. 1993. A 27mer oligonucleotide probe for the detection and measurement of the mRNA for uncoupling protein in brown adipose tissue of different species. *Comp. Biochem. Physiol.* B 104:125–31.

Brereton, A. 1995. Coercion-defence hypothesis: The evolution of primate sociality. *Folia Primatol.* 64:207–14.

Bromage, T. G., and M. C. Dean. 1985. Re-evaluation of the age at death of immature fossil hominids. *Nature* 317:525–27.

Bromham, L., A. Rambaut, and P. H. Harvey. 1996. Determinants of rate variation in mammalian DNA sequence evolution. *J. Mol. Evol.* 43:610–21.

Brommer, J. E. 2000. The evolution of fitness in life-history theory. *Biol. Rev.* 75: 377–404.

Bronikowski, A., and J. Altmann. 1996. Foraging in a variable environment: Weather patterns and the behavioral ecology of baboons. *Behav. Ecol. Sociobiol.* 39: 11–25.

Brooks, D. R., and D. A. McLennan. 1991. *Phylogeny, ecology, and behavior.* Chicago: University of Chicago Press.

Brown, D. 1988. Components of lifetime reproductive success. In *Reproductive success,* edited by T. H. Clutton-Brock, 439–53. Chicago: University of Chicago Press.

Bryant, J. P., F. S. Chapin, and D. R. Klein. 1983. Carbon/nutrient balance of boreal plants in relation to vertebrate herbivory. *Oikos* 40:357–68.

Buchanan, L. S. 2000. Brain size and evolution in *Papio. Am. J. Phys. Anthropol.* suppl. 30:114.

Bunn, H. T. 1981. Archaeological evidence for meat eating by Plio-Pleistocene hominids from Koobi Fora and Olduvai Gorge. *Nature* 291:574–77.

———. 1982. Meat-eating and human evolution: Studies on the diet and subsistence patterns of Plio-Pleistocene hominids in East Africa. Ph.D. thesis, University of California, Berkeley.

Bunn, H. T., and E. M. Kroll. 1986. Systematic butchery by Plio-Pleistocene hominids at Olduvai Gorge, Tanzania. *Curr. Anthropol.* 27:413–52.

Burghardt, G. M. 1998. The evolutionary origins of play revisited: Lessons from turtles. In *Animal Play,* edited by M. Bekoff and J. A. Byers, 1–26. New York: Cambridge University Press.

Burt, A. 1989. Comparative methods using phylogenetically independent contrasts. In *Oxford surveys in evolutionary biology,* vol. 6, edited by P. H. Harvey and L. Partridge, 33–53. Oxford: Oxford University Press.

Byers, J. A., and C. Walker. 1995. Refining the motor training hypothesis for the evolution of play. *Am. Nat.* 146:25–40.

Byrne, R. W. 1995a. Primate cognition. *Am. J. Primatol.* 37:127–41.

———. 1995b. *The thinking ape.* Oxford: Oxford University Press.

Byrne, R. W., and J. M. E. Byrne. 1993. Complex leaf gathering skills of mountain gorillas *(Gorilla g. beringei):* Variability and standardization. *Am. J. Primatol.* 31:241–61.

Byrne, R. W., and A. Whiten. 1988. *Machiavellian intelligence: Social expertise and the evolution of intellect in monkeys, apes and humans.* Oxford: Oxford University Press.

Calder, W. A. I. 1976. Aging in vertebrates: Allometric considerations of spleen size and life span. *Fed. Proc.* 35:96–97.

———. 1984. *Size, function, and life history.* Cambridge, MA: Harvard University Press.

Call, J., and M. Tomasello. 1999. A nonverbal theory of mind test: The performance of children and apes. *Child Dev.* 70:381–95.

Caro, T. M., D. W. Sellen, A. Parish, R. Frank, D. M. Brown, E. Voland, and M. Borgerhoff Mulder. 1995. Termination of reproduction in non-human and human female primates. *Int. J. Primatol.* 16:205–20.

Case, R. 1992. The role of the frontal lobes in the regulation of human development. *Brain Cogn.* 20:51–73.

Case, T. J. 1978. On the evolution and adaptive significance of postnatal growth rates in the terrestrial vertebrates. *Q. Rev. Biol.* 53:243–82.

Caswell, H. 1983. Phenotypic plasticity in life-history traits: Demographic effects and evolutionary consequences. *Am. Zool.* 23:35–46.

———. 2001. *Matrix population models.* Sunderland, MA: Sinauer Associates.

Cavigelli, S. A., and M. E. Pereira. 2000. Mating season aggression and fecal testosterone levels in male ring-tailed lemurs *(Lemur catta). Horm. Behav.* 37:246–55.

Cerling, T. E. 1992. Development of grasslands and savannas in East Africa during the Neogene. *Palaeogeogr. Palaeoclimatol. Palaeoecol.* 97:241–47.

Chapais, B., M. Girard, and G. Primi. 1991. Non-kin alliances and the stability of matrilineal dominance relations in Japanese macaques. *Anim. Behav.* 41:481–91.

Chapais, B., and S. Schulman. 1980. An evolutionary model of female dominance relations in primates. *J. Theor. Biol.* 82:47–89.

Chapman, C. A. 1990. Ecological constraints on group size in three species of Neotropical primates. *Folia Primatol.* 55:1–9.

Chapman, C. A., A. Gautier-Hion, J. F. Oates, and D. A. Onderdonk. 1999. African primate communities: Determinants of structure and threats to survival. In *Primate communities,* edited by J. F. Fleagle, C. H. Janson, and K. E. Reed, 1–37. Cambridge: Cambridge University Press.

Chapman, C. A., S. Walker, and L. Lefebvre. 1990. Reproductive strategies of primates: The influence of body size and diet on litter size. *Primates* 31:1–13.

Chapman, C. A., R. W. Wrangham, and L. J. Chapman. 1994. Indices of habitat-wide fruit abundance in tropical forests. *Biotropica* 26:160–71.

———. 1995. Ecological constraints on group size: An analysis of spider monkey and chimpanzee subgroups. *Behav. Ecol. Sociobiol.* 36:59–70.

Charlesworth, B. 1994. *Evolution in age-structured populations.* Cambridge: Cambridge University Press.

Charnov, E. L. 1991. Evolution of life history variation among female mammals. *Proc. Natl. Acad. Sci. USA* 88:1134–37.

——. 1993. *Life history invariants: Some explorations of symmetry in evolutionary ecology.* Oxford: Oxford University Press.

——. 1997. Trade-off-invariant rules for evolutionarily stable life histories. *Nature* 387:393–94.

——. 2001. Evolution of mammal life histories. *Evol. Ecol. Res.* 3:521–35.

Charnov, E. L., and D. Berrigan. 1991. Dimensionless numbers and the assembly rules for life histories. *Phil. Trans. R. Soc. Lond.* B 33:241–48.

——. 1993. Why do female primates have such long life spans and so few babies? *or* Life in the slow lane. *Evol. Anthropol.* 2:191–94.

Charnov, E. L., and W. M. Schaffer. 1973. Life history consequences of natural selection: Cole's result revisited. *Am. Nat.* 107:791–93.

Cheek, D. B. 1975. The fetus. In *Fetal and postnatal cellular growth,* edited by D. B. Cheek, 3–22. New York: Wiley.

Cheney, D. L., and R. M. Seyfarth. 1990. *How monkeys see the world: Inside the mind of another species.* Chicago: University of Chicago Press.

Cheney, D. L., R. M. Seyfarth, and B. B. Smuts. 1986. Social relationships and social cognition in non-human primates. *Science* 234:1361–66.

Cheney, D. L., and R. W. Wrangham. 1987. Predation. In *Primate societies,* edited by B. B. Smuts, D. L. Cheney, R. M. Seyfarth, R. W. Wrangham, and T. T. Struhsaker, 227–39. Chicago: University of Chicago Press.

Chism, J. 2000. Allocare patterns among cercopithecines. *Folia Primatol.* 71:55–66.

Chivers, D. J., B. A. Wood, and A. Bilsborough. 1984. *Food acquisition and processing in primates.* New York: Plenum Press.

Clark, A. 1978. Sex ratio and local resource competition in a prosimian primate. *Science* 20:163–65.

Clark, C. W., and M. Mangel. 1984. Foraging and flocking strategies—information in an uncertain environment. *Am. Nat.* 123:626–41.

Clegg, M., and L. C. Aiello. 1999. A comparison of the Nariokotome *Homo erectus* with juveniles from a modern human population. *Am. J. Phys. Anthropol.* 110:81–93.

Clutton-Brock, T. H. 1989. Mammalian mating systems. *Proc. R. Soc. Lond.* B 236:339–72.

——. 1991. The evolution of sex differences and the consequences of polygyny in mammals. In *The development and integration of behaviour,* edited by P. Bateson, 229–53. Cambridge: Cambridge University Press.

Clutton-Brock, T. H., S. D. Albon, and F. E. Guinness. 1985. Parental investment and sex differences in juvenile mortality in birds and mammals. *Nature* 313:131–33.

——. 1988. Reproductive success in male and female red deer. In *Reproductive success,* edited by T. H. Clutton-Brock, 325–43. Chicago: University of Chicago Press.

——. 1989. Fitness costs of gestation and lactation in wild mammals. *Nature* 337:260–62.

Clutton-Brock, T. H., D. Gaynor, G. M. McIlrath, A. D. C. Maccoll, R. Kansky, P. Chadwick, M. Manser, J. D. Skinner, and P. N. M. Brotherton. 1999. Predation, group size and mortality in a cooperative mongoose, *Suricata suricatta. J. Anim. Ecol.* 68:672–83.

Clutton-Brock, T. H., and P. H. Harvey. 1976. Evolutionary rules and primate societies. In *Growing points in ethology,* edited by P. Bateson and R. Hinde, 195–237. Cambridge: Cambridge University Press.

——. 1977. Primate ecology and social organization. *J. Zool.* 183:1–39.

――. 1978. Mammals, resources and reproductive strategies. *Nature* 273:191–95.

――. 1979. Comparison and adaptation. *Proc. R. Soc. Lond.* B 205:547–65.

――. 1997. Refuge use and predation risk in a desert baboon population. *Anim. Behav.* 54:241–53.

Clutton-Brock, T. H., P. H. Harvey, and B. Rudder. 1977. Sexual dimorphism, socionomic sex ratio and body weight in primates. *Nature* 269:191–95.

Clutton-Brock, T. H., and G. A. Parker. 1992. Potential reproductive rates and the operation of sexual selection. *Q. Rev. Biol.* 67:437–56.

――. 1995. Sexual coercion in animal societies. *Anim. Behav.* 49:1345–65.

Cole, L. C. 1954. The population consequences of life history phenomena. *Q. Rev. Biol.* 29:103–37.

Colinvaux, P. A. 1978. *Why big fierce animals are rare: An ecologist's perspective.* Princeton, NJ: Princeton University Press.

Collard, M., and B. Wood. 1999. Grades among the African early hominids. In *African biogeography, climate change, and human evolution,* edited by T. G. Bromage and S. Friedemann, 316–27. Oxford: Oxford University Press.

Combes, S. L., and J. Altmann. 2001. Status change during adulthood: Life-history by-product or kin selection based on reproductive value? *Proc. R. Soc. Lond.* B 268:1367–73.

Conklin-Brittain, N. L., R. W. Wrangham, and K. D. Hunt. 1998. Dietary response of chimpanzees and cercopithecines to seasonal variation in fruit abundance. II. Macronutrients. *Int. J. Primatol.* 19:971–98.

Count, E. W. 1947. Brain and body weight in man: Their antecedents in growth and evolution. *Ann. N.Y. Acad. Sci.* 66:993–1122.

Coursey, D. G. 1973. Hominid evolution and hypogeous plant foods. *Man* 8:634–35.

Couture, M. D., M. F. Ricks, and L. Housley. 1986. Foraging behavior of a contemporary northern Great Basin population. *J. Calif. Great Basin Anthropol.* 8:150–60.

Covert, H. H. 1986. Biology of Cenozoic primates. In *Comparative primate biology,* vol. 1, *Systematics, evolution and anatomy,* edited by D. R. Swindler and J. Erwin, 335–59. New York: Alan R. Liss.

Cowlishaw, G. 1994. Vulnerability to predation in baboon populations. *Behaviour* 131:293–304.

――. 1997. Trade-offs between foraging and predation risk determine habitat use in a desert baboon population. *Anim. Behav.* 53:667–86.

Cowlishaw, G., and R. I. M. Dunbar. 1991. Dominance rank and mating success in male primates. *Anim. Behav.* 41:1045–56.

――. 2000. *Primate conservation biology.* Chicago: University of Chicago Press.

Crockett, C. M., and J. F. Eisenberg. 1987. Howlers: Variation in group size and demography. In *Primate societies,* edited by B. B. Smuts, D. L. Cheney, R. M. Seyfarth, R. W. Wrangham, and T. T. Struhsaker, 54–68. Chicago: University of Chicago Press.

Crockett, C. M., and C. H. Janson. 2000. Infanticide in red howlers: Female group size, male membership, and a possible link to folivory. In *Infanticide by males and its implications,* edited by C. P. van Schaik and C. H. Janson, 75–98. Cambridge: Cambridge University Press.

Crockett, C. M., and T. R. Pope. 1993. Consequences of sex differences in dispersal for juvenile red howler monkeys. In *Juvenile primates: Life history, development, and behavior,* edited by M. E. Pereira and L. A. Fairbanks, 104–18. New York: Oxford University Press.

Crook, J. H. 1964. The evolution of social organisation and visual communication in the weaver birds (Ploceinae). *Behaviour* 10:1–178.

——. 1968. The adaptive significance of avian social organizations. *Symp. Zool. Soc. Lond.* 14:181–218.

——. 1970. The socioecology of primates. In *Social behaviour in birds and mammals,* edited by J. H. Crook, 103–66. London: Academic Press.

Crook, J. H., J. E. Ellis, and J. D. Goss-Custard. 1976. Mammalian social systems: Structure and function. *Anim. Behav.* 24:261–74.

Crook, J. H., and S. C. Gartlan. 1966. Evolution of primate societies. *Nature* 210: 1200–1203.

Crooks, K. R., M. A. Sanjayan, and D. F. Doak. 1998. New insights on cheetah conservation through demographic modeling. *Conserv. Biol.* 12:889–95.

Crowl, T. A., and A. P. Covich. 1990. Predator-induced life-history shifts in a freshwater snail. *Science* 324:58–60.

Cunningham, S. A., B. Summerhayes, and M. Westoby. 1999. Evolutionary divergences in leaf structure and chemistry, comparing rainfall and soil nutrient gradients. *Ecol. Monogr.* 69:569–88.

Curtis, D. J., and A. Zaramody. 1999. Social structure and seasonal variation in the behaviour of *Eulemur mongoz. Folia Primatol.* 70:79–96.

Cutler, R. G. 1976. Evolution of longevity in primates. *J. Hum. Evol.* 5:169–202.

Damuth, J. 1993. Cope's rule, the island rule, and the scaling of mammalian population density. *Nature* 365:748–50.

Dang, D.-C., G. Chaouat, A. Delmas, J. Hureau, G. Hidden, and J.-P. Lassau. 1992. Assessment of the factors responsible for variability of birth weight in the longtailed macaque *(Macaca fascicularis). Folia Primatol.* 59:149–56.

Darwin, C. R. 1871. *The descent of man and selection in relation to sex.* London: Murray.

Dausmann, K. H., J. U. Ganzhorn, and G. Heldmaier. 2000. Body temperature and metabolic rate of a hibernating primate in Madagascar: Preliminary results from a field study. In *Life in the cold. Eleventh international hibernation symposium,* edited by G. Heldmaier and M. Klingenspor, 41–47. Berlin, Heidelberg, New York: Springer Verlag.

Davies, N. B. 1991. Mating systems. In *Behavioural ecology,* edited by J. R. Krebs and N. B. Davies, 263–94. Oxford: Blackwell.

——. 2000. Multi-male breeding groups in birds: Ecological causes and social conflict. In *Primate males,* edited by P. M. Kappeler, 11–20. Cambridge: Cambridge University Press.

Deacon, T. W. 1990. Problems of ontogeny and phylogeny in brain-size evolution. *Int. J. Primatol.* 11:237–82.

——. 1997. *The symbolic species: The co-evolution of language and the brain.* New York: W. W. Norton.

——. 2000. Heterochrony in brain evolution: Cellular versus morphological analyses. In *Biology, brains, and behavior: The evolution of human development,* edited by S. T. Parker, J. Langer, and M. L. McKinney, 41–88. Santa Fe, NM: School of American Research Press.

Dean, C., M. G. Leakey, D. Reid, F. Schrenk, G. T. Schwartz, C. Stringer, and A. Walker. 2001. Growth processes in teeth distinguish modern humans from *Homo erectus* and earlier hominins. *Nature* 414:628–31.

Dean, M. C. 1987. Growth layers and incremental markings in hard tissues: A review of the literature and some preliminary observations about enamel structure in *Paranthropus boisei*. *J. Hum. Evol.* 16:157–72.

Deaner, R. O., and C. L. Nunn. 1999. How quickly do brains catch up with bodies? A comparative method for detecting evolutionary lag. *Proc. R. Soc. Lond.* B 266: 687–94.

Deaner, R. O., C. L. Nunn, and C. P. van Schaik. 2000. Comparative tests of primate cognition: Different scaling methods produce different results. *Brain Behav. Evol.* 55:44–52.

DeCasper, A. J., and M. J. Spence. 1986. Prenatal maternal speech influences newborns' perceptions of speech sounds. *Infant Behav. Dev.* 9:133–50.

Dekaban, A. S., and D. Sadowsky. 1978. Changes in brain weights during the span of human life: Relation of brain weights to body heights and body weights. *Ann. Neurol.* 4:345–56.

DeMenocal, P. B. 1995. Plio-Pleistocene African climate. *Science* 270:53–59.

Dennell, R., and W. Roebroeks. 1996. The earliest colonization of Europe: The short chronology revisited. *Antiquity* 70:535–42.

Derrickson, E. M. 1992. Comparative reproductive strategies of altricial and precocial eutherian mammals. *Funct. Ecol.* 6:57–65.

de Ruiter, J. R., and J. A. R. A. M. van Hooff. 1993. Male dominance rank and reproductive success in primate groups. *Primates* 34:513–23.

de Ruiter, J. R., J. A. R. A. M. van Hooff, and W. Scheffrahn. 1994. Social and genetic aspects of paternity in wild long-tailed macaques. *Behaviour* 129:203–24.

de Waal, F. B. M. 1989. Food sharing and reciprocal obligations among chimpanzees. *J. Hum. Evol.* 18:433–59.

———. 1991. Complementary methods and convergent evidence in the study of primate social cognition. *Behaviour* 118:297–320.

de Waal, F. B. M., and L. M. Luttrell. 1988. Mechanisms of social reciprocity in three primate species: Symmetrical relationship characteristics or cognition? *Ethol. Sociobiol.* 9:101–18.

de Waal, F. B. M., and A. van Roosmalen. 1979. Reconciliation and consolation among chimpanzees. *Behav. Ecol. Sociobiol.* 5:55–66.

Dewar, R. E., and J. R. Wallis. 1999. Geographical patterning in intra-annual rainfall variability in the tropics and near tropics: An L-moments approach. *J. Climatol.* 12:3457–66.

Dial, K. P., and J. M. Marzluff. 1988. Are the smallest organisms the most diverse? *Ecology* 69:1620–24.

Diamond, J. 1997. *Guns, germs, and steel: The fates of human societies.* New York: W. W. Norton.

Dieckmann, U. 1997. Can adaptive dynamics invade? *Trends Ecol. Evol.* 12:128–31.

Dietz, J., A. Baker, and D. Miglioretti. 1994. Seasonal variation in reproduction, juvenile growth, and adult body mass in golden lion tamarins *(Leontopithecus rosalia)*. *Am. J. Primatol.* 34:115–32.

Dirks, W. 1998. Histological reconstruction of dental development and age at death in a juvenile gibbon *(Hylobates lar)*. *J. Hum. Evol.* 35:411–25.

Dittus, W. 1998. Birth sex ratios in toque macaques and other mammals: Integrating the effects of maternal condition and competition. *Behav. Ecol. Sociobiol.* 44: 149–60.

Dixon, P., N. Friday, P. Ang, S. Heppell, and M. Kshatriya. 1997. Sensitivity analysis of structured-population models for management and conservation. In *Structured population models in marine, terrestrial and freshwater systems,* edited by S. Tuljapurkar and H. Caswell, 471–514. New York: Chapman and Hall.

Dobbing, J., and J. Sands. 1973. Quantitative growth and development of human brains. *Arch. Dis. Child.* 48:757–67.

———. 1979. Comparative aspects of the brain growth spurt. *Early Hum. Dev.* 3: 79–83.

Dobzhansky, T. 1962. *Mankind evolving: The evolution of the human species.* New Haven: Yale University Press.

Doebeli, M. 1995. Phenotypic variation, sexual reproduction and evolutionary population dynamics. *J. Evol. Biol.* 8:173–94.

Doebeli, M., and G. de Jong. 1999. Genetic variability in sensitivity to population density affects the dynamics of simple ecological models. *Theor. Popul. Biol.* 55:37–52.

Doebeli, M., and J. C. Koella. 1994. Sex and population dynamics. *Proc. R. Soc. Lond.* B 257:17–23.

Doebeli, M., and G. D. Ruxton. 1998. Stabilization through spatial pattern formation in metapopulations with long-range dispersal. *Proc. R. Soc. Lond.* B 265: 1325–32.

Dukas, R. 1998. Evolutionary ecology of learning. In *Cognitive ecology: The evolutionary ecology of information processing and decision making,* edited by R. Dukas, 129–74. Chicago: University of Chicago Press.

Dunbar, R. I. M. 1984. *Reproductive decisions: An economic analysis of gelada baboon social strategies.* Princeton, NJ: Princeton University Press.

———. 1987a. Demography and reproduction. In *Primate societies,* edited by B. B. Smuts, D. L. Cheney, R. M. Seyfarth, R. W. Wrangham, and T. T. Struhsaker, 240–49. Chicago: University of Chicago Press.

———. 1987b. Habitat quality, population dynamics, and group composition in colobus monkeys *(Colobus guereza). Int. J. Primatol.* 8:299–329.

———. 1988. *Primate social systems.* Ithaca, NY: Cornell University Press.

———. 1990. Environmental determinants of intraspecific variation in body weight in baboons (*Papio* ssp.). *J. Zool.* 220:157–69.

———. 1992. Neocortex size as a constraint on group size in primates. *J. Hum. Evol.* 22:469–93.

———. 1993. Coevolution of neocortex size, group size and language in humans. *Behav. Brain Sci.* 16:681–735.

———. 1995a. The mating system of callitrichid primates: I. Conditions for the coevolution of pair bonding and twinning. *Anim. Behav.* 50:1057–70.

———. 1995b. Neocortex size and group size in primates: A test of the hypothesis. *J. Hum. Evol.* 28:287–96.

———. 1998. The social brain hypothesis. *Evol. Anthropol.* 6:178–90.

Dunbar, R. I. M., and J. Bever. 1998. Neocortex size predicts group size in carnivores and some insectivores. *Ethology* 104:695–708.

Eaglen, R. H. 1985. Behavioral correlates of tooth eruption in Madagascar lemurs. *Am. J. Phys. Anthropol.* 66:307–15.

Early, J. D., and T. N. Headland. 1998. *Population dynamics of a Philippine rain forest people.* Gainesville: University of Florida Press.

Eberhard, W. G. 1996. *Female control: Sexual selection by cryptic female choice.* Princeton, NJ: Princeton University Press.

Economos, A. C. 1980a. Brain-life span conjecture: A re-evaluation of the evidence. *Gerontol.* 26:82–89.

———. 1980b. Taxonomic differences in the mammalian life span-body weight relationship and the problem of body weight. *Gerontol.* 26:90–98.

Edwards, C. P. 1993. Behavioral sex differences in children of diverse cultures: The case of nurturance to infants. In *Juvenile primates: Life history, development, and behavior,* edited by M. E. Pereira and L. A. Fairbanks, 327–338. New York: Oxford University Press.

Eisenberg, J. F. 1981. *The mammalian radiations: An analysis of trends in evolution, adaptation, and behavior.* Chicago: University of Chicago Press.

Elias, M. F., and K. W. Samonds. 1977. Protein and calorie malnutrition in infant *Cebus* monkeys: Growth and behavioral development during deprivation and rehabilitation. *Am. J. Clin. Nutr.* 30:355–66.

Ellis, R. S. 1920. Norms for some structural changes in the human cerebellum from birth to old age. *J. Comp. Neurol.* 30:229–52.

Elman, J. L., E. A. Bates, M. H. Johnson, A. Karmiloff-Smith, D. Parisi, and K. Plunkett. 1996. *Rethinking innateness: A connectionist perspective on development.* Cambridge, MA: MIT Press.

Emlen, S. T. 1994. Benefits, constraints and the evolution of the family. *Trends Ecol. Evol.* 9:282–84.

Emlen, S. T., and L. W. Oring. 1977. Ecology, sexual selection, and the evolution of mating systems. *Science* 197:215–23.

Emmons, L. H. 1987. Comparative feeding ecology of felids in a Neotropical rainforest. *Behav. Ecol. Sociobiol.* 20:271–83.

Etter, R. J. 1989. Life history variation in the intertidal snail *Nucella lapillus* across a wave-exposure gradient. *Ecology* 70:1857–76.

Evans, C. S., and R. W. Goy. 1968. Social behaviour and reproductive cycles in captive ringtailed lemurs *(Lemur catta). J. Zool.* 156:181–97.

Fagen, R. M. 1981. *Animal play behavior.* New York: Oxford University Press.

———. 1993. Primate juveniles and primate play. In *Juvenile primates: Life history, development, and behavior,* edited by M. E. Pereira and L. A. Fairbanks, 182–96. New York: Oxford University Press.

Fairbanks, L. A. 1988a. Mother-infant behavior in vervet monkeys: Response to failure of last pregnancy. *Behav. Ecol. Sociobiol.* 23:157–65.

———. 1988b. Vervet monkey grandmothers: Interactions with infant grandoffspring. *Int. J. Primatol.* 91:425–41.

———. 1990. Reciprocal benefits of allomothering for female vervet monkeys. *Anim. Behav.* 40:553–62.

———. 1993. Juvenile vervet monkeys: Establishing relationships and practicing skills for the future. In *Juvenile primates: Life history, development, and behavior,* edited by M. E. Pereira and L. A. Fairbanks, 211–27. New York: Oxford University Press.

———. 2000. The developmental timing of primate play: A neural selection model. In *Biology, brains, and behavior: The evolution of human development,* edited by S. T. Parker, J. Langer, and M. McKinney, 131–58. Santa Fe, NM: School of American Research Press.

Fairbanks, L. A., and M. T. McGuire. 1986. Age, reproductive value, and dominance-related behavior in vervet monkey females: Cross-generational influences on social relationships and reproduction. *Anim. Behav.* 34:1710–21.

Feibel, C. S., F. Brown, and I. McDougal. 1989. Stratigraphic context of fossil hominids from the Omo Group deposits: Northern Turkana Basin, Kenya and Ethiopia. *Am. J. Phys. Anthropol.* 78:595–622.

Feistner, A. T. C., and W. C. McGrew. 1989. Food sharing in primates: A critical review. In *Perspectives in primate biology*, vol. 3, edited by P. K. Seth and S. Seth, 21–36. New Delhi: Today and Tomorrow's Printers and Publishers.

Felsenstein, J. 1985. Phylogenies and the comparative method. *Am. Nat.* 125:1–15.

Ferrière, J., and J. Clobert. 1992. Evolutionarily stable age at first reproduction in a density-dependent model. *J. Theor. Biol.* 157:253–67.

Field, S. A., and G. Calbert. 1998. Patch defence in the parasitoid wasp *Trissolcus basalis:* When to begin fighting? *Behaviour* 135:629–42.

Fietz, J., and J. U. Ganzhorn. 1999. Feeding ecology of the hibernating primate *Cheirogaleus medius:* How does it get so fat? *Oecologia* 121:157–64.

Finch, C. E. 1997. Comparative perspectives on plasticity in human aging and life spans. In *Between Zeus and the salmon*, edited by K. W. Wachter and C. E. Finch, 245–68. Washington, D.C.: National Academy Press.

Finch, C. E., and R. M. Sapolsky. 1999. The evolution of Alzheimer disease, the reproductive schedule, and apoE isoforms. *Neurobiol. Aging* 20:407–28.

Finch, C. E., and R. E. Tanzi. 1997. Genetics of aging. *Science* 278:407–11.

Finlay, B. L., and R. B. Darlington. 1995. Linked regularities in the development and evolution of mammalian brains. *Science* 268:1578–84.

Fisher, R. A. 1930. *The genetical theory of natural selection.* Oxford: Clarendon Press.

Fleagle, J. G. 1978. Size distributions of living and fossil primate faunas. *Paleobiology* 4:67–76.

———. 1999. *Primate adaptation and evolution.* New York: Academic Press.

Fleagle, J. G., C. H. Janson, and K. E. Reed. 1999. *Primate communities.* Cambridge: Cambridge University Press.

Foley, R. A. 1995. Evolution and adaptive significance of hominid maternal behaviour. In *Motherhood in human and nonhuman primates*, edited by C. R. Pryce, R. D. Martin, and D. Skuse, 27–36. Basel, Switzerland: Karger.

Follett, B. K. 1984. The environment and reproduction. In *Reproduction in mammals*, vol. 4, *Reproductive fitness*, edited by C. R. Austin and R. V. Short, 103–32. Cambridge: Cambridge University Press.

Foster, R. B. 1982. Famine on Barro Colorado Island. In *The ecology of a tropical forest*, edited by E. G. Leigh Jr., A. S. Stanley, and D. M. Windsor, 201–12. Washington, D.C.: Smithsonian Institution Press.

French, A. R. 1982. Effects on temperature on the duration of arousal episodes during hibernation. *J. Appl. Physiol.* 52:216–20.

Friedenthal, H. 1910. Über die Gültigkeit der Massenwirkung für den Energieumsatz der lebendigen Substanz. *Zentralbl. Physiol.* 24:321–27.

Frith, C. 1996. Brain mechanisms for having a "theory of mind." *J. Psychopharmacol.* 10:9–16.

Froehlich, J. W., R. W. Thorington Jr., and J. S. Otis. 1981. The demography of howler monkeys *(Alouatta palliata)* on Barro Colorado Island, Panama. *Int. J. Primatol.* 2:207–36.

Gabunia, L., and A. Vekua. 1995. A Plio-Pleistocene hominid mandible from Dmanisi, East Georgia, Caucasus. *Nature* 373:509–12.

Gabunia, L., A. Vekua, D. Lordkipanidze, C. C. Swisher III, R. Ferring, A. Justus, M. Nioradze, M. Tvalchrelidze, S. C. Antón, G. Bosinski, O. Jöris, M.-A. de Lumley, G. Majsuradze, and A. Mouskhelishvili. 2000. Earliest Pleistocene hominid cranial remains from Dmanisi, Republic of Georgia: Taxonomy, geological setting, and age. *Science* 288:1019–25.

Gadgil, M., and W. H. Bossert. 1970. Life history consequences of natural selection. *Am. Nat.* 104:1–24.

Ganzhorn, J. U. 1992. Leaf chemistry and the biomass of folivorous primates in tropical forests. *Oecologia* 91:540–47.

———. 1999. Body mass, competition and the structure of primate communities. In *Primate communities,* edited by J. G. Fleagle, C. H. Janson, and K. E. Reed, 141–57. Cambridge: Cambridge University Press.

———. 2002. Distribution of folivorous lemurs in relation to seasonally varying food resources: Integrating quantitative and qualitative aspects of food characteristics. *Oecologia.* In press.

Ganzhorn, J. U., P. C. Wright, and J. Ratsimbazafy. 1999. Primate communities: Madagascar. In *Primate communities,* edited by J. G. Fleagle, C. H. Janson, and K. E. Reed, 75–89. Cambridge: Cambridge University Press.

Garber, P. A., and S. R. Leigh. 1997. Ontogenetic variation in small-bodied New World primates: Implications for patterns of reproduction and infant care. *Folia Primatol.* 68:1–22.

Gardezi, T. F., and J. da Silva. 1999. Diversity in relation to body size in mammals: A comparative study. *Am. Nat.* 153:110–23.

Gariépy, J.-L., M. H. Lewis, and R. B. Cairns. 1996. Genes, neurobiology, and aggression: Time frames and functions of social behaviors in adaptation. In *Aggression and violence: Neurobiological, biosocial and genetic perspectives,* edited by D. M. Stoff and R. B. Cairns, 41–63. New York: Lawrence Erlbaum.

Garland, T. Jr. 1992. Rate tests for phenotypic evolution using phylogenetically independent comparisons. *Am. Nat.* 140:509–19.

Garland, T. Jr., A. W. Dickerman, C. M. Janis, and J. A. Jones. 1993. Phylogenetic analysis of covariance by computer simulation. *Syst. Biol.* 42:265–92.

Garland, T. Jr., P. H. Harvey, and A. R. Ives. 1992. Procedures for the analysis of comparative data using phylogenetically independent contrasts. *Syst. Biol.* 41:18–32.

Garland, T. Jr., P. E. Midford, and A. R. Ives. 1999. An introduction to phylogenetically based statistical methods with a new method for confidence intervals on ancestral values. *Am. Zool.* 39:374–88.

Gawelczyk, A. 1998. *Allometries.* Krakow: Institute of Environmental Biology, Jagiellonian University.

Geffen, E., M. Gompper, J. L. Gittleman, H.-K. Luh, D. Macdonald, and R. Wayne. 1996. Size, life-history traits, and social organization in the Canidae: A reevaluation. *Am. Nat.* 147:140–60.

Gelvin, B. R., G. H. Albrecht, and J. M. A. Miller. 2000. Complex allometry of brain size scaling among mammals. *Am. J. Phys. Anthropol.* suppl. 30:157.

Genoud, M., R. D. Martin, and D. Glaser. 1997. Rate of metabolism in the smallest simian primate, the pygmy marmoset *(Cebuella pygmaea). Am. J. Primatol.* 41:229–45.

Gittleman, J. L. 1986a. Carnivore brain size, behavioral ecology, and phylogeny. *J. Mammal.* 67:23–36.

———. 1986b. Carnivore life history patterns: Allometric, phylogenetic, and ecological associations. *Am. Nat.* 127:744–71.

Gittleman, J. L., C. G. Anderson, M. Kot, and H.-K. Luh. 1996a. Comparative tests of evolutionary lability and rates using molecular phylogenies. In *New uses for new phylogenies,* edited by P. H. Harvey, A. J. L. Brown, J. Maynard Smith, and S. Nee, 289–307. Oxford: Oxford University Press.

———. 1996b. Phylogenetic lability and rates of evolution: A comparison of behavioral, morphological and life history traits. In *Phylogenies and the comparative method in animal behavior,* edited by E. P. Martins, 166–205. Oxford: Oxford University Press.

Gittleman, J. L., and A. Purvis. 1998. Body size and species richness in primates and carnivores. *Proc. R. Soc. Lond.* B 265:113–19.

Glander, K. E. 1980. Reproduction and population growth in free-ranging mantled howling monkeys. *Am. J. Phys. Anthropol.* 53:25–36.

Godfrey, L. R., W. L. Jungers, K. E. Reed, E. L. Simons, and P. S. Chatrath. 1997a. Subfossil lemurs: Inferences about past and present primate communities. In *Natural change and human impact in Madagascar,* edited by S. M. Goodman and B. Patterson, 218–56. Washington, D.C.: Smithsonian Institution Press.

Godfrey, L. R., W. L. Jungers, R. E. Wunderlich, and B. G. Richmond. 1997b. Reappraisal of the postcranium of *Hadropithecus* (Primates, Indroidea). *Am. J. Phys. Anthropol.* 103:529–56.

Godfrey, L. R., A. J. Petto, and M. R. Sutherland. 2002. Dental ontogeny and life-history strategies: The case of the giant extinct indroids of Madagascar. In *Reconstructing behavior in the primate fossil record,* edited by J. M. Plavcan, R. Kay, W. L. Jungers, and C. P. van Schaik, 113–57. New York: Kluwer Academic/ Plenum Press.

Godfrey, L. R., K. E. Samonds, W. L. Jungers, and M. R. Sutherland. 2001. Teeth, brains, and primate life histories. *Am. J. Phys. Anthropol.* 114:192–214.

Godwin, J., R. Sawby, R. R. Warner, D. Crews, and M. S. Grober. 2000. Hypothalamic arginine vasotocin mRNA abundance variation across sexes and with sex change in a coral reef fish. *Brain Behav. Evol.* 55:77–84.

Goldizen, A. W. 1987. Tamarins and marmosets: Communal care of offspring. In *Primate societies,* edited by B. B. Smuts, D. L. Cheney, R. M. Seyfarth, R. W. Wrangham, and T. T. Struhsaker, 34–43. Chicago: University of Chicago Press.

Goodall, J. 1986. *The chimpanzees of Gombe.* Cambridge, MA: Harvard University Press.

Gosden, R. G., R. I. M. Dunbar, D. Haig, E. Heyer, R. Mace, M. Milinski, G. Pichon, H. Richner, B. I. Strassmann, D. Thaler, C. Wedekind, and S. C. Stearns. 1998. Evolutionary interpretations of the diversity of reproductive health and disease. In *Evolution in health and disease,* edited by S. C. Stearns, 108–20. Oxford: Oxford University Press.

Gould, L., R. W. Sussman, and M. L. Sauther. 1999. Natural disasters and primate populations: The effects of a 2-year drought on a naturally occurring population of ring-tailed lemurs *(Lemur catta)* in southwestern Madagascar. *Int. J. Primatol.* 20:69–84.

Gould, S. J. 1977. *Ontogeny and phylogeny.* Cambridge, MA: Harvard University Press.

Grafen, A. 1989. The phylogenetic regression. *Phil. Trans. R. Soc. Lond.* B 326: 119–57.

Grant, A., and T. G. Benton. 2000. Elasticity analysis for density-dependent populations in stochastic environments. *Ecology* 81:680–93.

Grant, P. 1986. *Ecology and evolution of Darwin's finches.* Princeton, NJ: Princeton University Press.

Grob, B., L. A. Knapp, R. D. Martin, and G. Anzenberger. 1998. The major histocompatibility complex and mate choice: Inbreeding avoidance and selection of good genes. *Exp. Clin. Immunogenet.* 15:119–29.

Grober, M. S. 1998. Socially controlled sex change: Integrating ultimate and proximate levels of analysis. *Acta Ethol.* 1:3–17.

Gross, K., J. R. Lockwood III, C. C. Frost, and W. F. Morris. 1998. Modeling controlled burning and trampling reduction for conservation of *Hudsonia montana. Conserv. Biol.* 12:1291–1301.

Gursky, S. 2000. Effect of seasonality on the behavior of an insectivorous primate, *Tarsius spectrum. Int. J. Primatol.* 21:477–95.

Gust, D. A., and T. P. Gordon. 1991. Female rank instability in newly formed groups of familiar Sooty mangabeys *(Cercocebus torquatus atys). Primates* 32:465–71.

———. 1994. The absence of a matrilineally based dominance system in Sooty mangabeys, *Cercocebus torquatus atys. Anim. Behav.* 47:589–94.

Hakeem, A., G. R. Sandoval, M. Jones, and J. M. Allman. 1996. Brain and life span in primates. In *Handbook of the psychology of aging,* edited by J. E. Birren and K. W. Schaie, 78–104. New York: Academic Press.

Hall, R. L. 1982. *Sexual dimorphism in Homo sapiens: A question of size.* New York: Praeger.

Hall, T. C., K. H. Miller, and J. A. N. Corsellis. 1975. Variations in the human Purkinje cell population according to age and sex. *Neuropathol. Appl. Neurobiol.* 1:267–92.

Hamada, Y., S. Hayakawa, J. Suzuki, and S. Ohkura. 1999. Adolescent growth and development in Japanese macaques *(Macaca fuscata):* Punctuated adolescent growth spurt by season. *Primates* 40:439–52.

Hamilton, W. D., and J. Bulger. 1992. Facultative expression of behavioral differences between one-male and multi-male savanna baboon groups. *Am. J. Primatol.* 28:61–71.

Hamilton, W. D., and M. Zuk. 1982. Heritable true fitness and bright birds: A role for parasites? *Science* 218:384–87.

Hapke, A., D. Zinner, and H. Zischler. 2001. Mitochondrial DNA variation in Eritrean hamadryas baboons *(Papio hamadryas):* Life history influences population genetic structure. *Behav. Ecol. Sociobiol.* 50:483–92.

Happel, R. 1988. Seed-eating by West African cercopithecines, with reference to the possible evolution of bilophodont molars. *Am. J. Phys. Anthropol.* 75:303–27.

Harcourt, A. 1992. Coalitions and alliances: Are primates more complex than nonprimates? In *Coalitions and alliances in humans and other animals,* edited by A. Harcourt and F. B. M. de Waal, 445–71. Oxford: Oxford University Press.

Hare, B., J. Call, B. Agnetta, and M. Tomasello. 2000. Chimpanzees know what conspecifics see and do not see. *Anim. Behav.* 59:771–85.

Hart, R. W., and R. B. Setlow. 1974. Correlation between deoxyribonucleic acid excision-repair and life-span in a number of mammalian species. *Proc. Natl. Acad. Sci. USA* 71:2169–73.

Hartwig, W. C. 1996. Perinatal life history traits in New World monkeys. *Am. J. Primatol.* 40:99–130.

Harvey, P. H., and P. M. Bennett. 1983. Brain size, energetics, ecology and life history patterns. *Nature* 306:244–92.

Harvey, P. H., and T. H. Clutton-Brock. 1985. Life history variation in primates. *Evolution* 39:559–81.

Harvey, P. H., and A. E. Keymer. 1991. Comparing life histories using phylogenies. *Phil. Trans. R. Soc. Lond.* B 332:31–39.

Harvey, P. H., and J. R. Krebs. 1990. Comparing brains. *Science* 249:140–46.

Harvey, P. H., and G. M. Mace. 1982. Comparisons between taxa and adaptive trends: A problem of methodology. In *Current trends in sociobiology,* edited by King's College Sociobiology Group, Cambridge, 343–62. Cambridge: Cambridge University Press.

Harvey, P. H., R. D. Martin, and T. H. Clutton-Brock. 1987. Life histories in comparative perspective. In *Primate societies,* edited by B. B. Smuts, D. L. Cheney, R. M. Seyfarth, R. W. Wrangham, and T. T. Struhsaker, 181–96. Chicago: University of Chicago Press.

Harvey, P. H., and S. Nee. 1991. How to live like a mammal. *Nature* 350:23–24.

Harvey, P. H., and M. D. Pagel. 1991. *The comparative method in evolutionary biology.* Oxford: Oxford University Press.

Harvey, P. H., M. D. Pagel, and J. Rees. 1991. Mammalian metabolism and life histories. *Am. Nat.* 137:556–66.

Harvey, P. H., D. E. L. Promislow, and A. F. Read. 1989. Causes and correlates of life history differences among mammals. In *Comparative socioecology: The behavioral ecology of humans and other animals,* edited by V. Standen and R. A. Foley, 305–18. Oxford: Blackwell.

Harvey, P. H., and A. Purvis. 1991. Comparative methods for explaining adaptations. *Nature* 351:619–24.

———. 1999. Understanding the ecological and evolutionary reasons for life history variation: Mammals as a case study. In *Advanced ecological theory: Principles and applications,* edited by J. McGlade, 232–48. Oxford: Blackwell.

Harvey, P. H., and A. F. Read. 1988. How and why do mammalian life histories vary? In *Evolution of life histories: Patterns and process from mammals,* edited by M. S. Boyce, 213–32. New Haven: Yale University Press.

Harvey, P. H., A. F. Read, and D. E. L. Promislow. 1989. Life history variation in placental mammals: Unifying the data with theory. In *Oxford surveys in evolutionary biology,* vol. 6, edited by P. H. Harvey and L. Partridge, 13–31. Oxford: Oxford University Press.

Harvey, P. H., and R. M. Zammuto. 1985. Patterns of mortality and age at first reproduction in natural populations of mammals. *Nature* 315:319–20.

Harwerth, R. S., E. L. Smith III, G. C. Duncan, M. L. J. Crawford, and G. K. von Norden. 1986. Multiple sensitive periods in the development of the primate visual system. *Science* 232:235–37.

Hatley, T., and J. Kappelman. 1980. Bears, pigs, and Plio-Pleistocene hominids: A case for the exploitation of below-ground food resources. *Hum. Ecol.* 8:371–87.

Haug, H. 1985. Are neurons of the human cerebral cortex really lost during aging? A morphometric examination. In *Senile dementia of the Alzheimer type,* edited by J. Traber and W. H. Gispen, 150–63. New York: Springer-Verlag.

Hawkes, K. 1993. Why hunter-gatherers work: An ancient version of the problem of public goods. *Curr. Anthropol.* 34:341–61.

Hawkes, K., J. F. O'Connell, and N. G. Blurton Jones. 1989. Hardworking Hadza grandmothers. In *Comparative socioecology of mammals and man: The behavioral ecology of humans and other animals,* edited by V. Standen and R. A. Foley, 341–66. London: Blackwell.

———. 1991. Hunting income patterns among the Hadza: Big game, common goods, foraging goals, and the evolution of the human diet. *Phil. Trans. R. Soc. Lond.* B 334:243–51.

———. 1995. Hadza children's foraging: Juvenile dependency, social arrangements and mobility among hunter-gatherers. *Curr. Anthropol.* 36:688–700.

———. 1997. Hadza women's time allocation, offspring provisioning, and the evolution of post-menopausal life spans. *Curr. Anthropol.* 38:551–78.

———. 2000. Why do women have mid-life menopause? Grandmothering and the evolution of human longevity. In *Female reproductive aging,* edited by E. R. te Velde, P. L. Pearson, and F. J. Broekmans, 27–41. New York: Parthenon.

———. 2001a. Hadza meat sharing. *Evol. Hum. Behav.* 22:113–42.

———. 2001b. Hunting and nuclear families: Some lessons from the Hadza about men's work *Curr. Anthropol.* 42:681–709.

Hawkes, K., J. F. O'Connell, N. G. Blurton Jones, J. Alvarez, and E. L. Charnov. 1998. Grandmothering, menopause, and the evolution of human life histories. *Proc. Natl. Acad. Sci. USA* 95:1336–39.

Hayssen, V., A. van Tienhoven, and A. van Tienhoven. 1993. *Asdell's patterns of mammalian reproduction: A compendium of species-specific data.* Ithaca, NY: Cornell University Press.

Hebb, D. O. 1949. *The organization of behavior: A neuropsychological theory.* New York: Wiley.

Heinzelin, J. de, J. D. Clark, T. D. White, W. Hart, P. Renne, G. WoldeGabriel, Y. Beyene, and E. Vrba. 1999. Environment and behavior of 2.5 million-year-old Bouri hominids. *Science* 284:625–29.

Heldmaier, G. 1971. Zitterfreie Wärmebildung und Körpergrösse bei Säugetieren. *Z. vergl. Physiol.* 73:222–48.

———. 1989. Seasonal acclimatization of energy requirements in mammals: Functional significance of body weight control, hypothermia, torpor and hibernation. In *Energy transformation in cells and organisms,* edited by W. Wieser and E. Gnaiger, 130–39. Stuttgart: Georg Thieme Verlag.

Hemingway, C. A. 1996. Morphology and phenology of seeds and whole fruit eaten by Milne-Edwards' sifaka, *Propithecus diadema edwardsi,* in Ranomafana National Park, Madagascar. *Int. J. Primatol.* 17:637–59.

———. 1998. Selectivity and variability in the diet of Milne-Edwards' sifakas *(Propithecus diadema edwardsi):* Implications for folivory and seed-eating. *Int. J. Primatol.* 19:355–77.

Hennemann, W. W. 1983. Relationship among body mass, metabolic rate, and the intrinsic rate of natural increase in mammals. *Oecologia* 56:104–8.

Heppell, S. S., L. B. Crowder, and D. T. Crouse. 1996. Models to evaluate headstarting as a management tool for long-lived turtles. *Ecol. Appl.* 6:556–65.

Heppell, S. S., J. R. Walters, and L. B. Crowder. 1994. Evaluating management alternatives for red-cockaded woodpeckers: A modeling approach. *J. Wildl. Mgmt.* 58:479–87.

Heymann, E. W. 2000. The number of males in callitrichine groups and its implications for callitrichine social evolution. In *Primate males,* edited by P. M. Kappeler, 64–71. Cambridge: Cambridge University Press.

Heymann, E. W., and P. Soini. 1999. Offspring number in pygmy marmosets, *Cebuella pygmaea,* in relation to group size and the number of adult males. *Behav. Ecol. Sociobiol.* 46:400–404.

Heyning, J. E. 1997. Sperm whale phylogeny revisited: Analysis of the morphological evidence. *Mar. Mamm. Sci.* 13:596–613.

Hill, K. 1993. Life history theory and evolutionary anthropology. *Evol. Anthropol.* 2:78–88.

Hill, K., C. Boesch, J. Goodall, A. Pusey, J. Williams, and R. W. Wrangham. 2001. Mortality rates among wild chimpanzees. *J. Hum. Evol.* 39:1–14.

Hill, K., and A. M. Hurtado. 1991. The evolution of premature reproductive senescence and menopause in human females: An evaluation of the "grandmother" hypothesis. *Hum. Nat.* 2:313–50.

———. 1996. *Ache life history: The ecology and demography of a foraging people.* Hawthorne, NY: Aldine de Gruyter.

———. 1999. Packer and colleagues' model of menopause for humans. *Hum. Nat.* 10:199–204.

Hill, R. A., and R. I. M. Dunbar. 1998. An evaluation of the roles of predation rate and predation risk as selective pressures on primate grouping behaviour. *Behaviour* 135:411–30.

Hill, R. A., and P. C. Lee. 1998. Predation risk as an influence on group size in cercopithecoid primates: Implications for social structure. *J. Zool.* 245:447–56.

Hladik, C. M. 1977. A comparative study of the feeding strategies of two sympatric species of leaf monkeys: *Presbytis senex* and *P. entellus.* In *Primate ecology: Studies of feeding and ranging behaviour in lemurs, monkeys and apes,* edited by T. H. Clutton-Brock, 324–53. London, New York: Academic Press.

Hladik, C. M., D. J. Chivers, and P. Pasquet. 1999. On diet and gut size in nonhuman primates: Is there a relationship to brain size? *Curr. Anthropol.* 40:695–97.

Hofman, M. A. 1983. Energy metabolism, brain size and longevity in mammals. *Q. Rev. Biol.* 58:495–512.

Holekamp, K. E., and L. Smale. 1995. Rapid change in offspring sex ratios after clan fission in the spotted hyena. *Am. Nat.* 145:261–78.

Holliday, M. A. 1971. Metabolic rate and organ size during growth from infancy to maturity and during late gestation and early infancy. *Pediatrics* 47:169–79.

Holmes, D. J., and D. Sherry. 1997. Selected approaches to using individual variation for understanding mammalian life history evolution. *J. Mammal.* 78:311–19.

Holt, A. B., D. B. Cheek, D. Mellits, and D. E. Hill. 1975. Brain size and the relation of the primate to the nonprimate. In *Fetal and postnatal cellular growth,* edited by D. B. Cheek, 23–45. New York: Wiley.

Hood, L. C., and A. Jolly. 1995. Troop fission in female *Lemur catta* at Berenty Reserve, Madagascar. *Int. J. Primatol.* 16:997–1015.

Horrocks, J. A., and W. Hunte. 1983. Rank relations in vervet sisters: A critique of the role of reproductive value. *Am. Nat.* 122:417–21.

Houston, A. I., and J. M. McNamara. 1992. Phenotypic plasticity as a state-dependent life-history decision. *Evol. Ecol.* 6:243–53.

Howell, N. 1979. *Demography of the Dobe !Kung.* New York: Academic Press.

———. 1982. Village composition implied by a paleodemographic life table: The Libben site. *Am. J. Phys. Anthropol.* 59:263–69.

Hrdy, S. B. 1970. The care and exploitation of non_human primate infants by conspecifics other than the mother. In *Advances in the study of behaviour,* vol. 6, edited by J. S. Rosenblatt, R. A. Hinde, E. Shaw, and C. Beer, 251–383. New York: Academic Press.

———. 1979. Infanticide among animals: A review, classification, and examination of the implications for the reproductive strategies of females. *Ethol. Sociobiol.* 1:13–40.

Hrdy, S. B., and D. B. Hrdy. 1976. Hierarchical relations among female hanuman langurs (Primates: Colobinae, *Presbytis entellus*). *Science* 193:913–15.

Hrdy, S. B., and P. L. Whitten. 1987. Patterning of sexual activity. In *Primate societies*, edited by B. B. Smuts, D. L. Cheney, R. M. Seyfarth, R. W. Wrangham, and T. T. Struhsaker, 370–84. Chicago: University of Chicago Press.

Hudson, J. W. 1973. Torpidity in mammals. In *Comparative physiology of thermoregulation*, edited by G. C. Whittow, 97–165. London: Academic Press.

Huey, R. B. 1991. Physiological consequences of habitat selection. *Am. Nat.* 137: S91–115.

Hughes, R. N. 1993. *Diet selection*. Oxford: Blackwell.

Hulbert, A. J., and T. J. Dawson. 1974. Standard metabolism and body temperature of perameloid marsupials from different environments. *Comp. Biochem. Physiol.* A 47:583–90.

Hunter, M. D., and P. W. Price. 1992. Playing chutes and ladders: Heterogeneity and the relative roles of bottom-up and top-down forces in natural communities. *Ecology* 73:724–32.

Ims, R. A. 1990. On the adaptive value of reproductive synchrony as a predator-swamping strategy. *Am. Nat.* 136:485–98.

Isaac, G. L. 1978. The food sharing behavior of protohuman hominids. *Sci. Am.* 238:90–108.

———. 1997. *Koobi Fora research project: Plio-Pleistocene archaeology*, vol. 5. Oxford: Oxford University Press.

Isbell, L. A. 1991. Contest and scramble competition: Patterns of female aggression and ranging behavior among primates. *Behav. Ecol.* 2:145–55.

———. 1994. Predation on primates: Ecological patterns and evolutionary consequences. *Evol. Anthropol.* 3:61–71.

Jacob, S., M. K. McClintock, B. Zelano, and C. Ober. 2002. Paternally-inherited HLA alleles are associated with women's choice of male odor. *Nature Genet.* 30:175–79.

James, F. C. 1983. Environmental component of morphological differentiation in birds. *Science* 221:184–86.

Jansen, R. P. S., and K. de Boer. 1998. The bottleneck: Mitochondrial imperatives in oogenesis and ovarian follicular fate. *Mol. Cell. Endocrinol.* 145:81–88.

Janson, C. H. 1988. Intra_specific food competition and primate social structure: A synthesis. *Behaviour* 105:1–17.

———. 1990. Ecological consequences of individual spatial choice in foraging brown capuchin monkeys, *Cebus apella*. *Anim. Behav.* 40:922–34.

———. 1992. Evolutionary ecology of primate social structure. In *Evolutionary ecology and human behavior*, edited by E. A. Smith and B. Winterhalder, 95–130. New York: Aldine de Gruyter.

———. 1998. Testing the predation hypothesis for vertebrate sociality: Prospects and pitfalls. *Behaviour* 135:389–410.

Janson, C. H., and M. Goldsmith. 1995. Predicting group size in primates: Foraging costs and predation risk. *Behav. Ecol.* 6:326–36.

Janson, C. H., and C. P. van Schaik. 1993. Ecological risk aversion in juvenile primates: Slow and steady wins the race. In *Juvenile primates: Life history, develop-*

ment and behavior, edited by M. E. Pereira and L. A. Fairbanks, 57–74. Oxford: Oxford University Press.

——. 2000. The behavioral ecology of infanticide by males. In *Infanticide: Functional and theoretical considerations,* edited by C. P. van Schaik and C. H. Janson, 469–94. Cambridge: Cambridge University Press.

Janson, C. H., E. W. Stiles, and D. W. White. 1986. Selection of plant fruiting traits by brown capuchin monkeys: A multivariate approach. In *Frugivores and seed dispersal,* edited by A. Estrada and T. H. Fleming, 83–92. Dordrecht: Dr. W. Junk Publishers.

Jarman, P. J. 1974. The social organization of antelope in relation to their ecology. *Behaviour* 48:215–56.

——. 1983. Mating system and sexual dimorphism in large terrestrial mammalian herbivores. *Biol. Rev.* 58:485–520.

Jenkins, S. J. 1988. Use and abuse of demographic models of population growth. *Bull. Ecol. Soc. Am.* 69:201–7.

Jerison, H. J. 1973. *Evolution of the brain and intelligence.* New York: Academic Press.

Jernvall, J., and P. C. Wright. 1998. Diversity components of impending primate extinctions. *Proc. Natl. Acad. Sci. USA* 95:11279–83.

Joffe, T. H. 1997. Social pressures have selected for an extended juvenile period in primates. *J. Hum. Evol.* 32:593–605.

Joffe, T. H., and R. I. M. Dunbar. 1997. Visual and socio-cognitive information processing in primate brain evolution. *Proc. R. Soc. Lond.* B 264:1303–7.

Jolicoeur, P., G. Baron, and T. Cabana. 1988. Cross-sectional growth and decline of human stature and brain weight in 19th century Germany. *Growth Dev. Aging* 52:201–6.

Jolly, A. 1966. *Lemur behavior.* Chicago: University of Chicago Press.

——. 1998. Pair-bonding, female aggression and the evolution of lemur societies. *Folia Primatol.* 69:1–13.

Jolly, A., S. Caless, S. Cavigelli, L. Gould, A. Pitts, M. E. Pereira, R. E. Pride, H. D. Rabenandrasana, J. D. Walker, and T. Zafison. 2000. Infant killing, wounding, and predation in *Eulemur* and *Lemur. Int. J. Primatol.* 21:21–40.

Jolly, A., A. Dobson, H. M. Rasamimanana, J. D. Walker, S. O'Connor, M. Solberg, and V. Perel. 2002. Demography of *Lemur catta* at Berenty Reserve, Madagascar: Effects of troop size, habitat and rainfall. *Int. J. Primatol.* 23:327–53.

Jolly, A., and R. E. Pride. 1999. Troop histories and range inertia of *Lemur catta* at Berenty, Madagascar: A thirty year perspective. *Int. J. Primatol.* 20:359–73.

Jones, K. E. 1998. Evolution of bat life histories. Ph.D. thesis, University of Surrey.

Jones, K. E., and A. MacLarnon. 2001. Bat life histories: Testing models of mammalian life history evolution. *Evol. Ecol. Res.* 3:465–76.

Judge, D. S., and J. R. Carey. 2000. Post-reproductive life period predicted by primate patterns. *J. Gerontol. Biol. Sci.* 55A:B201–9.

Jungers, W. L., L. R. Godfrey, E. L. Simons, R. E. Wunderlich, B. G. Richmond, P. S. Chatrath, and B. Rakotosaminanana. 2002. Ecomorphology and behavior of giant extinct lemurs from Madagascar. In *Reconstructing behavior in the primate fossil record,* edited by J. M. Plavcan, R. F. Kay, W. L. Jungers, and C. P. van Schaik, 371–411. New York: Kluwer Academic/Plenum Press.

Kalbfleisch, J. D., and R. L. Prentice. 1980. *The statistical analysis of failure time data.* New York: Wiley.

Kamerich, E. 1999. *Guide to Maple.* New York: Springer Verlag.

Kaplan, H. 1997. The evolution of the human life course. In *Between Zeus and the salmon,* edited by K. W. Wachter and C. E. Finch, 175–211. Washington, D.C.: National Academy Press.

Kaplan, H., L. Hill, J. Lancaster, and A. M. Hurtado. 2000. A theory of human life history evolution: Diet, intelligence, and longevity. *Evol. Anthropol.* 9:156–85.

Kappeler, P. M. 1990. The evolution of sexual size dimorphism in prosimian primates. *Am. J. Primatol.* 21:201–14.

———. 1993. Reconciliation and post-conflict behavior in ringtailed lemurs, *Lemur catta* and redfronted lemurs, *Eulemur fulvus rufus. Anim. Behav.* 45:901–15.

———. 1995. Life history variation among nocturnal prosimians. In *Creatures of the dark: The nocturnal prosimians,* edited by L. Alterman, M. K. Izard, and G. A. Doyle, 75–92. New York: Plenum Press.

———. 1996. Causes and consequences of life history variation among strepsirhine primates. *Am. Nat.* 148:868–91.

———. 1997. Intrasexual selection and testis size in strepsirhine primates. *Behav. Ecol.* 8:10–19.

———. 1998a. Nests, tree holes, and the evolution of primate life histories. *Am. J. Primatol.* 46:7–33.

———. 1998b. To whom it may concern: The transmission and function of chemical signals in *Lemur catta. Behav. Ecol. Sociobiol.* 42:411–21.

———. 1999. Primate socioecology: New insights from males. *Naturwissenschaften* 86:18–29.

———. 2000. *Primate males.* Cambridge: Cambridge University Press.

Kappeler, P. M., and E. W. Heymann. 1996. Nonconvergence in the evolution of primate life history and socioecology. *Biol. J. Linn. Soc.* 59:297–326.

Kappelman, J. 1996. The evolution of body mass and relative brain size in fossil hominids. *J. Hum. Evol.* 30:243–76.

Kawai, M., R. I. M. Dunbar, H. Ohsawa, and U. Mori. 1983. Social organisation of gelada baboons: Social units and definitions. *Primates* 24:13–24.

Kawecki, T. J., and S. C. Stearns. 1993. The evolution of life histories in spatially heterogeneous environments: Optimal reaction norms revisited. *Evol. Ecol.* 7:155–74.

Kay, R. F. 1978. Molar structure and diet in extant Cercopithecidae. In *Development, function and evolution of teeth,* edited by P. M. Butler and K. A. Joysey, 309–39. London: Academic Press.

———. 1984. On the use of anatomical features to infer foraging behavior in extinct primates. In *Adaptations for foraging in nonhuman primates,* edited by P. S. Rodman and J. G. H. Cant, 21–53. New York: Columbia University Press.

Kay, R. F., R. H. Madden, C. P. van Schaik, and D. Higdon. 1997. Primate species richness is determined by plant productivity: Implications for conservation. *Proc. Natl. Acad. Sci.* 94:13023–27.

Kay, R. F., C. Ross, and B. A. Williams. 1997. Anthropoid origins. *Science* 275:797–804.

Kelley, J. 1999. Age at first molar emergence in *Afropithecus turkanensis. Am. J. Phys. Anthropol.* suppl. 28:167.

Kendall, B. E. 1998. Estimating the magnitude of environmental stochasticity in survivorship data. *Ecol. Appl.* 8:184–93.

Key, C. A., L. Aiello, and T. Molleson. 1994. Cranial suture closure and its implications for age estimation. *Int. J. Osteoarchaeol.* 4:193–207.

Keyfitz, N., and W. Flieger. 1968. *World population: An analysis of vital data.* Chicago: University of Chicago Press.

——. 1990. *World population growth and aging: Demographic trends in the late twentieth century.* Chicago: University of Chicago Press.

Kinderman, P., R. I. M. Dunbar, and R. Bentall. 1998. Theory-of-mind deficits and causal attributions. *Brit. J. Psychol.* 89:191–204.

King, S. J., L. R. Godfrey, and E. L Simons. 2001. Adaptive and phylogenetic significance of ontogenetic sequences in *Archaeolemur,* subfossil lemur from Madagascar. *J. Hum. Evol.* 41:545–76.

Kirkpatrick, R. C. 1999. Colobine diet and social organization. In *The nonhuman primates,* edited by P. Dolhinow and A. Fuentes, 93–105. Mountain View, CA: Mayfield Publishing Company.

Kirkwood, J. K. 1985. Patterns of growth in primates. *J. Zool.* 205:123–36.

Kirkwood, J. K., and K. Stathatos. 1992. *Biology, rearing, and care of young primates.* Oxford: Oxford University Press.

Kirkwood, T. B. L. 1977. Evolution of aging. *Nature* 270:301–2.

Klaus, S., L. Casteilla, F. Bouillaud, and D. Ricquier. 1991. The uncoupling protein UCP: A membraneous mitochondrial ion carrier exclusively expressed in brown adipose tissue. *Int. J. Biochem.* 23:791–801.

Kleiber, M. 1932. Body size and metabolism. *Hilgardia* 6:315–53.

Klein, R. G. 2000. Archaeology and the evolution of human behavior. *Evol. Anthropol.* 9:17–36.

Knott, C. D. 1998. Changes in orangutan caloric intake, energy balance, and ketones in response to fluctuating fruit availability. *Int. J. Primatol.* 19:1061–79.

Koyama, N. 1988. Mating behavior of ring-tailed lemurs *(Lemur catta)* at Berenty, Madagascar. *Primates* 29:163–75.

——. 1991. Troop division and inter-troop relationships of ring-tailed lemurs *(Lemur catta)* at Berenty, Madagascar. In *Primatology today,* edited by A. Ehara, O. Takenaka, and M. Iwamoto, 173–76. Amsterdam: Elsevier.

——. 1992. Some demographic data of ring-tailed lemurs *(Lemur catta)* at Berenty, Madagascar. In *Social structure of Madagascar higher vertebrates in relation to their adaptive radiation,* edited by S. Yamagishi, 10–16. Osaka: Osaka City University Press.

Kozlowski, J. 1993. Measuring fitness in life-history studies. *Trends Ecol. Evol.* 8:84–85.

Kozlowski, J., and J. Weiner. 1997. Interspecific allometries are byproducts of body size optimization. *Am. Nat.* 149:352–80.

Krakauer, D. C., and A. Mira. 1999. Mitochondria and germ-cell death. *Nature* 400:125–26.

Krishtalka, L., R. K. Stucky, and C. Beard. 1990. The earliest fossil evidence for sexual dimorphism in primates. *Proc. Natl. Acad. Sci. USA* 87:5223–26.

Kroll, E. M. 1994. Behavioral implications of Plio-Pleistocene archaeological site structure. *J. Hum. Evol.* 27:107–38.

Ku, H.-H., U. T. Brunk, and R. S. Sohal. 1993. Relationship between mitochondrial superoxide and hydrogen peroxide production and longevity of mammalian species. *Free Radic. Biol. Med.* 15:621–27.

Kudo, H., and R. I. M. Dunbar. 2001. The size and structure of primate social networks. *Anim. Behav.* 62:711–22.

Kurland, J. A., and J. D. Pearson. 1986. Ecological significance of hypometabolism in nonhuman primates: Allometry, adaptation, and deviant diets. *Am. J. Phys. Anthropol.* 71:445–57.

Kvarnemo, C., and I. Ahnesjö. 1996. The dynamics of operational sex ratios and competition for mates. *Trends Ecol. Evol.* 11:404–8.

Lambert, J. E. 1998. Primate digestion: Interactions among anatomy, physiology, and feeding ecology. *Evol. Anthropol.* 7:8–20.

Lan, D. 2001. Ecology and behaviour of black gibbons in Yunnan, China. Ph.D. thesis, University of Liverpool.

Lancaster, J. B., H. S. Kaplan, K. Hill, and A. M. Hurtado. 2000. The evolution of life history, intelligence and diet among chimpanzees and human foragers. In *Perspectives in ethology,* vol. 13: *Evolution, culture, and behavior,* edited by N. S. Thompson and F. Tonneau, 47–72. New York: Kluwer Academic/Plenum Publishers.

Lancaster, J. B., and C. S. Lancaster. 1983. Parental investment: The hominid adaptation. In *How humans adapt: A biocultural odyssey,* edited by D. J. Ortner, 33–65. Washington, D.C.: Smithsonian Institution Press.

Lande, R. 1982a. Elements of a quantitative genetic model of life history evolution. In *Evolution and genetics of life histories,* edited by H. Dingle and J. P. Hegmann, 21–29. New York: Springer-Verlag.

———. 1982b. A quantitative genetic theory of life history evolution. *Ecology* 63:607–15.

Lande, R., and S. J. Arnold. 1983. The measurement of selection on correlated characters. *Evolution* 37:1210–26.

Lawler, I. R., W. J. Foley, and B. M. Eschler. 2000. Foliar concentration of a single toxin creates habitat patchiness for a marsupial folivore. *Ecology* 81:1327–38.

Leakey, M. D. 1971. *Olduvai Gorge.* Vol. 3, *Excavations in beds I and II, 1960–1963.* Cambridge: Cambridge University Press.

Le Duc, R. G., W. F. Perrin, and A. E. Dizon. 1999. Phylogenetic relationships among the delphinid cetaceans based on full cytochrome *B* sequences. *Mar. Mamm. Sci.* 15:619–48.

Lee, E. T. 1992. *Statistical methods for survival data analysis.* New York: Wiley.

Lee, P. C. 1984. Ecological constraints on the social development of vervet monkeys. *Behaviour* 91:245–62.

———. 1987. Nutrition, fertility, and maternal investment in primates. *J. Zool.* 213:409–22.

———. 1996. The meaning of weaning: Growth, lactation, and life history. *Evol. Anthropol.* 5:87–96.

———. 1999. Comparative ecology of postnatal growth and weaning among haplorhine primates. In *Comparative primate socioecology,* edited by P. C. Lee, 111–36. Cambridge: Cambridge University Press.

Lee, P. C., and J. E. Bowman. 1995. Influence of ecology and energetics on primate mothers and infants. In *Motherhood in human and nonhuman primates,* edited by C. R. Pryce, R. D. Martin, and D. Skuse, 47–58. Basel: Karger.

Lee, P. C., P. Majluf, and I. J. Gordon. 1991. Growth, weaning and maternal investment from a comparative perspective. *J. Zool.* 225:99–114.

Lee, R. D. 1997. Intergeneration relations and the elderly. In *Between Zeus and the salmon,* edited by K. W. Wachter and C. E. Finch, 212–33. Washington, D.C.: National Academy Press.

Lee-Thorp, J., J. F. Thackery, and N. van der Merwe. 2000. The hunters and the hunted revisited. *J. Hum. Evol.* 39:1–14.

Lefebvre, L., A. Gaxiola, S. Dawson, S. Timmermans, L. Rosza, and P. Kabai. 1998. Feeding innovations and forebrain size in Australasian birds. *Behaviour* 135: 1077–97.

Lefebvre, L., P. Whittle, E. Lascaris, and A. Finkelstein. 1997. Feeding innovations and forebrain size in birds. *Anim. Behav.* 53:549–60.

Lefkovitch, L. P. 1965. The study of population growth in organisms grouped by stages. *Biometrics* 21:1–18.

Leibold, M. A. 1997. Species turnover and the regulation of trophic structure. *Annu. Rev. Ecol. Syst.* 28:467–94.

Leidy, L. E. 1999. Menopause in evolutionary perspective. In *Evolutionary medicine,* edited by W. R. Trevathan, E. O. Smith, and J. J. McKenna, 407–28. Oxford: Oxford University Press.

Leigh, S. R. 1992a. Cranial capacity evolution in *Homo erectus* and early *Homo sapiens. Am. J. Phys. Anthropol.* 87:1–13.

———. 1992b. Ontogeny and body size dimorphism in anthropoid primates. Ph.D. thesis, Northwestern University.

———. 1992c. Patterns of variation in the ontogeny of primate body size dimorphism. In *Ontogenetic perspectives on primate evolutionary biology,* edited by M. Ravosa and A. Gomez, 27–50. London: Academic Press.

———. 1994a. Ontogenetic correlates of diet in anthropoid primates. *Am. J. Phys. Anthropol.* 94:499–522.

———. 1994b. Relations between captive and non-captive weights in anthropoid primates. *Zoo Biol.* 13:21–43.

———. 1995. Socioecology and the ontogeny of sexual size dimorphism in anthropoid primates. *Am. J. Phys. Anthropol.* 97:339–56.

———. 1996. Evolution of human growth spurts. *Am. J. Phys. Anthropol.* 101:455–74.

———. 2000. Dynamics of canine development in mangabeys and baboons. *Am. J. Phys. Anthropol.* suppl. 30:209.

Leigh, S. R., and P. B. Park. 1998. Evolution of human growth prolongation. *Am. J. Phys. Anthropol.* 107:331–50.

Leigh, S. R., and B. T. Shea. 1995. Ontogeny and the evolution of adult body size dimorphism in apes. *Am. J. Primatol.* 36:37–60.

Leigh, S. R., and C. J. Terranova. 1998. Comparative perspectives on bimaturism, ontogeny, and dimorphism in lemurid primates. *Int. J. Primatol.* 19:723–49.

Leonard, W. R., and M. L. Robertson. 1992. Nutritional requirements and human evolution: A bioenergetics model. *Am. J. Hum. Biol.* 4:179–95.

———. 1994. Evolutionary perspectives on human nutrition: The influence of brain and body size on diet and metabolism. *Am. J. Hum. Biol.* 6:77–88.

Leslie, A. M. 1982. The perception of causality in infants. *Perception* 11:173–86.

Leslie, P. H. 1945. The use of matrices in certain population mathematics. *Biometrika* 33:183–212.

Leutenegger, W. 1979. Evolution of litter size in primates. *Am. Nat.* 114:525–31.

Leutenegger, W., and J. Cheverud. 1982. Correlates of sexual dimorphism in primates: Ecological and size variables. *Int. J. Primatol.* 3:387–402.

Leutenegger, W., and J. T. Kelly. 1977. Relationship of sexual dimorphism in canine size and body size to social, behavioral, and ecological correlates in anthropoid primates. *Primates* 18:117–36.

Levitsky, D. A., and B. J. Strupp. 1995. Malnutrition and the brain: Changing concepts, changing concerns. *J. Nutr.* 125: S2212–20.

Lewis, K. 2000. A comparative study of primate play behaviour: Implications for the study of cognition. *Folia Primatol.* 71:417–21.

Lima, S. L. 1987. Vigilance while feeding and its relation to the risk of predation. *J. Theor. Biol.* 24:303–16.

———. 1988. Initiation and termination of daily feeding in dark-eyed juncos—influences of predation risk and energy reserves. *Oikos* 53:3–11.

Lima, S. L., and L. M. Dill. 1990. Behavioral decisions made under the risk of predation—a review and prospectus. *Can. J. Zool.* 68:619–40.

Lima, S. L., and P. A. Zollner. 1996. Anti-predatory vigilance and the limits to collective detection: Visual and spatial separation between foragers. *Behav. Ecol. Sociobiol.* 38:355–63.

Lindstedt, S. L., and W. A. Calder. 1976. Body size and longevity in birds. *Condor* 78:91–94.

———. 1981. Body size, physiological time, and longevity of homeothermic animals. *Q. Rev. Biol.* 56:1–16.

Lindstrom, A. 1989. Finch flock size and risk of hawk predation at a migratory stopover site. *Auk* 106:225–32.

Linklater, W., E. Cameron, E. Minot, and K. Stafford. 1999. Stallion harassment and the mating system of horses. *Anim. Behav.* 58:295–306.

Lively, C. M. 1986a. Competition, comparative life histories, and maintenance of shell dimorphism in a barnacle. *Ecology* 67:858–64.

———. 1986b. Predator-induced shell dimorphism in the acorn barnacle. *Evolution* 40:232–42.

Lorenz, K. 1965. *Evolution and modification of behavior.* Chicago: University of Chicago Press.

Lovejoy, O. 1981. The origin of man. *Science* 211:341–50.

Lucas, P. W., and M. F. Teaford. 1994. Functional morphology of colobine teeth. In *Colobine monkeys: Their ecology, behavior, and evolution,* edited by A. G. Davies and J. F. Oates, 173–204. Cambridge: Cambridge University Press.

Lyman, R. L., and G. L. Fox. 1989. A critical evaluation of bone weathering as an indication of bone assemblage formation. *J. Archaeol. Sci.* 16:293–317.

MacArthur, R. H., and E. O. Wilson. 1967. *Theory of island biogeography.* Princeton, NJ: Princeton University Press.

Mace, G. M. 1979. The evolutionary ecology of small mammals. Ph.D. thesis, University of Sussex, Brighton.

Mace, R. 2000. Evolutionary ecology of human life history. *Anim. Behav.* 59:1–10.

Macho, G. A., and B. A. Wood. 1995. The role of time and timing in hominid dental evolution. *Evol. Anthropol.* 4:17–31.

MacLarnon, A. M. 1999. The comparative method: Principles and illustrations from primate socioecology. In *Comparative primate socioecology,* edited by P. C. Lee, 5–24. Cambridge: Cambridge University Press.

Macphail, E. M. 1987. The comparative psychology of intelligence. *Behav. Brain Sci.* 10:645–95.

Maggioncalda, A. N., N. M. Czekala, and R. M. Sapolsky. 2000. Growth hormone and thyroid stimulating hormone concentrations in captive male orangutans: Implications for understanding developmental arrest. *Am. J. Primatol.* 50:67–76.

Maggioncalda, A. N., R. M. Sapolsky, and N. M. Czekala. 1999. Reproductive hormone profiles in captive male orangutans: Implications for understanding developmental arrest. *Am. J. Phys. Anthropol.* 109:19–32.

Malinow, M. R., B. L. Pope, J. R. Depaoli, and S. Katz. 1968. Laboratory observations on living howlers. In *Biology of the howler monkey* (Alouatta caraya), edited by M. R. Malinow. *Bibl. Primatol.* 7:224–30.

Mallouk, R. S. 1975. Longevity in vertebrates is proportional to relative brain weight. *Fed. Proc.* 34:2102–3.

Mann, A. 1972. Hominid and cultural origins. *Man* 7:379–86.

Manocha, S. L. 1979. Physical growth and brain development of captive-bred male and female squirrel monkeys, *Saimiri sciureus. Experientia* 35:96–98.

Marean, C. W., L. M. Spencer, R. J. Blumenschine, and S. D. Capaldo. 1992. Captive hyaena bone choice and destruction, the schlepp effect, and Olduvai archaeofaunas. *J. Archaeol. Sci.* 19:101–21.

Margulis, S. W., J. Altmann, and C. Ober. 1993. Sex-biased lactational duration in a human population and its reproductive costs. *Behav. Ecol. Sociobiol.* 32:41–45.

Marino, L. 1996. What dolphins can tell us about primate evolution. *Evol. Anthropol.* 5:81–86.

———. 1997. The relationship between gestation length, encephalization, and body weight in odontocetes. *Mar. Mamm. Sci.* 13:133–38.

———. 1998. A comparison of encephalization between odontocete cetaceans and anthropoid primates. *Brain Behav. Evol.* 51:230–38.

Martin, R. A. 1992. Generic species richness and body mass in North American mammals: Support for the inverse relationship of body size and speciation rate. *Hist. Biol.* 6:73–90.

Martin, R. D. 1968. Reproduction and ontogeny in tree-shrews *(Tupaia belangeri)* with reference to their general behaviour and taxonomic relationships. *Z. Tierpsychol.* 25:409–532.

———. 1972. Behaviour and ecology of nocturnal prosimians. *Z. Tierpsychol.* Beiheft 9:43–89.

———. 1980. Sexual dimorphism and the evolution of higher primates. *Nature* 287:273–75.

———. 1981. Relative brain size and metabolic rate in terrestrial vertebrates. *Nature* 293:57–60.

———. 1983. *Human brain size in an ecological context.* (52nd James Arthur Lecture on the evolution of the brain). New York: American Museum of Natural History.

———. 1990. *Primate origins and evolution: A phylogenetic reconstruction.* London: Chapman and Hall.

———. 1995. Phylogenetic aspects of primate reproduction: The context of advanced maternal care. In *Motherhood in human and nonhuman primates,* edited by C. R. Pryce, R. D. Martin, and D. Skuse, 16–26. Basel: Karger.

———. 1996. Scaling of the mammalian brain: The maternal energy hypothesis. *News Physiol. Sci.* 11:149–56.

Martin, R. D., and P. H. Harvey. 1985. Brain size allometry, ontogeny and phylogeny. In *Size and scaling in primate biology,* edited by W. L. Jungers, 147–73. New York: Plenum Press.

Martin, R. D., and A. M. MacLarnon. 1985. Gestation period, neonatal size, and maternal investment in placental mammals. *Nature* 313:220–23.

———. 1988. Comparative quantitative studies of growth and reproduction. *Symp. Zool. Soc. Lond.* 60:39–80.

———. 1990. Reproductive patterns in primates and other mammals: The dichotomy between altricial and precocial offspring. In *Primate life history and evolution,* edited by C. J. de Rousseau, 47–79. New York: Wiley-Liss.

Martin, R. D., L. A. Willner, and A. Dettling. 1994. The evolution of sexual size dimorphism in primates. In *Differences between the sexes,* edited by S. Balahan, 159–200. Cambridge: Cambridge Univ. Press.

Martins, E. P. 1994. Estimating the rate of phenotypic evolution from comparative data. *Am. Nat.* 144:193–209.

Martins, E. P., and T. Garland Jr. 1991. Phylogenetic analysis of the correlated evolution of continuous characters: A simulation study. *Evolution* 45:534–57.

Martins, E. P., and T. F. Hansen. 1996. The statistical analysis of interspecific data: A review and evaluation of phylogenetic comparative methods. In *Phylogenies and the comparative method in animal behavior,* edited by E. P. Martins, 22–75. Oxford: Oxford University Press.

Martorell, R., and T. Gonzalez-Cossio. 1987. Maternal nutrition and birth weight. *Yrbk. Phys. Anthropol.* 30:195–220.

Marzluff, J. M., and K. P. Dial. 1991. Life history correlates of taxonomic diversity. *Ecology* 72:428–39.

Masoro, E. J. 1995. Dietary restriction. *Exp. Gerontol.* 30:291–98.

Matsuzawa, T. 1994. Field experiments on use of stone tools in the wild. In *Chimpanzee cultures,* edited by R. W. Wrangham, W. C. McGrew, F. B. M. de Waal, P. G. Heltne, and L. A. Marquardt, 351–70. Cambridge, MA: Harvard University Press.

Mayr, E. 1963. *Animal species and evolution.* Cambridge, MA: Harvard University Press.

McClintock, M. K., J. J. Anisko, and N. T. Adler. 1982. Group mating among Norway rats. II. The social dynamics of copulation: Competition, cooperation, and mate choice. *Anim. Behav.* 30:410–25.

McDonald, D. B. 1993. Demographic consequences of sexual selection in the longtailed manakin. *Behav. Ecol.* 4:297–309.

McDonald, D. B., and H. Caswell. 1993. Matrix methods for avian demography. In *Current ornithology,* edited by D. M. Power, 139–85. New York: Plenum Press.

McGrew, W. C. 1992. *Chimpanzee material culture: Implications for human evolution.* Cambridge: Cambridge University Press.

———. 1998. Culture in nonhuman primates? *Annu. Rev. Anthropol.* 27:301–28.

McHenry, H. M. 1994. Behavioral ecological implications of early hominid body size. *J. Hum. Evol.* 27:77–87.

McKey, D. B. 1978. Soils, vegetation and seed eating by black colobus monkeys. In *The ecology of arboreal folivores,* edited by G. G. Montgomery, 423–37. Washington, D.C.: Smithsonian Institution Press.

McKinney, M. L. 1998. The juvenilized ape myth—our "overdeveloped" brain. *BioScience* 48:109–16.

McKinney, M. L., and K. J. McNamara. 1991. *Heterochrony: The evolution of ontogeny.* New York: Plenum Press.

McNab, B. K. 1979. Climatic adaptation in the energetics of heteromyid rodents. *Comp. Biochem. Physiol.* A 62:813–20.

———. 1980. Food habits, energetics, and the population biology of mammals. *Am. Nat.* 116:106–24.

———. 1983. Energetics, body size, and the limits to endothermy. *J. Zool.* 199:1–29.

McNab, B. K., and J. F. Eisenberg. 1989. Brain size and its relation to the rate of metabolism in mammals. *Am. Nat.* 133:157–67.

McNab, B. K., and P. C. Wright. 1987. Temperature regulation and oxygen consumption in the Philippine tarsier *Tarsius syrichta. Physiol. Zool.* 60:596–600.

McNamara, J. M., and A. I. Houston. 1996. State-dependent life histories. *Nature* 380:215–21.

Medawar, P. B. 1952. *An unsolved problem of biology.* London: Lewis.

Melnick, D. J., and M. C. Pearl. 1987. Cercopithecines in multimale groups: Genetic diversity and population structure. In *Primate societies,* edited by B. B. Smuts, D. L. Cheney, R. M. Seyfarth, R. W. Wrangham, and T. T. Struhsaker, 121–34. Chicago: University of Chicago Press.

Meyers, D. M., and P. C. Wright. 1993. Resource tracking: Food availability and *Propithecus* seasonal reproduction. In *Lemur social systems and their ecological basis,* edited by P. M. Kappeler and J. U. Ganzhorn, 179–92. New York: Plenum Press.

Millar, J. S. 1977. Adaptive features of mammalian reproduction. *Evolution* 31: 370–86.

Millar, J. S., and R. M. Zammuto. 1983. Life histories of mammals: An analysis of life tables. *Ecology* 64:631–35.

Miller, J. M. A. 1997. A hierarchical analysis of primate brain size and body size: Patterns of morphometric variation. Ph.D. thesis, University of Southern California.

Miller, M. N., and J. A. Byers. 1998. Sparring as play in young pronghorn males. In *Animal play,* edited by M. Bekoff and J. A. Byers, 141–60. New York: Cambridge University Press.

Mills, L. S., D. F. Doak, and M. J. Wisdom. 1999. Reliability of conservation actions based on elasticity analysis of matrix models. *Conserv. Biol.* 13:815–29.

Mitani, J. C., J. Gros-Louis, and A. Richards. 1996. Sexual dimorphism, the operational sex ratio, and the intensity of male competition in polygynous primates. *Am. Nat.* 147:966–80.

Mitani, J. C., and D. P. Watts. 1997. The evolution of non-maternal caretaking among anthropoid primates: Do helpers help? *Behav. Ecol. Sociobiol.* 40:213–20.

Mitter, C., B. Farrell, and B. Wiegmann. 1988. The phylogenetic study of adaptive zones: Has phytophagy promoted insect diversification? *Am. Nat.* 132:107–28.

Morbeck, M. E. 1997. Reading life history in teeth, bones, and fossils. In *The evolving female: A life-history perspective,* edited by M. E. Morbeck, A. Galloway, and A. L. Zihlman, 117–31. Princeton, NJ: Princeton University Press.

Morland, H. S. 1990. Parental behavior and infant development in ruffed lemurs *(Varecia variegata)* in a northeast Madagascar rainforest. *Am. J. Primatol.* 20: 253–65.

———. 1993. Seasonal behavioral variation and its relationship to thermoregulation in ruffed lemurs *(Varecia variegata variegata).* In *Lemur social systems and their ecological basis,* edited by P. M. Kappeler and J. U. Ganzhorn, 193–203. New York: Plenum Press.

Morris, W. F., and D. F. Doak. 2002. *Quantitative Conservation Biology: Theory and Practice of Population Viability Analysis.* Sunderland, MA: Sinauer Associates.

Müller, E. F. 1983. Thermoregulation and energy budget of prosimians. *Bonn. Zool. Beitr.* 34:29–71.

———. 1985. Basal metabolic rates in primates: The possible role of phylogenetic and ecological factors. *Comp. Biochem. Physiol.* A 81:707–11.

Muller, H. J. 1964. The relation of recombination to mutational advance. *Mutat. Res.* 1:2–9.

Myers, N., R. Mittermeier, C. Mittermeier, G. da Fonseca, and J. Kent. 2000. Biodiversity hotspots for conservation priorities. *Nature* 403:853–58.

Nagy, K. A., C. Meienberger, S. D. Bradshaw, and R. D. Wooller. 1995. Field metabolic rate of a small marsupial mammal, the honey possum *(Tarsipes rostratus).* *J. Mammal.* 76:862–66.

Nagy, T. R., B. A. Gower, and M. H. Stetson. 1995. Photoperiod effects on body mass, body composition, growth hormone, and thyroid hormones in male collared lemmings *(Dicrostonys groenlandicus).* *Can. J. Zool.* 72:1726–34.

Nee, S., A. F. Read, and P. H. Harvey. 1996. Why phylogenies are necessary for comparative analysis. In *Phylogenies and the comparative method in animal behaviour,* edited by E. P. Martins, 399–411. New York: Oxford University Press.

Needham, J. 1933. On the dissociability of the fundamental processes in ontogenesis. *Biol. Rev.* 8:180–223.

Newman, R. 1992. Adaptive plasticity in amphibian metamorphosis. *BioScience* 42:671–78.

Nezu, A., S. Kimura, S. Uehara, T. Kobayashi, M. Tanaka, and K. Saito. 1997. Magnetic stimulation of motor cortex in children: Maturity of corticospinal pathway and problem of clinical application. *Brain Dev.* 19:176–80.

Nicolson, N. A. 1987. Infants, mothers and other females. In *Primate societies,* edited by B. B. Smuts, D. L. Cheney, R. M. Seyfarth, R. W. Wrangham, and T. T. Struhsaker, 330–42. Chicago: University of Chicago Press.

Noordwijk, M. A. van, C. K. Hemelrijk, L. A. M. Herremans, and E. H. M. Sterck. 1993. Spatial position and behavioral sex differences in juvenile long-tailed macaques. In *Juvenile primates: Life history, development, and behavior,* edited by M. E. Pereira and L. A. Fairbanks, 77–85. New York: Oxford University Press.

Noordwijk, M. A. van, and C. P. van Schaik. 1999. The effects of dominance rank and group size on female lifetime reproductive success in wild long-tailed macaques, *Macaca fascicularis. Primates* 40:105–30.

———. 2000. Reproductive patterns in eutherian mammals: Adaptations against infanticide? In *Infanticide by males and its implications,* edited by C. P. van Schaik and C. H. Janson, 322–60. Cambridge: Cambridge University Press.

Nowak, R. M. 1991. *Walker's mammals of the world,* vol. 2. Baltimore, MD: The Johns Hopkins University Press.

Nowicki, S., S. Peters, and J. Podos. 1998. Song learning, early nutrition and sexual selection in songbirds. *Am. Zool.* 38:179–90.

Nunn, C. L. 1995. A simulation test of Smith's degrees of freedom correction for comparative studies. *Am. J. Phys. Anthropol.* 98:355–67.

Nunn, C. L., and R. A. Barton. 2001. Comparative methods for studying primate adaptation and allometry. *Evol. Anthropol.* 10:81–98.

Nunn, C. L., and M. E. Pereira. 2000. Group histories and offspring sex ratios in ring-tailed lemurs *(Lemur catta). Behav. Ecol. Sociobiol.* 48:18–28.

Nunn, C. L., and C. P. van Schaik. 2002. A comparative approach to reconstructing the socioecology of extinct primates. In *Reconstructing behavior in the primate fossil record,* edited by J. M. Plavcan, R. F. Kay, W. L. Jungers, and C. P. van Schaik, 159–216. New York: Kluwer Academic/Plenum Press.

Oates, J. F. 1987. Food distribution and foraging behavior. In *Primate societies,* edited by B. B. Smuts, D. L. Cheney, R. M. Seyfarth, R. W. Wrangham, and T. T. Struhsaker, 197–209. Chicago: University of Chicago Press.

Oates, J. F., G. H. Whitesides, A. G. Davies, P. G. Waterman, S. M. Green, G. L. Dasilva, and S. Mole. 1990. Determinants of variation in tropical forest primate biomass: New evidence from West Africa. *Ecology* 71:328–43.

Ober, C., S. Elias, D. D. Kostyu, and W. W. Hauck. 1992. Decreased fecundability in Hutterite couples sharing HLA-DR. *Am. J. Hum. Genet.* 50:6–14.

Ober, C., L. R. Weitkamp, N. Cox, H. Dytch, D. Kostyu, and S. Elias. 1997. HLA and mate choice in humans. *Am. J. Hum. Genet.* 61:497–505.

O'Connell, J. F., K. Hawkes, and N. G. Blurton Jones. 1988. Hadza scavenging: Implications for Plio-Pleistocene hominid subsistence. *Curr. Anthropol.* 29:356–63.

——. 1999. Grandmothering and the evolution of *Homo erectus. J. Hum. Evol.* 36:461–85.

——. 2000. A critical look at the role of carnivory in early human evolution. *J. Hum. Evol.* 38:A23–24.

O'Connell, J. F., P. K. Latz, and P. Barnett. 1983. Traditional and modern uses of native plants among the Alyawara of Central Australia. *Econ. Bot.* 37:83–112.

O'Connell, S. 1995. Theory of mind in chimpanzees. Ph.D. thesis, University of Liverpool.

Oftedal, O. T. 1984. Milk composition, milk yield and energy output at peak lactation: A comparative review. *Symp. Zool. Soc. Lond.* 51:33–85.

——. 1991. The nutritional consequences of foraging in primates: The relationship of nutrient intake to nutrient requirements. *Phil. Trans. R. Soc. Lond.* B 334: 161–70.

Olivier, E., S. A. Edgley, J. Armand, and R. N. Lemon. 1997. An electrophysiological study of the postnatal development of the corticospinal system in the macaque monkey. *J. Neurosci.* 17:267–76.

Oliver, J. S. 1994. Estimates of hominid and carnivore involvement in the FLK Zinjanthropus fossil assemblage: Some socioecological implications. *J. Hum. Evol.* 27:267–94.

Olshansky, S. J., B. A. Carnes, and D. Grahn. 1998. Confronting the boundaries of human longevity. *Am. Sci.* 86:52–61.

Orr, W. C., and R. S. Sohal. 1994. Extension of life-span by overexpression of superoxide-dismutase and catalase in *Drosophila melanogaster. Science* 263:1128–30.

Ortmann, S., G. Heldmaier, J. Schmid, and J. U. Ganzhorn. 1997. Spontaneous daily torpor in Malagasy mouse lemurs. *Naturwissenschaften* 84:28–32.

Ostfeld, R. S., and F. Keesing. 2000. Pulsed resources and community dynamics of consumers in terrestrial ecosystems. *Trends Ecol. Evol.* 15:232–37.

Overdorff, D. J. 1993. Ecological and reproductive correlates to range use in redbellied lemurs *(Eulemur rubriventer)* and rufous lemurs *(Eulemur fulvus rufus).* In *Lemur social systems and their ecological basis,* edited by P. M. Kappeler and J. U. Ganzhorn, 167–78. New York: Plenum Press.

Overdorff, D. J., and S. G. Strait. 1998. Seed handling by three prosimian primates in Southeastern Madagascar: Implications for seed dispersal. *Am. J. Primatol.* 45: 69–82.

Owens, I. P. F., and P. M. Bennett. 1995. Ancient ecological diversification explains life-history variation among living birds. *Proc. R. Soc. Lond.* B 261:227–32.

Owens, I. P. F., P. M. Bennett, and P. H. Harvey. 1999. Species richness among birds: Body size, life history, sexual selection or ecology? *Proc. R. Soc. Lond.* B 266: 933–39.

Packer, C., and P. A. Abrams. 1990. Should cooperative groups be more vigilant than selfish groups? *J. Theor. Biol.* 142:341–57.

Packer, C., L. Herbst, A. E. Pusey, J. D. Bygott, J. P. Hanby, S. J. Cairns, and M. Borgerhoff Mulder. 1988. Reproductive success of lions. In *Reproductive success,* edited by T. H. Clutton-Brock, 363–83. Chicago: University of Chicago Press.

Packer, C., M. Tatar, and A. Collins. 1998. Reproductive cessation in female mammals. *Nature* 392:807–11.

Pagel, M. D. 1992. A method for the analysis of comparative data. *J. Theor. Biol.* 156:431–32.

———. 1993. Seeking the evolutionary regression coefficient: An analysis of what comparative methods measure. *J. Theor. Biol.* 164:191–203.

———. 1999. Inferring the historical patterns of biological evolution. *Nature* 401: 877–84.

Pagel, M. D., and P. H. Harvey. 1988a. How mammals produce large-brained offspring. *Evolution* 42:948–57.

———. 1988b. The taxon level problem in the evolution of mammalian brain size: Facts and artifacts. *Am. Nat.* 32:344–59.

———. 1989. Taxonomic differences in the scaling of brain and body weight among mammals. *Science* 244:1589–93.

———. 1993. Evolution of the juvenile period in mammals. In *Juvenile primates: Life history, development, and behavior,* edited by M. E. Pereira and L. A. Fairbanks, 28–37. New York: Oxford University Press.

Pagès, E., and A. Petter-Rousseaux. 1980. Annual variations in the circadian activity rhythms of five sympatric species of nocturnal prosimians in captivity. In *Nocturnal Malagasy primates,* edited by P. Charles-Dominique, H. M. Cooper, A. Hladik, C. M. Hladik, G. F. Pariente, A. Petter-Rousseaux, A. Schilling, and J. J. Petter, 153–67. New York: Academic Press.

Paine, R. R. 1997. *Integrating archaeological demography: Multidisciplinary approaches to prehistoric population.* Occasional Paper 24, Center for Archaeological Investigations. Carbondale: Southern Illinois University.

Palombit, R. A. 1999. Infanticide and the evolution of pair bonds in nonhuman primates. *Evol. Anthropol.* 7:117–29.

Palombit, R. A., R. M. Seyfarth, and D. L. Cheney. 1997. The adaptive value of "friendships" to female baboons: Experimental and observational evidence. *Anim. Behav.* 54:599–614.

Parker, K. 2000. The behavioral neurobiology of affiliation and paternal care in *Microtus pennsylvanicus* (meadow vole). Ph.D. thesis, University of Michigan.

Parker, S. T., and K. R. Gibson. 1977. Object manipulation, tool use and sensorimotor intelligence as feeding adaptations in cebus monkeys and great apes. *J. Hum. Evol.* 6:623–41.

Parker, S. T., J. Langer, and M. McKinney. 2000. *Biology, brains, and behavior: The evolution of human development.* Santa Fe, NM: School of American Research Press.

Parker, S. T., and M. L. McKinney. 1999. *Origins of intelligence: The evolution of cognitive development in monkeys, apes, and humans.* Baltimore, MD: The Johns Hopkins University Press.

Parnham, M. J. 1998. Brain size, ecology and neural specialization in insectivores. M.Sc. thesis, University of Durham.

Partridge, L., and P. H. Harvey. 1988. The ecological context of life history evolution. *Science* 241:1449–55.

Passingham, R. E. 1975a. The brain and intelligence. *Brain Behav. Evol.* 11:1–15.

———. 1975b. Changes in size and organisation of the brain of man and his ancestors. *Brain Behav. Evol.* 11:73–90.

Paul, A., S. Preuschoft, and C. P. van Schaik. 2000. The other side of the coin: Infanticide and the evolution of affiliative male-infant interactions in Old World primates. In *Infanticide by males and its implications,* edited by C. P. van Schaik and C. H. Janson, 269–92. Cambridge: Cambridge University Press.

Paus, T., A. Zijdenbos, K. Worsley, D. L. Collins, J. Blumenthal, J. N. Giedd, J. L. Rapoport, and A. C. Evans. 1999. Structural maturation of neural pathways in children and adolescents: In vivo study. *Science* 283:1908–11.

Pavelka, M. S. M., and L. M. Fedigan. 1991. Menopause: A comparative life history perspective. *Yrbk. Phys. Anthropol.* 34:13–38.

———. 1999. Reproductive termination in female Japanese monkeys: A comparative life history perspective. *Am. J. Phys. Anthropol.* 109:455–64.

Pawlowski, B. P., C. B. Lowen, and R. I. M. Dunbar. 1998. Neocortex size, social skills and mating success in primates. *Behaviour* 135:357–68.

Peacor, S. C., and E. E. Werner. 2001. Predator effects on an assemblage of consumers through induced changes in consumer foraging behavior. *Ecology* 81: 1998–2010.

Pearl, R. 1928. *The rate of living: Being an account of some experimental studies on the biology of life duration.* New York: Alfred Knopf.

Peccei, J. S. 1995. The origin and evolution of menopause: The altriciality-lifespan hypothesis. *Ethol. Sociobiol.* 16:425–49.

Peeling, A. N., and J. L. Smart. 1994. Review of literature showing that undernutrition affects the growth rate of all processes in the brain to the same extent. *Metab. Brain Dis.* 9:33–42.

Pellis, S. M., and V. C. Pellis. 1998. Structure-function interface in the analysis of play fighting. In *Animal play,* edited by M. Bekoff and J. A. Byers, 115–40. New York: Cambridge University Press.

Pereira, M. E. 1984. Age changes and sex differences in the social behavior of juvenile yellow baboons *(Papio cynocephalus).* Ph.D. thesis, University of Chicago.

———. 1988a. Agonistic interactions of juvenile savanna baboons. I. Fundamental features. *Ethology* 79:195–217.

———. 1988b. Effects of age and sex on intra-group spacing behaviour in juvenile savannah baboons *(Papio cynocephalus cynocephalus). Anim. Behav.* 36:184–204.

———. 1991. Asynchrony within estrous synchrony among ringtailed lemurs (Primates: Lemuridae). *Physiol. Behav.* 49:47–52.

———. 1992. The development of dominance relations before puberty in cercopithecine societies. In *Aggression and peacefulness in humans and other primates,* edited by J. Silverberg and P. Gray, 117–49. New York: Oxford University Press.

———. 1993a. Agonistic interaction, dominance relation, and ontogenetic trajectories in ringtailed lemurs. In *Juvenile primates: Life history, development, and behavior,* edited by M. E. Pereira and L. A. Fairbanks, 285–305. New York: Oxford University Press.

———. 1993b. Juvenility in animals. In *Juvenile primates: Life history, development, and behavior,* edited by M. E. Pereira and L. A. Fairbanks, 17–27. New York: Oxford University Press.

———. 1993c. Seasonal adjustment of growth rate and adult body weight in ringtailed lemurs. In *Lemur social systems and their ecological basis,* edited by P. M. Kappeler and J. U. Ganzhorn, 205–21. New York: Plenum Press.

———. 1995. Development and social dominance among group-living primates. *Am. J. Primatol.* 37:143–75.

Pereira, M. E., and J. Altmann. 1985. Development of social behavior in free-living nonhuman primates. In *Nonhuman primate models for human growth and development,* edited by E. S. Watts, 217–309. New York: Alan R. Liss.

Pereira, M. E., T. H. Clutton-Brock, and P. M. Kappeler. 2000. Understanding male primates. In *Primate males,* edited by P. M. Kappeler, 271–77. Cambridge: Cambridge University Press.

Pereira, M. E., and L. A. Fairbanks. 1993a. *Juvenile primates: Life history, development, and behavior.* New York: Oxford University Press.

———. 1993b. What are juvenile primates all about? In *Juvenile primates: Life history, development, and behavior,* edited by M. E. Pereira and L. A. Fairbanks, 3–12. New York: Oxford University Press.

Pereira, M. E., and P. M. Kappeler. 1997. Divergent systems of agonistic behaviour in lemurid primates. *Behaviour* 134:225–74.

Pereira, M. E., A. Klepper, and E. L. Simons. 1987. Tactics of care for young infants by forest-living ruffed lemurs *(Varecia variegata variegata):* Ground nests, parking, and biparental guarding. *Am. J. Primatol.* 13:129–44.

Pereira, M. E., and C. A. McGlynn. 1997. Special relationships instead of female dominance for redfronted lemurs, *Eulemur fulvus rufus. Am. J. Primatol.* 43: 239–58.

Pereira, M. E., and C. M. Pond. 1995. Organization of white adipose tissue in Lemuridae. *Am. J. Primatol.* 35:1–13.

Pereira, M. E., M. L. Seeligson, and J. M. Macedonia. 1988. The behavioral repertoire of the black-and-white ruffed lemur, *Varecia variegata variegata* (Primates: Lemuridae). *Folia Primatol.* 51:1–32.

Pereira, M. E., R. A. Strohecker, S. A. Cavigelli, C. L. Hughes, and D. D. Pearson. 1999. Metabolic strategy and social behavior in Lemuridae. In *New directions in lemur studies,* edited by B. Rakotosamimanana, H. Rasamimanana, J. U. Ganzhorn, and S. M. Goodman, 93–118. New York: Plenum Press.

Pereira, M. E., and M. L. Weiss. 1991. Female mate choice, male migration, and the threat of infanticide in ringtailed lemurs. *Behav. Ecol. Sociobiol.* 28:141–52.

Peres, C. A. 1997. Effects of habitat quality and hunting pressure on arboreal folivore densities in Neotropical forests: A case study of howler monkeys *(Alouatta* spp.). *Folia Primatol.* 68:199–222.

Perret, M. 1990. Influence of social factors on sex ratio at birth, maternal investment and young survival in a prosimian primate. *Behav. Ecol. Sociobiol.* 27:447–54.

———. 1992. Environmental and social determinants of sexual function in the male lesser mouse lemur *(Microcebus murinus). Folia Primatol.* 59:1–25.

Perret, M., and F. Aujard. 2001. Regulation by photoperiod of seasonal changes in body mass and reproductive function in gray mouse lemurs *(Microcebus murinus):* Differential responses by sex. *Int. J. Primatol.* 22:5–24.

Peters, A., C. Sethares, and M. B. Moss. 1998. The effects of aging on layer 1 in area 46 of prefrontal cortex in the rhesus monkey. *Cereb. Cortex* 8:671–84.

Pfister, C. A. 1998. Patterns of variance in stage-structured populations: Evolutionary predictions and ecological implications. *Proc. Natl. Acad. Sci. USA* 95: 213–18.

Piaget, J. 1980. *Adaptation and intelligence: Organic selection and phenocopy.* Chicago: University of Chicago Press.

Pianka, E. R. 1970. On "r" and "K" selection. *Am. Nat.* 104:592–97.

Pinker, S. 1994. *The language instinct.* New York: W. Morrow.

Plavcan, J. M., and C. P. van Schaik. 1997. Interpreting hominid behavior on the basis of sexual dimorphism. *J. Hum. Evol.* 32:345–74.

Plavcan, J. M., C. P. van Schaik, and P. M. Kappeler. 1995. Competition, coalitions and canine size in primates. *J. Hum. Evol.* 28:245–76.

Plummer, T. W., L. C. Bishop, P. Ditchfield, and J. Hicks. 1999. Research on Late Pliocene Oldowan sites at Kanjera South, Kenya. *J. Hum. Evol.* 36:151–70.

Poirier, F. E., and E. O. Smith. 1974. Socializing functions of primate play. *Am. Zool.* 14:275–87.

Pond, C. M. 1998. *The fats of life.* Cambridge: Cambridge University Press.

Pook, A. G. 1978. A comparison between the reproduction and parental behaviour of the Geoldi's monkey *(Callimico goeldi)* and the true marmosets (Callitrichidea). In *Biology and behaviour of marmosets,* edited by H. Rothe, H. J. Wolters, and J. P. Hearn, 1–14. Göttingen: Eigenverlag Hartmut Rothe.

Portmann, A. 1939. Nesthocker und Nestflüchter als Entwicklungszustände von verschiedener Wertigkeit bei Vögeln und Säugern. *Rev. Suisse Zool.* 46:385–90.

Potts, R. 1988. *Early hominid activities at Olduvai.* Hawthorne, NY: Aldine de Gruyter.

Potts, R., and P. Shipman. 1981. Cutmarks made by stone tools on bones from Olduvai Gorge. *Nature* 291:577–80.

Prader, A., J. M. Tanner, and G. A. van Harnack. 1963. Catch-up growth following illness or starvation. *J. Pediatr.* 62:646–59.

Pratap, R. 1998. *Getting started with MATLAB 5: A quick introduction for scientists and engineers.* Oxford: Oxford University Press.

Prentice, A. M., and R. G. Whitehead. 1987. The energetics of human reproduction. *Symp. Zool. Soc. Lond.* 75:275–304.

Price, T. 1997. Correlated evolution and independent contrasts. *Phil. Trans. R. Soc. Lond.* B 352:519–29.

Promislow, D. E. L. 1991. Senescence in natural populations of mammals: A comparative study. *Evolution* 45:1869–87.

———. 1996. Using comparative approaches to integrate behavior and population biology. In *Phylogenies and the comparative method in animal behavior,* edited by E. P. Martins, 288–323. Oxford: Oxford University Press.

Promislow, D. E. L., and P. H. Harvey. 1990. Living fast and dying young: A comparative analysis of life-history variation among mammals. *J. Zool.* 220:417–37.

Prothero, J., and K. D. Jürgens. 1987. Scaling of animal lifespan in mammals: A review. In *Evolution of longevity in animals: A comparative approach,* edited by A. D. Woodhead and K. H. Thompson, 49–74. New York: Plenum Press.

Przybylo, R., B. C. Sheldon, and J. Merilä. 2000. Climatic effects on breeding and morphology: Evidence for phenotypic plasticity. *J. Anim. Ecol.* 69:395–403.

Pulliam, H. R., and T. Caraco. 1984. Social living: Is there an optimal group size. In *Behavioural ecology: An evolutionary approach,* edited by J. R. Krebs and N. B. Davies, 122–47. Oxford: Blackwell.

Pulliam, H. R., G. H. Pyke, and T. Caraco. 1982. The scanning behavior of juncos: A game_theoretical approach. *J. Theor. Biol.* 95:89–103.

Purvis, A. 1995. A composite estimate of primate phylogeny. *Phil. Trans. R. Soc. Lond.* B 348:405–21.

———. 1996. Using interspecific phylogenies to test macroevolutionary hypotheses. In *New uses for new phylogenies,* edited by P. H. Harvey, A. J. Leigh Brown, J. Maynard Smith, and S. Nee, 153–68. Oxford: Oxford University Press.

Purvis, A., J. L. Gittleman, G. Cowlishaw, and G. M. Mace. 2000. Predicting extinction risk in declining species. *Proc. R. Soc. Lond.* B 267:1947–52.

Purvis, A., J. L. Gittleman, and H. K. Luh. 1994. Truth or consequences: Effects of phylogenetic accuracy on two comparative methods. *J. Theor. Biol.* 167:293–300.

Purvis, A., and P. H. Harvey. 1995. Mammalian life history evolution: A comparative test of Charnov's model. *J. Zool.* 237:259–83.

———. 1996. Miniature mammals: Life history strategies and macro evolution. *Symp. Zool. Soc. Lond.* 69:159–74.

———. 1997. The right size for a mammal. *Nature* 386:332–33.

Purvis, A., and A. Hector. 2000. Getting the measure of biodiversity. *Nature* 405:212–19.

Purvis, A., S. Nee, and P. H. Harvey. 1995. Macroevolutionary inferences from primate phylogeny. *Proc. R. Soc. Lond.* B 260:329–33.

Purvis, A., and A. Rambaut. 1995. Comparative analysis by independent contrasts (CAIC): An Apple Macintosh application for analysing comparative data. *Comput. Appl. Biosci.* 11:247–51.

Purvis, A., and A. J. Webster. 1999. Phylogenetically independent comparisons and primate phylogeny. In *Comparative primate socioecology,* edited by P. C. Lee, 44–70. Cambridge: Cambridge University Press.

Pusey, A., and C. Packer. 1994. Non-offspring nursing in social carnivores. *Behav. Ecol.* 5:362–74.

Pusey, A., J. Williams, and J. Goodall. 1997. The influence of dominance rank on the reproductive success of female chimpanzees. *Science* 277:828–31.

Quartz, S. R., and T. J. Sejnowski. 1997. The neural basis of cognitive development: A constructivist manifesto. *Behav. Brain Sci.* 20:537–96.

Quiatt, D. 1979. Aunts and mothers: Adaptive implications of allomaternal behaviour of non_human primates. *Am. Anthropol.* 81:310–19.

Rajpurohit, L., V. Sommer, and S. Mohnot. 1995. Wanderers between harems and bachelor bands: Male hanuman langurs *(Presbytis entellus)* at Jodhpur in Rajasthan. *Behaviour* 132:255–99.

Rannala, B. H., and C. R. Brown. 1994. Relatedness and conflict over optimal group size. *Trends Ecol. Evol.* 9:117–19.

Ranta, E., V. Kaitala, S. Alaja, and D. Tesar. 2000. Nonlinear dynamics and the evolution of semelparous and iteroparous reproductive strategies. *Am. Nat.* 155: 294–300.

Rasamimanana, H., and E. Rafidinarivo. 1993. Feeding behavior of *Lemur catta* females in relation to their physiological state. In *Lemur social systems and their ecological basis,* edited by P. M. Kappeler and J. U. Ganzhorn, 123–33. New York: Plenum Press.

Raunkiaer, C. 1934. *The life forms of plants and statistical plant geography.* Oxford: Clarendon Press.

Razafindraibe, H., D. Montagnon, and Y. Rumpler. 1997. Phylogenetic relationships among Indriidae (Primates, Strepsirhini) inferred from highly repeated DNA band patterns. *C. R. Acad. Sci. III* 320:469–75.

Read, A. F., and P. H. Harvey. 1989. Life history differences among the eutherian radiations. *J. Zool.* 219:329–53.

Reavis, R. H., and M. S. Grober. 1999. An integrative approach to sex change: Social, behavioural and neurochemical changes in *Lythrypnus dalli* (Pisces). *Acta Ethol.* 2:51–60.

372 REFERENCES

Reed, D. J., G. T. Schwartz, C. Dean, and M. S. Chandrasekera. 1998. A histological reconstruction of dental development in the common chimpanzee, *Pan troglodytes. J. Hum. Evol.* 35:427–48.

Reed, K. E. 1997. Early hominid evolution and ecological changes through the African Plio-Pleistocene. *J. Hum. Evol.* 32:289–322.

Rees, M. 1995. EC-PC comparative analyses? *J. Ecol.* 83:891–93.

Rettig, N. 1978. Breeding behavior of the harpy eagle *(Harpia harpyja). Auk* 95:629–43.

Reznick, D. N. 1996. Life history evolution in guppies: A model system for the empirical study of adaptation. *Neth. J. Zool.* 46:172–90.

Reznick, D. N., H. Bryga, and J. A. Endler. 1990. Experimentally-induced life-history evolution in a natural population. *Nature* 346:357–59.

Rice, D. W. 1998. *Marine mammals of the world: Systematics and distribution.* Special Publication 4. Lawrence, KS: The Society for Marine Mammalogy.

Rice, W. R. 1989. Analyzing tables of statistical tests. *Evolution* 43:223–25.

Richard, A. F., and R. E. Dewar. 1991. Lemur ecology. *Annu. Rev. Ecol. Syst.* 22:145–75.

Richard, A. F., R. E. Dewar, M. Schwartz, and J. Ratsirarson. 2000. Mass change, environmental variability and female fertility in wild *Propithecus verreauxi. J. Hum. Evol.* 39:381–91.

Richard, A. F., P. Rakotomanga, and M. Schwartz. 1993. Dispersal by *Propithecus verreauxi* at Beza Mahafaly, Madagascar: 1984–1991. *Am. J. Primatol.* 30:1–20.

Ricklefs, R. E. 1979. Adaptation, constraint, and compromise in avian postnatal development. *Biol. Rev.* 54:269–90.

———. 1983. Comparative avian demography. In *Current ornithology,* vol. 1, edited by R. F. Johnston, 1–32. New York: Plenum Press.

———. 1996. Phylogeny and ecology. *Trends Ecol. Evol.* 11:229–30.

———. 1998. Evolutionary theories of aging: Confirmation of a fundamental prediction, with implications for the genetic basis and evolution of life span. *Am. Nat.* 152:24–44.

Ricklefs, R. E., and J. M. Starck. 1996. Applications of phylogenetically independent contrasts: A mixed progress report. *Oikos* 77:167–72.

———. 1998. The evolution of the developmental mode in birds. In *Avian growth and development: Evolution within the altricial-precocial spectrum,* edited by J. M. Starck and R. E. Ricklefs, 366–80. New York: Oxford University Press.

Riddel, W. I., and K. G. Corl. 1977. Comparative investigation of the relationship between cerebral indices and learning abilities. *Brain Behav. Evol.* 14:385–98.

Ridley, M. 1992. Darwin sound on comparative method. *Trends Ecol. Evol.* 7:37.

Robinson, J. G., and K. Redford. 1986. Body size, diet, and population density of Neotropical forest mammals. *Am. Nat.* 128:665–80.

Robinson, S. K. 1994. Habitat selection and foraging ecology of raptors in Amazonian Peru. *Biotropica* 26:443–58.

Roff, D. A. 1992. *The evolution of life histories.* New York: Chapman and Hall.

Rogers, A. 1993. Why menopause? *Evol. Ecol.* 7:406–20.

Rogers, M., C. S. Feibel, and J. W. K. Harris. 1994. Changing patterns of land use by Plio-Pleistocene hominids in the Lake Turkana Basin. *J. Hum. Evol.* 27:139–58.

Rose, L., and F. Marshall. 1996. Meat eating, hominid sociality, and home bases revisited. *Curr. Anthropol.* 37:307–38.

Rose, M. R. 1991. *Evolutionary biology of aging.* New York: Oxford University Press.

Ross, C. 1988. The intrinsic rate of natural increase and reproductive effort in primates. *J. Zool.* 214:199–219.

———. 1991. Life history patterns of New World monkeys. *Int. J. Primatol.* 12: 481–502.

———. 1992a. Basal metabolic rate, body weight and diet in primates: An evaluation of the evidence. *Folia Primatol.* 58:7–23.

———. 1992b. Environmental correlates of the intrinsic rate of natural increase in primates. *Oecologia* 90:383–90.

———. 1998. Primate life histories. *Evol. Anthropol.* 6:54–63.

———. 2001. Park or ride? The evolution of infant carrying in primates. *Int. J. Primatol.* 22:749–71.

Ross, C., and K. E. Jones. 1999. Socioecology and the evolution of primate reproductive rates. In *Comparative primate socioecology,* edited by P. C. Lee, 73–110. Cambridge: Cambridge University Press.

Ross, C., and A. M. MacLarnon. 1995. Ecological and social correlates of maternal expenditure on infant growth in haplorhine primates. In *Motherhood in human and nonhuman primates,* edited by C. R. Pryce, R. D. Martin, and D. Skuse, 37–46. Basel: Karger.

———. 2000. The evolution of non-maternal care in anthropoid primates: A test of the hypotheses. *Folia Primatol.* 71:93–113.

Rowlett, R. M. 1999. Comment on R. W. Wrangham, J. H. Jones, G. Laden, D. Pilbeam, and N. L. Conklin-Brittain. The raw and the stolen: Cooking and the ecology of human origins. *Curr. Anthropol.* 40:584–85.

Rubenstein, D. I. 1993. On the evolution of juvenile life-styles in mammals. In *Juvenile primates: Life history, development, and behavior,* edited by M. E. Pereira and L. A. Fairbanks, 38–56. New York: Oxford University Press.

Rubenstein, D. I., and R. W. Wrangham. 1986. *Ecological aspects of social evolution: Birds and mammals.* Princeton, NJ: Princeton University Press.

Rubner, M. 1908. *Das Problem der Lebensdauer und seine Beziehungen zu Wachstum und Ernährung.* München: Verlag R. Oldenbourg.

Ruff, C. B. 1991. *Aging and osteoporosis in native Americans from Pecos Pueblo, New Mexico: Behavioral and biomechanical effects.* New York: Garland.

Ruff, C. B., E. Trinkhaus, and T. W. Holliday. 1997. Body mass and encephalization in Pleistocene *Homo. Nature* 387:173–76.

Rumbaugh, D. M., E. S. Savage-Rumbaugh, and D. A. Washburn. 1996. Toward a new outlook on primate learning and behavior: Complex learning and emergent processes in comparative perspective. *Jpn. Psychol. Res.* 38:113–25.

Sacher, G. A. 1959. Relation of lifespan to brain weight and body weight in mammals. In *The lifespan of animals. CIBA foundation colloquia on ageing,* vol. 5, edited by G. E. W. Wolstenholme and M. O'Connor, 115–33. Boston: Little, Brown.

———. 1975. Maturation and longevity in relation to cranial capacity in hominid evolution. In *Primate functional morphology and evolution,* edited by R. H. Tuttle, 417–41. Mouton: The Hague.

———. 1978. Longevity, aging, and death: An evolutionary perspective. *Gerontologist* 18:112–19.

Sacher, G. A., and E. F. Staffeldt. 1974. Relation of gestation time to brain weight for placental mammals: Implications for the theory of vertebrate growth. *Am. Nat.* 108:593–615.

Sade, D. S., K. Cushing, J. Cushing, J. Dunaif, A. Figueroa, J. R. Kaplan, C. Lauer, D. Rhodes, and J. Schneider. 1976. Population dynamics in relation to social structure on Cayo Santiago. *Yrbk. Phys. Anthropol.* 20:253–62.

Saether, B., and I. Gordon. 1994. The adaptive significance of reproductive strategies in ungulates. *Proc. R. Soc. Lond.* B 256:263–68.

Samonds, K. E., L. R. Godfrey, W. L. Jungers, and L. B. Martin. 1999. Primate dental development and the reconstruction of life history strategies in subfossil lemurs. *Am. J. Phys. Anthropol.* suppl. 28:238–39.

Sanchez, S., F. Pelaez, C. Gil-Burman, and W. Kaumanns. 1999. Costs of infant-carrying in the cotton-top tamarin *(Saguinus oedipus). Am. J. Primatol.* 48: 99–111.

Sauermann, U., P. Nürnberg, F. Bercovitch, J. Berard, A. Trefilov, A. Widdig, M. Kessler, J. Schmidtke, and M. Krawczak. 2001. Increased reproductive success of MHC class II heterozygous males among free-ranging rhesus macaques. *Hum. Genet.* 108:249–54.

Sauther, M. L. 1991. Reproductive behavior of free-ranging *Lemur catta* at Beza Mahafaly Special Reserve, Madagascar. *Am. J. Phys. Anthropol.* 84:463–77.

———. 1993. Resource competition in wild populations of ringtailed lemurs *(Lemur catta):* Implications for female dominance. In *Lemur social systems and their ecological basis,* edited by P. M. Kappeler and J. U. Ganzhorn, 135–52. New York: Plenum Press.

———. 1998. Interplay of phenology and reproduction in ring-tailed lemurs: Implications for ring-tailed lemur conservation. *Folia Primatol.* 69:309–20.

Sawaguchi, T., and H. Kudo. 1990. Neocortical development and social structure in primates. *Primates* 31:283–89.

Schaefer, S. A., and G. V. Lauder. 1996. Testing historical hypotheses of morphological change: Biomechanical decoupling in loricarioid catfishes. *Evolution* 50: 1661–75.

Schaffer, W. M. 1974. Selection for optimal life histories: The effects of age structure. *Ecology* 5:291–303.

Scheibel, A. B. 1996. Structural and functional changes in the aging brain. In *Handbook of the psychology of aging,* edited by J. E. Birren and K. W. Schaie, 105–28. New York: Academic Press.

Schew, W. A., and R. E. Ricklefs. 1998. Developmental plasticity. In *Avian growth and development: Evolution within the altricial-precocial spectrum,* edited by J. M. Starck and R. E. Ricklefs, 288–304. Oxford: Oxford University Press.

Schlichting, C. D. 1986. The evolution of adaptive plasticity in plants. *Annu. Rev. Ecol. Syst.* 17:667–93.

Schlichting, C. D., and M. Pigliucci. 1998. *Phenotypic evolution: A reaction norm perspective.* Sunderland, MA: Sinauer Association.

Schlund, W., F. Scharfe, and J. U. Ganzhorn. 2002. Long-term comparison of food availability and reproduction in the edible dormouse *(Glis glis). Mamm. Biol.* 67:1–14.

Schmid, J. 2000. Daily torpor in the gray mouse lemur *(Microcebus murinus)* in Madagascar: Energetic consequences and biological significance. *Oecologia* 123: 175–83.

Schmid, J., and P. M. Kappeler. 1998. Fluctuating sexual dimorphism and differential hibernation by sex in a primate, the gray mouse lemur *(Microcebus murinus). Behav. Ecol. Sociobiol.* 43:125–32.

Schmid, J., T. Ruf, and G. Heldmaier. 2000. Metabolism and temperature regulation during daily torpor in the smallest primate, the pygmy mouse lemur *(Microcebus myoxinus)* in Madagascar. *J. Comp. Physiol.* B 170:59–68.

Schmid, J., and J. R. Speakman. 2000. Daily energy expenditure of the grey mouse lemur *(Microcebus murinus)*: A small primate that uses torpor. *J. Comp. Physiol.* B 170:633–41.

Schmidt-Nielsen, K. 1984. *Scaling: Why is animal size so important?* Cambridge: Cambridge University Press.

———. 1997. *Animal physiology: Adaptation and environment.* Cambridge: Cambridge University Press.

Schmitz, O. J., P. A. Hamback, and A. P. Beckerman. 2000. Trophic cascades in terrestrial systems: A review of the effects of carnivore removals on plants. *Am. Nat.* 155:141–53.

Schneider, J. 1999. Delayed oviposition: A female strategy to counter infanticide by males? *Behav. Ecol.* 10:567–71.

Schoeninger, M. J., H. T. Bunn, S. S. Murrray, and J. A. Marlett. 2001. Composition of tubers used by Hadza foragers of Tanzania. *J. Food Comp. Anal.* 14:15–25.

Schrire, C., ed. 1984. *Past and present in hunter gatherer studies.* Orlando, FL: Academic Press.

Schulman, S. R., and B. Chapais. 1980. Reproductive value and rank relations among macaque sisters. *Am. Nat.* 115:580–93.

Schwartz, G. T. 2000. Taxonomic and functional aspects of the patterning of enamel thickness distribution in extant large_bodied hominoids. *Am. J. Phys. Anthropol.* 111:221–44.

Semaw, S., P. Renne, J. W. K. Harris, C. S. Feibel, R. L. Bernor, N. Fessena, and K. Mowbray. 1997. 2.5-million-year-old stone tools from Gona, Ethiopia. *Nature* 385:333–36.

Semendeferi, K., H. Damasio, R. Frank, and G. W. van Hoesen. 1997. The evolution of the frontal lobes: A volumetric analysis based on three-dimensional reconstructions of magnetic resonance scans of human and ape brains. *J. Hum. Evol.* 32:375–88.

Shanley, D. P., and T. B. L. Kirkwood. 2001. Evolution of the human menopause. *Bioessays* 23:282–87.

Shapiro, S., E. R. Schlesinger, and R. E. L. Nesbitt Jr. 1968. *Infant, perinatal, maternal, and childhood mortality in the United States.* Cambridge, MA: Harvard University Press.

Shea, B. T. 1990. Dynamic morphology: Growth, life history, and ecology in primate evolution. In *Primate life history and evolution,* edited by C. J. de Rousseau, 325–52. New York: Wiley-Liss.

Sherman, P. W., and M. L. Morton. 1984. Demography of Belding's ground squirrels. *Ecology* 65:1617–28.

Shine, R. 1989. Ecological causes for the evolution of sexual dimorphism: A review of the evidence. *Q. Rev. Biol.* 64:419–61.

Shipman, P., and A. Walker. 1989. The costs of becoming a predator. *J. Hum. Evol.* 18:373–92.

Shoemaker, W. J., and F. E. Bloom. 1977. Effect of undernutrition on brain morphology. In *Nutrition and the brain,* vol. 2, edited by R. J. Wurtman and J. J. Wurtman, 147–92. New York: Raven Press.

Shuchat, A., and F. Shultz. 2000. *The joy of mathematica.* New York: Harcourt/Academic Press.

Silk, J. B. 1983. Local resource competition and facultative adjustment of sex ratios in relation to competitive abilities. *Am. Nat.* 121:56–66.

——. 1986. Eating for two: Behavioural and environmental correlates of gestation length. *Int. J. Primatol.* 7:583–602.

Silk, J. B., A. Samuels, and P. S. Rodman. 1981. The influence of kinship, rank, and sex on affiliation and aggression between adult female and immature bonnet macaques *(Macaca radiata). Behaviour* 78:111–37.

Siviy, S. M. 1998. Neurobiological substrates of play behavior: Glimpses into the structure and function of mammalian playfulness. In *Animal play,* edited by M. Bekoff and J. A. Byers, 221–42. New York: Cambridge University Press.

Smart, J. L. 1991. Critical periods in brain development. In *The childhood environment and adult disease,* edited by G. R. Bock and J. Whelan, 109–28. CIBA Foundation symposium 156. Chichester, UK: Wiley.

Smith, B. H. 1989. Dental development as a measure of life history in primates. *Evolution* 4:683–88.

——. 1991a. Age of weaning approximates age of emergence of the first permanent molar in nonhuman primates. *Am. J. Phys. Anthropol.* suppl. 12:163–64.

——. 1991b. Dental development and the evolution of life history in Hominidae. *Am. J. Phys. Anthropol.* 86:157–74.

——. 1992. Life history and the evolution of human maturation. *Evol. Anthropol.* 1:134–42.

——. 1993. The physiological age of KNM-WT 15000. In *The Nariokotome Homo erectus Skeleton,* edited by A. Walker and R. Leakey, 195–220 and 438–441. Cambridge, MA: Harvard University Press.

Smith, B. H., T. L. Crummett, and K. L. Brandt. 1994. Ages of eruption of primate teeth: A compendium for aging individuals and comparing life histories. *Yrbk. Phys. Anthropol.* 37:177–231.

Smith, B. H., and R. L. Tompkins. 1995. Toward a life history of the Hominidae. *Annu. Rev. Anthropol.* 24:257–79.

Smith, C. C., and S. D. Fretwell. 1974. The optimal balance between the size and number of offspring. *Am. Nat.* 108:499–506.

Smith, D. W. E. 1993. *Human longevity.* New York: Oxford University Press.

Smith, K. 2000. Paternal kin matter: The distribution of social behavior among wild, adult female baboons. Ph.D. dissertation, University of Chicago.

Smith, R. J. 1994. Degrees of freedom in interspecific allometry: An adjustment for the effects of phylogenetic constraint. *Am. J. Phys. Anthropol.* 93:95–107.

——. 1996. Biology and body size in human evolution. *Curr. Anthropol.* 37:451–81.

——. 1999. Statistics of sexual size dimorphism. *J. Hum. Evol.* 36:423–59.

Smith, R. J., and W. L. Jungers. 1997. Body mass in comparative primatology. *J. Hum. Evol.* 32:523–59.

Smith, R. J., and S. R. Leigh. 1998. Sexual dimorphism in primate neonatal body mass. *J. Hum. Evol.* 34:173–201.

Smith-Gill, S. J. 1983. Developmental plasticity: Developmental conversion versus phenotypic modulation. *Am. Zool.* 23:47–55.

Smuts, B. B. 1985. *Sex and friendship in baboons.* Cambridge, MA: Harvard University Press.

Smuts, B. B., and R. W. Smuts. 1993. Male aggression and sexual coercion of females in nonhuman primates and other mammals: Evidence and theoretical implications. *Adv. Stud. Behav.* 22:1–63.

Sohal, R. S., B. H. Sohal, and U. T. Brunk. 1990. Relationship between antioxidant defenses and longevity in different mammalian species. *Mech. Ageing Dev.* 53: 217–27.

Sohal, R. S., B. H. Sohal, and W. C. Orr. 1995. Mitochondrial superoxide and hydrogen-peroxide generation, protein oxidative damage, and longevity in different species of flies. *Free Radic. Biol. Med.* 19:499–504.

Spencer, L. M. 1997. Dietary adaptations of Plio-Pleistocene Bovidae: Implications for hominid habitat use. *J. Hum. Evol.* 32:201–28.

Spitzer, G. 1972. Jahreszeitliche Aspekte der Biologie der Bartmeise *(Panarus biarmicus). J. Ornithol.* 113:241–75.

Stacey, P. B. 1986. Group size and foraging efficiency in yellow baboons. *Behav. Ecol. Sociobiol.* 18:175–87.

Stahl, W. R. 1962. Similarity and dimensional methods in biology. *Science* 137: 205–12.

Stammbach, E. 1978. On social differentiation in groups of captive female hamadryas baboons. *Behaviour* 67:322–38.

Stamps, J. A. 1993. Sexual size dimorphism in species with asymptotic growth after maturity. *Biol. J. Linn. Soc.* 50:123–45.

Standen, V., and R. A. Foley. 1989. *Comparative socioecology: The behavioral ecology of humans and other animals.* Oxford: Blackwell.

Stanford, C. B. 1992. Costs and benefits of allomothering in wild capped langurs *(Presbytis pileata). Behav. Ecol. Sociobiol.* 30:29–34.

———. 1995. The influence of chimpanzee predation on group size and anti-predator behaviour in red colobus monkeys. *Anim. Behav.* 49:577–87.

Starck, J. M. 1999a. Phenotypic flexibility of the avian gizzard: Rapid, reversible, and repeated changes of organ size in response to changes in dietary fiber content. *J. Exp. Biol.* 202:3171–79.

———. 1999b. Structural flexibility of the gastro-intestinal tract of vertebrates: Implications for evolutionary morphology. *Zool. Anz. Jena* 238:87–101.

Starck, J. M., and E. Kloss. 1995. Structural responses of Japanese quail intestine to different diets. *Dtsch. Tierärztl. Wochenschr.* 102:146–50.

Starck, J. M., and R. E. Ricklefs. 1998. Variation, constraint, and phylogeny: Comparative analysis of variation in growth. In *Avian growth and development: Evolution within the altricial-precocial spectrum,* edited by J. M. Starck and R. E. Ricklefs, 247–65. New York: Oxford University Press.

Stearns, S. C. 1976. Life-history tactics: A review of the ideas. *Q. Rev. Biol.* 51:3–47.

———. 1977. The evolution of life history traits: A critique of the theory and a review of the data. *Annu. Rev. Ecol. Syst.* 8:145–71.

———. 1983. The influence of size and phylogeny on patterns of covariation among life-history traits in the mammals. *Oikos* 41:173–87.

———. 1987. The selection arena hypothesis. In *The evolution of sex and its consequences,* edited by S. C. Stearns, 299–311. Basel: Birkhäuser Verlag.

———. 1989a. The evolutionary significance of phenotypic plasticity. *BioScience* 39: 436–45.

———. 1989b. Trade-offs in life history evolution. *Funct. Ecol.* 3:259–68.

———. 1992. *The evolution of life histories.* Oxford: Oxford University Press.

———. 2000. Life history evolution: Successes, limitations, and prospects. *Naturwissenschaften* 87:476–86.

Stearns, S. C., G. de Jong, and B. Newman. 1991. The effects of phenotypic plasticity on genetic correlations. *Trends Ecol. Evol.* 6:122–26.

Stearns, S. C., and J. Koella. 1986. The evolution of phenotypic plasticity in life-history traits: Predictions of reaction norms for age and size at maturity. *Evolution* 40:893–913.

Steenbeck, R. 2000. Infanticide by males and female choice in wild Thomas's langurs. In *Infanticide by males and its implications,* edited by C. P. van Schaik and C. H. Janson, 153–77. Cambridge: Cambridge University Press.

Stephan, H., G. Baron, and H. D. Frahm. 1991. *Comparative brain research in mammals.* Vol. 1, *Insectivores.* New York: Springer.

Stephan, H., H. D. Frahm, and G. Baron. 1981. New and revised data on volumes of brain structures in insectivores and primates. *Folia Primatol.* 35:1–29.

Sterck, E. H. M. 1999. Variation in langur social organization in relation to the socioecological model, human habitat alteration, and phylogenetic constraints. *Primates* 40:199–213.

Sterck, E. H. M., and A. Korstjens. 2000. Female dispersal and infanticide avoidance in primates. In *Infanticide by males and its implications,* edited by C. P. van Schaik and C. H. Janson, 293–321. Cambridge: Cambridge University Press.

Sterck, E. H. M., D. P. Watts, and C. P. van Schaik. 1997. The evolution of female social relationships in nonhuman primates. *Behav. Ecol. Sociobiol.* 41:291–309.

Stewart, C. A. 1918. Changes in the relative weights of the various parts of the brain in normal and underfed albino rats at different stages. *J. Exp. Zool.* 25:301–53.

Struhsaker, T. T., and M. Leakey. 1990. Prey selectivity by crowned hawk-eagles on monkeys in the Kibale Forest, Uganda. *Behav. Ecol. Sociobiol.* 26:435–43.

Sutherland, W. J., A. Grafen, and P. H. Harvey. 1986. Life history correlations and demography. *Nature* 320:88.

Suwa, G., T. D. White, and F. C. Howell. 1996. Mandibular postcanine dentition from the Shungura Formation, Ethiopia: Crown morphology, taxonomic allocations, and Plio-Pleistocene hominid evolution. *Am. J. Phys. Anthropol.* 101:247–82.

Symons, D. 1978. *Play and aggression.* New York: Columbia University Press.

Tan, C. L. 1999a. Group composition, home range size, and diet of three sympatric bamboo lemur species (Genus *Hapalemur*) in Ranomafana National Park, Madagascar. *Int. J. Primatol.* 20:547–66.

———1999b. Life history and infant rearing strategies of three *Hapalemur* species. *Primate Report* 54-1:33.

Tanaka, T., K. Tokuda, and S. Kotera. 1970. Effects of infant loss on the interbirth interval of Japanese monkeys. *Primates* 11:113–17.

Tanner, J. M. 1986. Growth as a target-seeking function: Catch-up and catch-down growth in man. In *Human growth: A comprehensive treatise,* vol. 1, *Developmental biology, prenatal growth,* edited by F. Falkner and J. M. Tanner, 167–79. New York: Plenum Press.

Tardieu, C. 1998. Short adolescence in early hominids: Infantile and adolescent growth of the early human femur. *Am. J. Phys. Anthropol.* 107:163–78.

Tardif, S. D. 1994. Relative energetic cost of infant care in small bodied Neotropical primates and its relation to infant-care patterns. *Am. J. Primatol.* 34:133–43.

Taylor, L. L. 1986. Kinship, dominance and social organization in a semi-free-ranging group of ringtailed lemurs *(Lemur catta).* Ph.D. thesis, Washington University.

Taylor, L. L., and R. W. Sussman. 1985. A preliminary study of kinship and social organization in a semi-free-ranging group of *Lemur catta. Int. J. Primatol.* 6:601–14.

Taylor, R. J. 1979. The value of clumping to prey when detectability increases with group size. *Am. Nat.* 113:299–301.

Teather, K., and P. Weatherhead. 1994. Allometry, adaptation, and the growth and development of sexually dimorphic birds. *Oikos* 71:515–25.

Terborgh, J. 1983. *Five New World primates.* Princeton, NJ: Princeton University Press.

———. 1992. *Diversity and the tropical rain forest.* New York: Scientific American Library.

Terborgh, J., and C. H. Janson. 1986. The socioecology of primate groups. *Annu. Rev. Ecol. Syst.* 17:111–35.

Terranova, C., and B. Coffman. 1997. Body weight of wild and captive lemurs. *Zoo Biol.* 16:17–30.

Thoms, A. 1989. The Northern roots of hunter-gatherer intensification: Camas and the Pacific Northwest. Ph.D. thesis, Washington State University, Pullman.

Tilden, C. C., and O. T. Oftedal. 1995. The bioenergetics of reproduction in prosimian primates: Is it related to female dominance? In *Creatures of the dark: The nocturnal primates,* edited by L. Alterman, G. A. Doyle, and M. K. Izard, 119–31. New York: Plenum Press.

———. 1997. Milk composition reflects pattern of maternal care in prosimian primates. *Am. J. Primatol.* 41:195–211.

Timmermans, P. J. A., and J. M. H. Vossen. 1996. The influence of rearing conditions on maternal behaviour in cymomolgus macaques *(Macaca fascicularis). Int. J. Primatol.* 17:259–76.

Timmermans, S., L. Lefebvre, D. Boire, and P. Basu. 2000. Relative size of the hyperstriatum ventrale is the best predictor of feeding innovation rate in birds. *Brain Behav. Evol.* 56:196–203.

Tomasello, M. 2000. Culture and cognitive development. *Curr. Dir. Psychol. Sci.* 9:37–40.

Tomasello, M., and J. Call. 1997. *Primate cognition.* Oxford: Oxford University Press.

Tomasi, T. E., and A. M. Stribling. 1996. Thyroid function in the thirteen-lined ground squirrel. In *Adaptations to the cold,* edited by F. Geiser, A. J. Hulbert, and S. C. Nicol, 263–69. Armidale: University of New England Press.

Torvik, A., S. Torp, and C. F. Lindboe. 1986. Atrophy of the cerebellar vermis in ageing: A morphometric and histologic study. *J. Neurol. Sci.* 76:283–94.

Trayhurn, P. 1993. Species distribution of brown adipose tissue: Characterization of adipose tissue from uncoupling protein and its mRNA. In *Life in the cold: Ecological, physiological and molecular mechanisms,* edited by C. Carey, G. L. Florant, A. B. Wunder, and B. Horwitz, 361–68. Boulder, CO: Westview Press.

Treherne, J. E., and W. A. Foster. 1982. Group size and anti-predator strategies in a marine insect. *Anim. Behav.* 32:536–42.

Treves, A. 1998. Primate social systems: Conspecific threat and coercion-defense hypotheses. *Folia Primatol.* 69:81–88.

———. 1999. Has predation shaped the social system of arboreal primates? *Int. J. Primatol.* 20:35–53.

Trinkaus, E. 1995. Neanderthal mortality patterns. *J. Archaeol. Sci.* 22:121–42.

Trivers, R. 1985. *Social evolution.* Menlo Park, CA: Benjamin Cummings.

Trut, L. N. 1996. Sex ratio in silver foxes: Effects of domestication and the star gene. *Theor. Appl. Genet.* 92:109–15.

———. 1999. Early canid domestication: The farm-fox experiment. *Am. Sci.* 87:160–69.

Tschudin, A. 1999. Relative neocortex size and its correlates in dolphins: Comparisons with humans and implications for mental evolution. Ph.D. thesis, University of Natal.

Tuomi, J., J. Agrell, and T. Mappes. 1997. On the evolutionary stability of female infanticide. *Behav. Ecol. Sociobiol.* 40:227–33.

Twente, J. W., J. A. Twente, and R. M. Moy. 1977. Regulation of arousal from hibernation by temperature in three species of *Citellus. J. Appl. Physiol.* 42:191–95.

van Buskirk, J., S. A. McCollum, and E. F. Werner. 1997. Natural selection for environmentally induced phenotypes in tadpoles. *Evolution* 51:1983–92.

Vandermeer, J. 1978. Choosing category size in a stage projection matrix. *Oecologia* 32:79–84.

van Groenendael, J., H. de Kroon, and H. Caswell. 1988. Projection matrices in population biology. *Trends Ecol. Evol.* 3:264–69.

van Groenendael, J., H. de Kroon, S. Kalisz, and S. Tuljapurkar. 1994. Loop analysis: Evaluating life history pathways in population projection matrices. *Ecology* 75:2410–15.

van Schaik, C. P. 1983. Why are diurnal primates living in groups? *Behaviour* 87: 120–44.

———. 1989. The ecology of social relationships amongst female primates. In *Comparative socioecology: The behavioral ecology of humans and other animals,* edited by V. Standen and R. A. Foley, 195–218. Oxford: Blackwell.

———. 1996. Social evolution in primates: The role of ecological factors and male behaviour. *Proc. Brit. Acad.* 88:9–31.

———. 2000a. Infanticide by male primates: The sexual selection hypothesis revisited. In *Infanticide by males and its implications,* edited by C. P. van Schaik and C. H. Janson, 27–60. Cambridge: Cambridge University Press.

———. 2000b. Vulnerability to infanticide by males: Patterns among mammals. In *Infanticide by males and its implications,* edited by C. P. van Schaik and C. H. Janson, 61–71. Cambridge: Cambridge University Press.

van Schaik, C. P., and R. O. Deaner. 2002. Life history and cognitive evolution in primates. In *Animal social complexity,* edited by F. B. M. de Waal and P. L. Tyack. Cambridge, MA: Harvard University Press. In press.

van Schaik, C. P., R. O. Deaner, and M. Merrill. 1999. The conditions for tool use in primates: Implications for the evolution of material culture. *J. Hum. Evol.* 36: 719–41.

van Schaik, C. P., J. K. Hodges, and C. L. Nunn. 2000. Paternity confusion and the ovarian cycle of female primates. In *Infanticide by males and its implications,* edited by C. P. van Schaik and C. H. Janson, 361–87. Cambridge: Cambridge University Press.

van Schaik, C. P., and J. A. R. A. M. van Hooff. 1983. On the ultimate causes of primate social systems. *Behaviour* 85:91–117.

van Schaik, C. P., and M. Hörstermann. 1994. Predation risk and the number of adult males in a primate group: A comparative test. *Behav. Ecol. Sociobiol.* 35:261–72.

van Schaik, C. P., and S. B. Hrdy. 1991. Intensity of local resource competition shapes the relationship between maternal rank and sex ratios at birth in cercopithecine primates. *Am. Nat.* 138:1555–62.

van Schaik, C. P., and C. H. Janson. 2000. *Infanticide by males and its implications.* Cambridge: Cambridge University Press.

van Schaik, C. P., and P. M. Kappeler. 1993. Life history and activity period as determinants of lemur social systems. In *Lemur social systems and their ecological ba-*

sis, edited by P. M. Kappeler and J. U. Ganzhorn, 241–60. New York: Plenum Press.

———. 1996. The social systems of gregarious lemurs: Lack of convergence with anthropoids due to evolutionary disequilibrium? *Ethology* 102:915–41.

———. 1997. Infanticide risk and the evolution of male-female association in primates. *Proc. R. Soc. Lond.* B 264:1687–94.

van Schaik, C. P., M. A. van Noordwijk, R. J. de Boer, and I. den Tonkelaar. 1983. The effect of group size on time budgets and social behavior in wild long-tailed macaques *(Macaca fascicularis)*. *Behav. Ecol. Sociobiol.* 13:173–81.

van Schaik, C. P., M. A. van Noordwijk, and C. L. Nunn. 1999. Sex and social evolution in primates. In *Comparative primate socioecology*, edited by P. C. Lee, 204–40. Cambridge: Cambridge University Press.

van Schaik, C. P., J. W. Terborgh, and S. J. Wright. 1993. The phenology of tropical forests: Adaptive significance and consequences for primary consumers. *Annu. Rev. Ecol. Syst.* 24:353–77.

van Tienderen, P. H. 1995. Life cycle trade-offs in matrix population models. *Ecology* 76:2482–89.

———. 2000. Elasticities and the link between demographic and evolutionary dynamics. *Ecology* 81:666–79.

Van Valen, L. 1973a. Body size and numbers of plants and animals. *Evolution* 27:27–35.

———. 1973b. A new evolutionary law. *Evol. Theory* 1:1–30.

Via, S., R. Gomulkiewicz, G. de Jong, S. M. Scheiner, C. D. Schlichting, and P. H. van Tienderen. 1995. Adaptive phenotypic plasticity: Consensus and controversy. *Trends Ecol. Evol.* 10:212–17.

Via, S., and R. Lande. 1985. Genotype-environment interaction and the evolution of phenotypic plasticity. *Evolution* 39:505–22.

———. 1998. Complexity matters. *Science* 279:1158–59.

Vincent, A. 1985. Plant foods in savannah environments: A preliminary report of tubers eaten by the Hadza of northern Tanzania. *World Archaeol.* 17:131–48.

Wagner, G. P. 1996. Homologues, natural kinds and the evolution of modularity. *Am. Zool.* 36:36–43.

Walker, P. L., J. R. Johnson, and P. M. Lambert. 1988. Age and sex bias in the preservation of human skeletal remains. *Am. J. Phys. Anthropol.* 76:183–88.

Wandsnider, L. A. 1997. The roasted and the boiled: Food composition and heat treatment with special emphasis on pit-hearth cooking. *J. Anthropol. Archaeol.* 16:1–48.

Washburn, S. L., and C. S. Lancaster. 1968. The evolution of hunting. In *Man the hunter*, edited by R. B. Lee and I. DeVore, 293–303. Chicago: Aldine de Gruyter.

Watts, D. P. 2000. Causes and consequences of variation in male mountain gorilla life histories and group membership. In *Primate males*, edited by P. M. Kappeler, 169–79. Cambridge: Cambridge University Press.

Watts, D. P., and A. E. Pusey. 1993. Behavior of juvenile and adolescent great apes. In *Juvenile primates: Life history, development, and behavior*, edited by M. E. Pereira and L. A. Fairbanks, 148–67. New York: Oxford University Press.

Wcislo, W. T. 1989. Behavioral environments and evolutionary change. *Annu. Rev. Ecol. Syst.* 20:137–69.

Webster, A. J., and A. Purvis. 2002. Ancestral states and evolutionary rates of continuous characters. In *Morphology, shape and phylogeny*, edited by N. MacLeod and P. L. Forey, 247–68. London: Taylor & Francis.

Wedekind, C. 1994. Mate choice and maternal selection for specific parasite resistances before, during and after fertilization. *Phil. Trans. R. Soc. Lond.* B 346: 303–11.

Weiss, K. M. 1973. Demographic models for anthropology. *Soc. Am. Archaeol. Mem.* 27:1–86.

Wellborn, G. A., D. K. Skelly, and E. E. Werner. 1996. Mechanisms creating community structure across a freshwater habitat gradient. *Annu. Rev. Ecol. Syst.* 27:337–63.

Werner, E. E., and B. R. Anholt. 1993. Ecological consequences of the trade-off between growth and mortality rates mediated by foraging activity. *Am. Nat.* 142: 242–72.

———. 1996. Predator-induced behavioral indirect effects: Consequences to competitive interactions in anuran larvae. *Ecology* 77:157–69.

West, G. B., J. H. Brown, and B. J. Enquist. 1997. A general model for the origin of allometric scaling laws in biology. *Science* 276:122–26.

West-Eberhard, M. J. 1989. Phenotypic plasticity and the origins of diversity. *Annu. Rev. Ecol. Syst.* 20:249–78.

Westendorp, R. G. J., and T. B. L. Kirkwood. 1998. Human longevity at the cost of reproductive success. *Nature* 396:743–46.

Western, D. 1979. Size, life history and ecology in mammals. *Afr. J. Ecol.* 17: 185–204.

Western, D., and J. Ssemakula. 1982. Life history patterns in birds and mammals and their evolutionary interpretation. *Oecologia* 54:281–90.

White, F. J. 1998. Seasonality and socioecology: The importance of variation in fruit abundance to bonobo society. *Int. J. Primatol.* 19:1013–27.

White, T. D. 1995. African omnivores: Global climatic change and Plio-Pleistocene hominids and suids. In *Paleoclimate and evolution, with emphasis on human origins,* edited by E. Vrba, G. Denton, T. Partridge, and L. Buckle, 369–84. New Haven: Yale University Press.

Whitehead, H., and J. Mann. 2000. Female reproductive strategies of cetaceans: Life histories and calf care. In *Cetacean societies: Field studies of dolphins and whales,* edited by J. Mann, R. C. Connor, P. L. Tyack, and H. Whitehead, 219–46. Chicago: University of Chicago Press.

Whiten, A., R. W. Byrne, R. A. Barton, P. G. Waterman, and S. P. Henzie. 1992. Dietary and foraging strategies of baboons. *Phil. Trans. R. Soc. Lond.* B 334: 187–97.

Whiten, A., J. Goodall, W. C. McGrew, T. Nishida, V. Reynolds, Y. Sugiyama, C. Tutin, R. W. Wrangham, and C. Boesch. 1999. Cultures in chimpanzees. *Nature* 399:682–85.

Widemo, F., and I. Owens. 1999. Size and stability of vertebrate leks. *Anim. Behav.* 58:1217–21.

Wilkinson, G. S., D. C. Presgraves, and L. Crymes. 1998. Male eye span in stalk-eyed flies indicates genetic quality by meiotic drive suppression. *Nature* 391:276–79.

Williams, G. C. 1957. Pleiotropy, natural selection, and the evolution of senescence. *Evolution* 11:398–411.

———. 1966. *Adaptation and natural selection: A critique of some current biological thought.* Princeton, NJ: Princeton University Press.

———. 1992. *Natural selection: Domains, levels, and challenges.* Oxford: Oxford University Press.

Winick, M., and A. Noble. 1966. Cellular response in rats during malnutrition at various ages. *J. Nutr.* 89:300–306.

Wisdom, M. J., L. S. Mills, and D. F. Doak. 2000. Life-stage simulation analysis: Estimating vital rate effects on population growth for species conservation. *Ecology* 81:628–41.

Wise, D. H. 1976. Variable rates of maturation of the spider, *Neriene radiata (Linyphia marginata). Am. Midl. Nat.* 96:66–75.

Wolf, E. R. 1982. *Europe and the people without history.* Berkeley: University of California Press.

Wolfe, L. D. 1984. Female rank and reproductive success among Ariyashiyama B Japanese macaques *(Macaca fuscata). Int. J. Primatol.* 5: 133–43.

———. 1986. Reproductive biology of rhesus and Japanese macaques. *Primates* 27:95–101.

Wolff, J. O. 1993. Why are female small mammals territorial? *Oikos* 68:364–70.

———. 1997. Population regulation in mammals: An evolutionary perspective. *J. Anim. Ecol.* 66:1–13.

Wood, B. A. 1992. Origin and evolution of the genus *Homo. Nature* 355:783–90.

Wood, B. A., and M. Collard. 1999a. The changing face of genus *Homo. Evol. Anthropol.* 8:195–207.

———. 1999b. The human genus. *Science* 284:65–71.

Wood, H. T. 2000. The energetics of weaning: Testing primate models of body and brain size ontogeny and evolution. *Am. J. Phys. Anthropol.* suppl. 30:326–27.

Woodburn, J. 1968. An introduction to Hadza ecology. In *Man the hunter,* edited by R. B. Lee and I. DeVore, 49–55. Chicago: Aldine de Gruyter.

Wrangham, R. W. 1980a. An ecological model of female-bonded primate groups. *Behaviour* 75:262–300.

———. 1980b. On the evolution of ape social systems. *Soc. Sci. Inform.* 18:335–68.

———. 1987. Evolution of social structure. In *Primate societies,* edited by B. B. Smuts, D. L. Cheney, R. M. Seyfarth, R. W. Wrangham, and T. T. Struhsaker, 282–97. Chicago: University of Chicago Press.

Wrangham, R. W., N. L. Conklin-Brittain, and K. D. Hunt. 1998. Dietary response of chimpanzees and cercopithecines to seasonal variation in fruit abundance. I. Antifeedants. *Int. J. Primatol.* 19:949–70.

Wrangham, R. W., J. L. Gittleman, and C. A. Chapman. 1993. Constraints on group size in primates and carnivores: Population density and day-range as assays of exploitation competition. *Behav. Ecol. Sociobiol.* 32:199–209.

Wrangham, R. W., J. H. Jones, G. Laden, D. Pilbeam, and N. L. Conklin-Brittain. 1999. The raw and the stolen: Cooking and the ecology of human origins. *Curr. Anthropol.* 40:567–94.

Wright, L. E., and H. P. Schwarz. 1998. Stable carbon and oxygen isotopes in human tooth enamel: Identifying breast feeding and weaning in prehistory. *Am. J. Phys. Anthropol.* 106:1–18.

Wright, P. C. 1990. Patterns of paternal care in primates. *Int. J. Primatol.* 11:89–102.

———. 1995. Demography and life history of free-ranging *Propithecus diadema edwardsi* in Ranomafana National Park, Madagascar. *Int. J. Primatol.* 16:835–54.

———. 1997. Behavioral and ecological comparisons of Neotropical and Malagasy primates. In *New World primates: Ecology, evolution, and behavior,* edited by W. G. Kinzey, 127–41. New York: Aldine de Gruyter.

———. 1999. Lemur traits and Madagascar ecology: Coping with an island environment. *Yrbk. Phys. Anthropol.* 42:31–72.

Wunder, B. A., and R. D. Gettinger. 1996. Effects of body mass and temperature acclimation on the nonshivering thermogenic response of small mammals. In *Adaptations to the cold,* edited by F. Geiser, A. J. Hulbert, and S. C. Nicol, 131–39. Armidale: University of New England Press.

Yamasaki, K., G. K. Beauchamp, J. Bard, and E. A. Boyse. 1990. Single MHC mutations alter urine odor constitution in mice. In *Chemical signals in vertebrates,* edited by D. W. MacDonald, D. Müller-Schwarze, and S. E. Natynczuk, 255–59. Oxford: Oxford University Press.

Yamashita, N. 1998a. Functional dental correlates of food properties in five Malagasy lemur species. *Am. J. Phys. Anthropol.* 106:169–88.

———. 1998b. Molar morphology and variation in two Malagasy lemur families (Lemuridae and Indriidae). *J. Hum. Evol.* 35:137–62.

Yeager, C. P., S. C. Silver, and E. S. Dierenfeld. 1997. Mineral and phytochemical influences on foliage selection by proboscis monkey *(Nasalis larvatus). Am. J. Primatol.* 41:117–28.

Zilles, K., and G. Rehkämper. 1988. The brain, with special reference to the telencephalon. In *Orang-utan biology,* edited by J. H. Schwartz, 157–76. Oxford: Oxford University Press.

Subject Index

Species Index

Page references to tables and figures appear in bold italicized type.